THE ARTERIAL CHEMORECEPTORS

ADVANCES IN EXPERIMENTAL MEDICINE AND BIOLOGY

Editorial Board:

NATHAN BACK, *State University of New York at Buffalo*
IRUN R. COHEN, *The Weizmann Institute of Science*
DAVID KRITCHEVSKY, *Wistar Institute*
ABEL LAJTHA, *N.S. Kline Institute for Psychiatric Research*
RODOLFO PAOLETTI, *University of Milan*

Recent Volumes in this Series

Volume 572
RETINAL DEGENERATIVE DISEASES
Edited by Joe Hollyfield, Robert Anderson, and Matthew LaVail

Volume 573
EARLY LIFE ORIGINS OF HEALTH AND DISEASE
Edited by Marelyn Wintour-Coghlan and Julie Owens

Volume 574
LIVER AND PANCREATIC DISEASES MANAGEMENT
Edited by Nagy A. Habib and Ruben Canelo

Volume 575
DIPEPTIDYL AMINOPEPTIDASES: BASIC SCIENCE AND CLINICAL APPLICATIONS
Edited by Uwe Lendeckel, Ute Bank, and Dirk Reinhold

Volume 576
N-ACETYLASPARTATE: A UNIQUE NEURONAL MOLECULE IN THE CENTRAL NERVOUS SYSTEM
Edited by John R. Moffett, Suzannah B. Tieman, Daniel R. Weinberger, Joseph T. Coyle and Aryan M.A. Namboodiri

Volume 577
EARLY LIFE ORIGINS OF HEALTH AND DISEASE
Edited by E. Marelyn Wintour and Julie A. Owens

Volume 578
OXYGEN TRANSPORT TO TISSUE XXVII
Edited by Giuseppe Cicco, Duane Bruley, Marco Ferrari, and David K. Harrison

Volume 579
IMMUNE MECHANISMS IN INFLAMMATORY BOWEL DISEASE
Edited by Richard S. Blumberg

Volume 580
THE ARTERIAL CHEMORECEPTORS
Edited by Yoshiaki Hayashida, Constancio Gonzalez, and Hisatake Kondo

---

A Continuation Order Plan is available for this series. A continuation order will bring delivery of each new volume immediately upon publication. Volumes are billed only upon actual shipment. For further information please contact the publisher.

THE ARTERIAL CHEMORECEPTORS

Edited by

Yoshiaki Hayashida
International Buddhist University, Osaka, Japan

Constancio Gonzalez
University of Valladolid, Valladolid, Spain

and

Hisatake Kondo
Tohoku University, Sendai, Japan

ISBN-10 0-387-31310-9 (HB)
ISBN-13 978-0-387-31310-8 (HB)
ISBN-10 0-387-31311-7 (e-book)
ISBN-13 978-0-387-31311-5 (e-book)

© 2006 Springer

All rights reserved. This work may not be translated or copied in whole or in part without the written permission of the publisher (Springer Science + Business Media, Inc., 233 Spring Street, New York, NY 10013, USA), except for brief excerpts in connection with reviews or scholarly analysis. Use in connection with any form of information storage and retrieval, electronic adaptation, computer software, or by similar or dissimilar methodology now known or hereafter developed is forbidden.

The use in this publication of trade names, trademarks, service marks and similar terms, even if they are not identified as such, is not to be taken as an expression of opinion as to whether or not they are subject to proprietary rights.

Printed in the Netherlands.

9 8 7 6 5 4 3 2 1

springer.com

Contents

Preface xiii

A Tribute to Professor Autar Singh Paintal...1
    Ravi K. and Vijayan V.K.

Structure of Chemoreceptors

Immunolocalization of Tandem Pore Domain K^+ Channels
in the Rat Carotid Body...9
    Yamamoto Y. and Taniguchi K.

Neuroglobin, a New Oxygen Binding Protein is Present in the
    Carotid Body and Increases after Chronic Intermittent Hypoxia...........15
    Di Giulio C., Bianchi G., Cacchio M., Artese L.,
    Piccirilli M., Verratti V., Valerio R., Iturriaga R.

Hypoxia-Inducible Factor (HIF)-1α and Endothelin-1 Expression
in the Rat Carotid Body during Intermittent Hypoxia............................21
    Lam S-Y., Tipoe G.L., Liong E.C., Fung M-L.

Expression of HIF-2α and HIF-3α in the Rat Carotid Body
in Chronic Hypoxia 29
    Lam S-Y., Liong E.C., Tipoe G.L., Fung M-L.

Modulation of Gene Expression in Subfamilies of TASK K^+
Channels by Chronic Hyperoxia Exposure in Rat Carotid Body.............37
    Kim I., Donnelly D.F., Carroll J.L.

Postnatal Changes in Gene Expression of Subfamilies
of TASK K^+ Channels in Rat Carotid Body...43
    Kim I., Kim J.H., Carroll J.L.

Morphological Changes in the Rat Carotid Body
in Acclimatization and Deacclimatization to Hypoxia..............................49
    Matsuda H., Hirakawa H., Oikawa S., Hayashida Y., Kusakabe T

Effect of Carbon Dioxide on the Structure of the Carotid Body:
A Comparison between Normoxic and Hypoxic Conditions....................55
    Kusakabe T., Hirakawa H., Oikawa S., Matsuda H., Hayashida Y.

S-Nitrosoglutathione (SNOG) Accumulates Hypoxia Inducible
Factor-1α in Main Pulmonary Artery Endothelial Cells
but not in Micro Pulmonary Vessel Endothelial Cells.............................63
    Fujiuchi S., Yamazaki Y., Fujita Y., Nishigaki Y., Takeda A.,
    Yamamoto Y., Fijikane T., Shimizu T., Osanai S.,
    Takahashi T., Kikuchi K.

Changes in Antioxidant Protein SP-22 of Chipmunk Carotid
Bodies during the Hibernation Season 73
    Fukuhara K., Wu Y., Nanri H., Ikeda M., Hayashida Y.,
    Yoshizaki K., Ohtomo K.

Potential Role of Mitochondria in Hypoxia Sensing by
Adrenomedullary Chromaffin Cells..79
    Buttigieg J., Zhang M., Thompson R., Nurse C.

Localization of Ca/Calmodulin-Dependent Protein Kinase I
in the Carotid Body Chief Cells and the Ganglionic Small
Intensely Fluorescent (SIF) Cells of Adult Rats 87
    Hoshi H., Sakagami H., Owada Y., Kondo H.

Developmental Aspects of Chemoreceptors

Dual Origins of the Mouse Carotid Body Revealed by Targeted
Disruption of *Hoxa3* and *Mash1* 93
    Kameda Y.

Genetic Regulation of Chemoreceptor Development in DBA/2J
and A/J Strains of Mice 99
    Balbir A., Okumura M., Schofield B., Coram J.,
    Tankersley C.G., Fitzgerald R.S., O'Donnell C.P.,
    Shirahata M.

Genetic Influence on Carotid Body Structure in DBA/2J
and A/J Strains of Mice 105
    Yamaguchi S., Balbir A., Okumura M., Schofield B.,
    Coram J., Tankersley C.G., Fitzgerald R.S.,
    O'Donnell C.P., Shirahata M.

The Effect of Hyperoxia on Reactive Oxygen Species (ROS)
In Petrosal and Nodose Ganglion Neurons during
Development (Using Organotypic Slices) 111
    Kwak D.J., Kwak S.D., Gauda E.B.

Carotid Body Volume in Three-Weeks-Old Rats Having
an Episode of Neonatal Anoxia 115
    Saiki C., Makino M., Matsumoto S.

The Effect of Development on the Pattern of A1
and A2a-Adenosine Receptor Gene and Protein Expression
in Rat Peripheral Arterial Chemoreceptors 121
    Gauda E.B., Cooper R.Z., Donnelly D.F., Mason A.,
    McLemore G.L.

A Comparative Study of the Hypoxic Secretory Response
between Neonatal Adrenal Medulla and Adult Carotid Body
from the Rat 131
    Rico A.J., Fernandcz S.P., Prieto-Lloret J., Gomez-Niño A.,
    Gonzalez C., Rigual R.

Molecular Biology of Chemoreceptors

In Search of the Acute Oxygen Sensor: Functional Proteomics
and   Acute Regulation of Large-Conductance, Calcium-
Activated Potassium Channels by Hemeoxygenase-2 137
  Kemp P.J., Peers C., Riccardi D., Iles D.E., Mason H.S.,
  Wootton P., Williams S.E.

Does AMP-activated Protein Kinase Couple Inhibition
of Mitochondrial Oxydative Phosphorylation by Hypoxia
to Pulmonary Artery Constriction? 147
  Evans A.M., Mustard K.J.W., Wyatt C.N., Dipp M.,
  Kinnear N.P., Hardie D.G.

Function of NADPH Oxidase and Signaling by Reactive
Oxygen Species in Rat Carotid Body Type I Cells 155
  He L., Dinger B., Gonzalez C., Obeso A., Fidone S.

Hypoxemia and Attenuated Hypoxic Ventilatory Responses
in Mice Lacking Heme Oxygenese-2: Evidence for a Novel Role
of Heme Oxygenese-2 as an Oxygen Sensor 161
  Zhang Y., Furuyama K., Adachi T., Ishikawa K.,
  Matsumoto H., Masuda T., Ogawa K., Takeda K.,
  Yoshizawa M., Ogawa H., Maruyama Y., Hida W.,
  Shibahara S.

Regulation of a TASK-like Potassium Channel
in Rat Carotid Body Type I Cells by ATP 167
  Varas R. and Buckler K.J.

Accumulation of Radiolabeled *N*-Oleoyl-Dopamine
in the Rat Carotid Body 173
  Pokorski M., Zajac D., Kapuściński A., Matysiak Z.,
  Czarnocki Z.

Profiles for ATP and Adenosine Release at the Carotid Body
in Response to O_2 Concentrations 179
   Conde S.V. and Monteiro E.C.

Biophysics of Ionic Channels in Chemoreceptors

Hypoxic Regulation of Ca^{2+} Signalling in Astrocytes and
Endothelial Cells 185
   Peers C., Aley P.K., Boyle J.P., Porter K.E., Pearson H.A.
   Smith I.F., Kemp P.J.

Does AMP-activated Protein Kinase Couple Hypoxic Inhibition
of Oxydative Phosphorylation to Carotid Body Excitation? 191
   Wyatt C.N., Kumar P., Aley P., Peers C., Hardie D.G.,
   Evans A.M.

Mitochondrial ROS Production Initiates $A\beta_{1-40}$-Mediated
Up-Regulation of L-Type Ca^{2+} Channels during
Chronic Hypoxia 197
   Fearon I.M., Brown S.T., Hudasek K., Scragg J.L.,
   Boyle J.P., Peers C.

Acute Hypoxic Regulation of Recombinant THIK-1 Stably
Expressed in HEK293 Cells 203
   Fearon I.M., Campanucci V.A., Brown S.T., Hudasek K.,
   O'Kelly I.M., Nurse C.A.

Differential Expression of Oxygen Sensitivity
in Voltage-Dependent K Channels in Inbred Strains of Mice 209
   Otsubo T., Yamaguchi S., Okumura M., Shirahata M.

An Overview on the Homeostasis of Ca^{2+} in Chemoreceptor
Cells of the Rabbit and Rat Carotid Bodies 215
   Conde S.V., Caceres A.I., Vicario I., Rocher A., Obeso A.,
   Gonzalez C.

Central Integration and Systemic Effects of Chemoreflex

Midbrain Neurotransmitters in Acute Hypoxic Ventilatory
Response 223
    Kazemi H.

Chronic Intermittent Hypoxia Enhances Carotid Body
Chemosensory Responses to Acute Hypoxia 227
    Iturriaga R., Rey S., Alcayaga J., Del Rio R.

The Cell-Vessel Architecture Model for the Central Respiratory
Chemoreceptor 233
    Okada M., Kuwana S., Oyamada Y., Chen Z.

Loop Gain of Respiratory Control upon Reduced Activity
of Carbonic Anhydrase or Na^+/H^+ Exchange 239
    Kiwull-Schöne H., Teppema L., Wiemann M., Kiwull P.

Adrenaline Increases Carotid Body CO_2 Sensitivity:
An *in vivo* Study 245
    Maskell P.D., Rusius C.J., Whitehead K.J., Kumar P.

Peripheral Chemoreceptor Activity on Exercise-Induced
Hyperpnea in Human 251
    Osanai S., Takahashi T., Nakao S., Takahashi M., Nakano H.,
    Kikuchi K.

Effects of Low-Dose Methazolamide on the Control of
Breathing in Cats 257
    Bijl J.H.L., Mousavi Gourabi B., Dahan A., Teppema L.J.

Stimulus Interaction between Hypoxia and Hypercapnia in
the Human Peripheral Chemoreceptors 263
    Takahashi T., Osanai S., Nakao S., Takahashi M.,
    Nakano H., Ohsaki Y., Kikuchi K.

Gene Expression and Signaling Pathways by Extracellular
Acidification 267
    Shimokawa N., Londoño M., Koibuchi N.

Hypoxic Modulation of the Cholinergic System
in the Cat Carotid Glomus Cell 275
    Mendoza J.A., Chang I., Shirahata M.

Mechanisms of Chemoreceptions

Are There "CO_2 Sensors" in the Lung?..281
    Lee L.Y., Lin R.L., Ho C.Y., Gu Q., Hong J.L.

Nitric Oxide in Brain Glucose Retention after Carotid Body
Receptors Stimulation with Cyanide in Rats...293
    Montero S.A., Cadenas J.L., Lemus M.,
    Roces De Álvarez-Buylla E., Álvarez-Buylla R.

Pulmonary Nociceptors are Potentially Connected with
Neuroepithelial Bodies 301
    Yu J., Lin S.X., Zhang J.W., Walker J.F.

Modulators of Cat Carotid Body Chemotransduction..............................307
    Fitzgerald R.S., Shirahata M., Chang I., Balbir A.

Identification and Characterization of Hypoxia Sensitive Kvα
Subunits in Pulmonary Neuroepithelial Bodies 313
    Fu X.W. and Cutz E.

Voltage-Dependent K Channels in Mouse Glomus Cells are
Modulated by Acetylcholine 319
    Otsubo T., Yamaguchi S., Shirahata M.

Modification of the Glutathione Redox Environment and
Chemoreceptor Cell Responses 325
    Gómez-Niño A., Agapito M.T., Obeso A., González C.

Carotid Body Transmitters Actions on Rabbit Petrosal
Ganglion *in Vitro* 331
    Alcayaga J., Soto C.R., Vargas R.V., Ortiz F.C., Arroyo J.,
    Iturriaga R.

Potassium Channels in the Central Control of Breathing 339
    Oyamada Y., Yamaguchi K., Murai M., Ishizaka A., Okada Y.

Role of Endothelin-1 on the Enhanced Carotid Body Activity
Induced by Chronic Intermittent Hypoxia 345
    Rey S., Del Rio R., Iturriaga R.

Concluding Remarks 351
    Gonzalez C.

Index 361

Preface

In the general assembly of International Society for Arterial Chemoreception (ISAC) at the XIV Meeting of ISAC in Philadelphia (June, 1999) the membership decided that tentatively the XVI Meeting of the Society will be held in Japan. At the conclusion of the XV ISAC Meeting held in Lyon (France) in November 2002, Hisatake Kondo took of the torch as president of ISAC, and took the responsibility of organizing the XVI Meeting of the Society in Sendai (Japan).

This book contains the Proceedings of XVI ISAC Meeting held at Miyagi Zao Royal Hotel in the suburb of Sendai, Japan, from May 9 to 12, 2005. Hisatake Kondo counted with a group of colleague co-organizers from different Japanese institutions, including Yoshiaki Hayashida from Osaka, Katsuaki Yoshizaki from Akita, Yoko Kameda from Kanagawa, Tatsumi Kusakabe from Tokyo, Yuji Owada from Sendai and Hiroyuki Sakagami also from Sendai. The Scientific Committee was formed by some ISAC members from abroad. Professors Helmut Acker (Germany), Carlos Eyzaguirre (USA), Salvatore Fidone (USA), Robert Fitzgerald (USA), Constancio Gonzalez (Spain), Sukhamay Lahiri (USA), Jean Marc Pequignot (France) and Patricio Zapata (Chile) helped with their advice to take decisions on specific aspects of the scientific programme. Recommendations given by Dr. Prem Kumar (UK), acting treasurer of ISAC, were invaluable to solve last minute contingencies.

In the XVI ISAC Meeting essentially all areas on Arterial Chemoreceptors were covered in the presentations and compiled in this volume. There were presentations on the structure and developmental aspects of the carotid body chemoreceptors, on the molecular biology and biophysical aspects of the ion channels expressed in chemoreceptor cells, on the neurotransmitters and their receptors expressed in the carotid body, on the central integration of the carotid body generated activity and on the systemic effects of the chemoreceptor reflexes. Some important studies on central chemoreceptors, on neuroepithelial bodies and other lung receptors, on hypoxic pulmonary vasoconstriction and on oxygen sensing in endothelial cells widened the scope and enriched the meeting. Probably the areas generating more enthusiastic discussions dealt with the mechanisms of chemoreception. Particularly animated were the discussions on the papers dealing with significance of different potassium channels in the hypoxic activation of chemoreceptor cells and with the role of reactive oxygen species as triggers or modulators of hypoxic transduction cascade.

The Arterial Chemoreceptors meetings have a history of over half a century, and shows alteration of generations. Dr. A. S. Paintal passed away on December 21, 2004. He was an excellent sensory physiologist. His skill and patience to record from single C-fibres allowed him to describe for the first time the J-receptors and to characterize many sensory receptors in thoracic and abdominal viscera. He also contributed to enrich the field of Arterial Chemoreception with his studies in the aortic and carotid bodies. A tribute to the memory of the late Paintal was offered and presented in this volume by K. Ravi and V. K. Vijayan at VP Chest Institute, University of Delhi.

The Heymans-De Castro-Neil Awards for young investigators were given to C. Wyatt (Scotland), T. Otsubo (USA) and R. Varas (United Kingdom). ISAC wishes the awardees a fruitful development of their current research projects and successful scientific careers.

At the business meeting the next Symposium was decided to be held in Valladolid, Spain, in 2008, with Constancio Gonzalez as the president. The following Symposium will be held in 2011 in Ontario, Canada with Colin Nurse as its president. During the assembly, membership discussed the future scope of ISAC in order to attract a more biomedical scientists. A dilemma emerged: the interest in widening the scope of the Arterial Chemoreception to get closer the oxygen (erythropoietin, hypoxic pulmonary vasoconstriction, central nervous system) and acid (central chemoreceptors) sensing fields, and the risk of losing our identity as ISAC in the diversity. There was no conclusion, but an agreement was reached to give the issue into the hand of C. Gonzalez, the president of both ISAC and the next meeting.

The Symposium was supported by funds from Japan Society for the Promotion of Science (JSPS) International meeting series, Sankyo Foundation for Life Science Research Promotion, Tokyo, Japan and Asaoka Eye Clinic Foundation, Hamamatsu, Japan. We are grateful to them all.

Finally, we are grateful to the participants who visited Sendai-Zao and contributed to the success of the Symposium. We are particularly grateful to Mr. Mike van den Bosch and Miss Marie Johnson of Springer for their expert management of the production of this volume.

The Editors,
Yoshiaki Hayashida (Osaka, Japan)
Constancio Gonzalez (Valladolid, Spain)
Hisatake Kondo (Sendai Japan)

A TRIBUTE TO PROFESSOR AUTAR SINGH PAINTAL (1925-2004)

K. RAVI AND V.K. VIJAYAN

Vallabhbhai Patel Chest Institute, University of Delhi, Delhi 110007, India

Great men are the true men in whom nature has prospered. They are not extra-ordinary – they are in the true order, what they are ought to be. They reached the summit by doing their jobs in hand with everything they had of energy, enthusiasm and hard work. More importantly, they made every man who came in contact with them feel great. Autar Singh Paintal who passed away on December 21, 2004 in Delhi was indeed such a great man who will be remembered not only for his contributions in Physiology but also for his acts of altruism.

Born on September 24, 1925 in the ruby-mining town of Mogok in Burma (now Myanmar) where his father Dr. Man Singh was a physician in the British Medical Services, after his initial schooling, Paintal was forced to move out of Burma to Lahore (presently in Pakistan) to complete his matriculation. He then did his intermediate examination of the Panjab University from Forman Christian College and subsequently obtained admission for doing MBBS at King George's Medical College, Lucknow, India. During his stay at the Medical College (1943-1948), he won several awards and was given the coveted HEWITT Gold Medal for being the best graduate student of his batch. Thus, it appears that Paintal was born great.

There are two things to aim at in life. First, to get what you want and second, to enjoy it after that. For being what he was, even though a career in clinical medicine was available to him, he surprised many when he decided to pursue M.D in Physiology. For his thesis work, he selected the topic "Electrical resistance of the skin in normals and psychotics" which required instrumentation that would record the events fast. Realising that not much technical help was coming forth, Paintal built his own apparatus from the ex-war disposal junk and collected some valuable data from these subjects (Paintal, 1951). He introduced a new index which came to be known as "Paintal index" for evaluation of galvanic skin responses which was used by clinicians successfully to diagnose psychosis (Elliot and Singer, 1953). Even to-day, it forms the basis of the lie detector test deployed for crime detection universally. After a short stint as lecturer in the Physiology Department of King George's Medical college, he proceeded to work for his Ph.D degree with Prof. David Whitteridge in the Physiology Department of the Medical School in Edinburgh on a Rockefeller Fellowship. This gesture on the part of the Foundation was unprecedented as this Fellowship is normally given for carrying out research in the USA. From the spectrum of discoveries which he made since then, it is apt to say that Paintal got what he wanted and he enjoyed what he got, for only the wisest of mankind achieve the second.

While in UK, Paintal was assigned the task of measuring the conduction velocities of single fibres of the vagus for which the cat used to be placed in a box and steam generated from an immersion heater was passed into it so that the nerve filaments would not dry up. Paintal felt that this procedure was cumbersome since proper viewing of the fibres was not possible as steam condensed on the glass plates. To avoid this discomfort, he started dissecting the filaments under a layer of liquid paraffin (Paintal, 1953a). Even though it was a clear departure from the then convention, the usage of liquid paraffin revolutionized studies on sensory physiology. While continuing his studies on vagal afferent fibres, he successfully demonstrated that injection of chemicals into the circulation could be used as a technique to discover 'silent' sensory receptors of the viscera (Paintal, 1954).

When we are guided by the light of reason, we must let our minds be bold. Paintal was such a man of courage who was full of confidence. He had the audacity to demonstrate that the pulmonary vascular receptors which his mentor Whitteridge believed were located in the small branches of the pulmonary artery, actually ended in the right and left atria of the heart. In the process, he used 'punctate' stimulation as a method for localizing these receptors in an 'open-chested' preparation. Thus, he was not only unconventional but unorthodox also. His Professor had the strength of character to accept his pupil's findings gracefully. More importantly, it was he who gave wide publicity for this discovery. Thus the 'atrial volume receptors' (Paintal, 1953b) which have a role to play in the body fluid volume regulation became conspicuous.

After obtaining his Ph.D., he returned to India to work as a Technical Officer of the Defence Laboratories in Kanpur before taking up the post of Assistant Director at V.P. Chest Institute where he made several discoveries for which he is famous globally. From 1956-58, he was invited as Visiting Professor at Albert Einstein College of Medicine, New-York, USA, University of Utah, Salt Lake City, USA and University of Gottingen, Germany. He was then offered the position of Professor of Physiology at AIIMS where he spent 6 years from 1958-64. It was during this period that he received his D.Sc., degree from the University of Edinburgh in 1960. In 1964, he returned to V.P. Chest Institute as the Director and stayed there till his retirement in 1990. During 1986-1991 he also guided the destiny of Medical Research in India as the Director General of Indian Council of Medical Research. Even after retirement and until his death, he continued his research at the Centre for Visceral Mechanisms housed at V.P. Chest Institute and investigated the sensory mechanisms which caused breathlessness and limited muscular performance especially in soldiers who were posted at high altitude.

During the years 1952—1960, he discovered several sensory receptors in the viscera. These include the ventricular pressure receptors, the gastric stretch receptors, the mucosal mechanoreceptors of the intestines (Paintal, 1973, for review) and the pressure pain receptors of muscles (Paintal, 1960). With C.C.

Hunt, he coined the term "fusimotor" for the nerve fibres to intrafusal muscle of muscle spindles (Hunt and Paintal, 1958). His studies on nerve conduction and differential cooling of the nerve for blockage of nerve conduction (Paintal, 1965) gave Physiologists a technique to differentiate reflexes mediated by myelinated afferents from those mediated by non-myelinated ones. He also demonstrated that the Head's Paradoxical reflex was an artefact (Paintal, 1966).

1. DISCOVERY OF TYPE J RECEPTORS

Paintal is best known for the discovery of the juxtapulmonary capillary or type J receptors. Ever since Adrian (1933) recorded afferent activity from the vagus nerve, the search for receptors which accounted for 'Hering-Breuer deflation reflex' continued. Even though Knowlton and Larrabee (1946) and later on Widdicombe (1954) reported a group of vagal nerve endings which were stimulated by deflation, their activity showed rapid adaptation. In his preliminary studies performed in UK, Paintal came across a few units, with conduction velocities in the range 2-10 m/sec, responding to deflation (Paintal, 1953a). He continued these studies while working at V.P. Chest Institute. He identified several of these endings which responded specifically to suction of air from the lungs. They were connected mostly to non-myelinated afferents. He called them 'specific pulmonary deflation receptors' (Paintal, 1955). Realising that deflation was a weak and artificial stimulus, he later demonstrated that these endings were stimulated within 1.5 to 2.5 sec following the injection of the chemical, phenyl diguanide, into the pulmonary circulation and in 0.3 sec following the insufflation of halothane. Systemic injection of phenyl diguanide did not stimulate these endings. Additionally, occlusion of left atrio-ventricular junction, injection of alloxan and inhalation of chlorine gas stimulated them. Based upon his electrophysiological findings, he predicted that these endings must be located at the interstitial space between the pulmonary capillary and alveolus and called them 'juxtapulmonary capillary' or 'type J' receptors (Paintal, 1969). Indeed, histological investigations using electron microscope (Meyrick and Reid, 1971) reveal the presence of sensory nerve endings in this region. Thus Paintal had the vision, the art of seeing things invisible. He also proposed that the natural stimulus for these receptors was pulmonary congestion. Thus, conditions which caused an increase in pulmonary capillary pressure would cause an increase in pulmonary interstitial fluid which in turn would stimulate the type J receptors (Paintal, 1973).

1.1 Physiological Mechanism

After establishing their pathophysiological significance, Paintal tried to explore the physiological function of the type J receptors for he firmly believed that nature would not have designed something to respond only to an extreme situation. He visualized that the type J receptors must be stimulated by exercise at sea-level. During exercise, the venous return increases which will result in an increase in cardiac output. There will be an increase in pulmonary capillary pressure, an increase in interstitial fluid volume and consequently an increase in type J receptor activity. To prove this point, experiments were performed in cats in whom the left pulmonary artery was occluded and the entire cardiac output was diverted into the right lung. This procedure resulted in marked activation of the type J receptors. Since the increase in pulmonary blood flow observed was similar to that seen during moderate exercise, he considered it logical to conclude that the type J receptors were stimulated by moderate exercise at sea-level also (Anand and Paintal, 1980).

1.2 Sensations from Type J Receptors

The type J receptors of man are stimulated by intravenous injections of lobeline. Besides producing the known respiratory effects namely apnea/tachypnea, type J receptor stimulation in man produces certain sensations notably choking and pressure in the throat and upper chest. Dry cough occurred following intensification of the sensations (Paintal, 1995). Similar sensations have been reported by people who suffer from high altitude pulmonary edema. Most of the subjects with high altitude pulmonary edema who have breathlessness as a major symptom report of having pressure and choking sensations in the throat and upper chest. Paintal proposed that such sensations generated by the type J receptors must accompany the sensation of breathlessness (Paintal, 1995).

1.3 J Reflex

Besides producing tachypnea/apnea, bronchoconstriction and bradycardia, J receptor stimulation causes inhibition of somatic muscles (Deshpande and Devanandan, 1970). The last viscerosomatic response has been termed as 'J Reflex' by Paintal (Paintal, 1970). According to him, this reflex acts as a feedback mechanism to limit exercise. Additionally, it functions as a protective reflex for the lungs also. This reflex is initiated by muscular exercise which operates the muscle pump with the result, there will be an increase in the venous return, an increase in cardiac output and excessive filling of the pulmonary capillaries. Consequently, there will be an increase in pulmonary capillary pressure and an increase in volume of the interstitial fluid which will lead to stimulation of type J receptors (Paintal, 1970). Experimental studies demonstrated that type J receptor stimulation caused inhibition of somatic muscle activity through cerebral pathways in the caudate nucleus of the basal ganglia (Kalia 1969). Thus, there would be termination of exercise itself. According to Paintal, in high altitude, when one gets the feeling of unusual weakness of the leg muscles along with the sensations of breathlessness, one must realize that his lungs are already congested. He warned that if these signals from the type J receptors were ignored, and if one continued to climb, there was the grave danger of precipitating pulmonary edema (Paintal, 1986). In the Bhopal gas tragedy of 1984, most of the victims died of pulmonary edema. Those who survived, complained of muscle weakness along with cough and breathlessness. Since methyl isocyanate gas produced pulmonary edema, Paintal explained that the muscle weakness must be due to the 'J Reflex' as there was no evidence of any neurological lesions (Paintal, 1986). In fact, the muscle weakness receded gradually in these patients when the interstitial edema became less and less.

Since type J receptors produce dry cough, breathlessness and muscle weakness, Paintal was of the view that these symptoms could be used for assessment of certain pathophysiological conditions pertaining to the heart and lungs. For instance, in patients with left ventricular failure, the appearance of dry cough might indicate the onset of interstitial edema. The presence of muscle weakness would support this deduction and these symptoms would help in the early diagnosis and treatment of these patients (Paintal, 1986).

2. CONTRIBUTIONS TO CHEMORECEPTION

Arterial chemoreceptors, the sensory receptors which respond to a decrease in the partial pressure of oxygen in the arterial blood, comprise of the carotid and aortic bodies. Ever since their discovery in 1927, there has been a continuing debate on some of the stimuli to these receptors and the mechanisms behind sensory transduction. There is general agreement that both these groups of receptors are stimulated by ventilation with hypoxic gas mixtures which lead to a fall in the arterial oxygen tension, a reduction in the oxygen content of arterial blood as in anemia or carbon monoxide poisoning and a reduction in the rate of oxygen carried by arterial blood as in hypotension. However, there is some disagreement on the role of carbon dioxide as a natural stimulus to these receptors. Paintal was the chief proponent of the theory that carbon dioxide, within physiological levels of increase, do not activate the aortic chemoreceptors (Paintal and Riley, 1966). Whatever excitation hypercapnia had on these receptors was due to local vasoconstriction in the aortic body mediated by increased sympathetic outflow (Anand and Paintal, 1988). He did almost all his investigations on the aortic chemoreceptors since he was of the opinion that in the studies on carotid chemoreceptors one cannot discount the contribution from the external environment which will have a higher partial pressure of oxygen than blood (Paintal, 1968). Continuing his studies on aortic chemoreceptors, he showed that the actual stimulus for these receptors is local Po_2 which will depend upon arterial Po_2, oxygen content and blood flow (Anand and Paintal, 1990).

2.1 Mechanical Hypothesis of Chemoreceptor Stimulation

Paintal was of the firm opinion that in nature there existed three ways of transmission of neural information namely i) electrical, along nerve fibres, ii) chemical, at synapses and iii) mechanical, at sensory receptors and there was the involvement of physico-chemical process in each one of them (Paintal, 1976a). He held the view that all sensory receptors including chemoreceptors were basically mechanoreceptors (Paintal, 1976b).

For the mechanical process of chemoreceptor stimulation, he collected evidence initially by the process of exclusion. Ever since Landgren and Neil (1951) reported that the carotid chemoreceptors were stimulated by haemorrhage, it was considered by many chemoreceptor physiologists that the stimulation must be due to accumulation of some metabolites in the vicinity of the receptors. This finding led to the genesis of 'transmitter theory' for chemoreception. It has been proposed that hypoxic stimulus causes depolarization of type I glomus cells; there is calcium entry and secretion of neurotransmitters. The neurotransmitters released gain access to the nerve terminal, depolarise it and there is the production and propagation of action potentials (Eyzaguirre and Abudara, 1999, for review). Such a view is actively pursued by several investigators all over the world.

But, Paintal was never afraid of waging battles. He believed that victory does not belong to the majority always - it is to the vigilant, the active, the brave. Contrary to the popular belief, he proposed the mechanical transmission hypothesis for chemoreception. Working with aortic chemoreceptors, he demonstrated that their activity fell in about 2.5 min after circulating arrest (Paintal, 1967). If there was the involvement of a metabolite, a transmitter, the activity should have continued for a longer time since there was no blood to wash away the metabolite/transmitter. Furthermore, there was no additional discharge in these units on subsequent ventilation of these animals with N_2 (Paintal, 1967). As hypoxia did not depolarize the endings directly, he proposed that their generator region must be stimulated by mechanical deformation (Paintal, 1976b). He showed that compression of the area in which these receptors were located led to their stimulation (Paintal, 1976). According to the scheme proposed by him, the type I cells consumed oxygen at a very high rate and this consumption, depending upon the oxygen availability determined by the oxygen content of the arterial blood, regulated the local Po_2 at the Po_2 sensor which is the type II cell. During hypoxia, when the local Po_2 falls, there would be mechanical deformation (a change in the shape or size) of the type II cell which will affect the contact points with the generator region of the nerve ending leading to the production of generator potential (Paintal, 1976, Anand and Paintal, 2000).

This unifying concept of the receptor mechanism though considered very interesting and challenging, was not welcomed by many. However, it must be remembered that even though this concept was proposed nearly forty years ago, it has withstood the test of time and has not been disproved till date.

3. ACADEMIC HONOURS

His contributions came to be described as having opened a new era in Physiology with Cornellie Heymans and Eric Neil coining the terms "Pre-Paintal" versus "Post-Paintal" while referring to the impact of his discoveries. Pursuing a scientific career is a great game. Some play it for the sheer love of it while others do it for attaining prizes. Paintal was passionate about his work and his dedication towards his chosen field won the appreciation of many. Awards and honours followed naturally. He was elected to the Fellowship of the Royal Society of Edinburgh in 1966, followed by an election to the National Academy of Medical Sciences, and the Indian National Science Academy. In 1981, he was elected to the Royal Society (U.K.) and was the first Indian medical scientist to be so honoured. An honorary membership of the Physiological Society (U.K). and the American Physiological Society followed soon after as did an Honorary Fellowship of the Royal College of Physicians. His outstanding scientific contributions won him several National Awards and Honours viz: Dr.B.C.Roy award (1973), Medical Council Silver Jubilee research Award (1979). Barclay

Medal (1982), Rameswar Birla National Award (1982), First Jawaharlal Award in Science (1983), Acharya J.C.Bose Medal (1985),Silver Jubilee Award, AIIMS (1986), C.V.Raman Award (1995), Jawaharlal Nehru Birth Centenary Award (2002) with the President of India bestowing on him the coveted honour of "Padma Vibhushan" in 1986. He was elected as a member of the International Council of Physiological Sciences in 1997 and re-elected for another term up to 2005 - which was an honour not only for him but also for India.

Throughout his career, Paintal set very high standards which were difficult to emulate. He was uncompromising and went to any extent to prove his point. But, inside him, he had a heart full of compassion. He believed that 'the best portion of a good man's life is his little, nameless, unremembered acts of kindness and love' shown towards his friends and colleagues. At the Patel Chest Institute, where he worked as its Director for a very long time, the popular saying used to be 'Confess your sins to Him, you will be forgiven'. While one could get away for not being punctual, for not keeping up the work place tidy, one could never get away for scientific fraudulence. To prevent it, he formed along with like-minded scientists, a Society of Scientific Values in India and served as its first President. This Society, the first of its kind in the world, aimed to promote integrity, objectivity and ethical values in the pursuit of science. He was of the opinion that scientific meetings should be held in the university/academic environments only and not in the lavish environment of five star hotels. At a time when heads of departments who had little time to spend in the laboratories but were very happy to add their names when the work got published, Paintal was very different. He never took any credit unless he contributed significantly. Even though the physiological research of to-day is becoming more and more molecular with fewer and fewer scientists opting for a career in integrative physiology, Paintal stuck to classical physiology and proved that fundamental discoveries have an application in solving human problems. It is indeed an honour to have known him and to have worked under him.

REFERENCES

Adrian E.D. Afferent impulses in the vagus and their effect on respiration. J Physiol 1933; 79:332-358.

Anand A., Paintal A.S. Reflex effects following selective stimulation of J receptors in the cat. J Physiol 1980; 299:553-572.

Anand A., Paintal A.S. The influence of the sympathetic outflow on aortic chemoreceptors of the cat during hypoxia and hypercapnia. J Physiol 1988; 395:215-231.

Anand A., Paintal A.S. "How real is the relation of arterial Po_2 to chemoreceptor activity?" In *Arterial Chemoreceptors*, C. Eyzaguirre., S.J. Fidone., R.S. Fitzgerald., S. Lahiri., D.M. McDonald, ed. New York: Springer Verlag, 1990.

Anand A., Paintal A.S. The present status of mechanical hypothesis for chemoreceptor stimulation. Adv Exp Med Biol 2000; 475:411-418.

Deshpande S.S., Devanandan M.S. Reflex inhibition of monosynaptic reflexes by stimulation of type J pulmonary endings. J Physiol 1970; 206:345-357.

Elliott D.N., Singer E.G. The Paintal index as an indicator of skin-resistance changes to emotional stimuli. J Exp Psychol 1953; 45:429-430.

Eyzaguire C., Abudara V. Carotid body glomus cells: chemical secretion and transmission (modulation) across cell nerve ending junctions. Resp Physiol 1999; 115:135-149.

Hunt C.C., Paintal A.S. Spinal reflex regulation of fusimotor neurons. J Physiol 1958; 143;195-212.

Kalia M. Cerebral pathways in reflex muscular inhibition from type J pulmonary receptors. J Physiol 1969; 204:92-93p.

Knowlton G.C., Larrabee M.G. A unitary analysis of pulmonary volume receptors. Am J Physiol 1946; 147:100-114.

Landgren S., Neil E. Chemoreceptor impulse activity following haemorrhage. Acta Physiol Scand 1951; 23:158-167.

Meyrick B., Reid, L. Nerves in rat intra-acinar alveoli: an electron microscopic study. Resp Physiol 1971; 11:367-377.

Paintal A.S. A comparison of the galvanic skin responses of normals and psychotics. J Exp Psychol 1951; 41:425-428.

Paintal A.S. The conduction velocities of respiratory and cardiovascular afferent fibres in the vagus nerve. J Physiol 1953a; 121:182-190.

Paintal A.S. A study of right and left atrial receptors. J Physiol 1953b; 120:596-610.

Paintal A.S. The responses of gastric stretch receptors and certain other abdominal and thoracic vagal receptors to some drugs. J Physiol 1954; 126:271-285.

Paintal A.S. Impulses in vagal afferent fibres from specific pulmonary deflation receptors. The responses of these receptors to phenyl diguanide, potato starch, 5-hydroxytryptamine and nicotine and their role in respiratory and cardiovascular reflexes. Quart J Exp Physiol 1955; 40:89-111.

Paintal A.S. Functional analysis of Group III afferent fibres of mammalian muscles. J Physiol 1960; 152:250-270.

Paintal A.S. Block of conduction in mammalian myelinated fibres by low temperatures. J Physiol 1965; 180:1-19.

Paintal A.S. Re-evaluation of respiratory reflexes. Quart J Exp Physiol 1966; 51:151-163.

Paintal A.S. Mechanism of stimulation of aortic chemoreceptors by natural stimuli and chemical substances. J Physiol 1967; 189:63-84.

Paintal A.S. The possible influence of the external environment on the responses of chemoreceptors. In *Arterial Chemoreceptors* R.W. Torrance, ed. Oxford, Blackwell, 1968.

Paintal A.S. Mechanism of stimulation of type J pulmonary receptors. J Physiol 1969; 203:511-532.

Paintal A.S. The mechanism of excitation of type J receptors and the J reflex. In *Ciba Found. Symp. Breathing: Hering-Breuer Centenary Symposium*, R. Porter, ed. London: Churchill, 1970.

Paintal A.S. Vagal sensory receptors and their reflex effects. Physiol Rev 1973; 53:159-227.

Paintal A.S. Natural and paranatural stimulation of sensory receptors. In *Sensory Functions of the Skin*, Y. Zotterman, ed. Pergamon Press – Oxford and New York, 1976a.

Paintal A.S. Mechanical transmission of sensory information at chemoreceptors. In *Morphology and Mechanisms of Chemoreceptors*, A.S. Paintal, ed. Navchetan Press, New Delhi, 1976b.

Paintal A.S. The significance of dry cough, breathlessness and muscle weakness. Indian J Tuberculosis 1986; 33:51-55.

Paintal A.S. Some recent advances in studies on J receptors. In *Control of the Cardiovascular and Respiratory Systems in Health and Disease*, C.T. Kappagoda and M.P. Kaufman, ed. Plenum Press, New York and London, 1995.

Paintal A.S., Riley, R.L. Responses of aortic chemoreceptors. J Appl Physiol 1966; 21:543-548.

Immunolocalization of Tandem Pore Domain K$^+$ Channels in the Rat Carotid Body

YOSHIO YAMAMOTO AND KAZUYUKI TANIGUCHI

Laboratory of Veterinary Anatomy, Faculty of Agriculture, Iwate University, 18-8, Ueda 3-chome, Morioka, Iwate 020-8550, JAPAN

1. INTRODUCTION

Tandem pore domain K$^+$ channels can be divided into six subfamilies; TWIK, TASK, TREK, TALK, THIK and TRESK (Patel and Lazdunski, 2004). Its subfamily consists of several subunits. These channel subunits. have 4 transmembrane segments and 2 pore domains. These channels are ubiquitously expressed in the body including central and peripheral nervous systems. Of these channels, TASK (TWIK-related acid-sensitive K$^+$ channels) family including TASK-1, TASK-2 and TASK-3 are inhibited by extracellular low pH (Lesage, 2003). In addition, it has been shown that TASK-1 and TASK-3 are closed by hypoxia (Hartness et al., 2001; Lewis et al., 2001). Thus, these channels are one of the candidates for oxygen and/or CO_2/H^+ sensor in chemosensory cells. In the isolated glomus cells of the carotid body, Buckler et al. (2000) found TASK-like current with electrophysiological method and expression of TASK-1 mRNA. On the other hand, TREK (TWIK-related K$^+$ channels), comprises three subunits, TREK-1, TREK-2 and TRAAK (Kim, 2003). These channels are regulated by polyunsaturated fatty acid, cellular volume, intracellular pH and general anesthetics. It has been suggested that they play an important role in potent neuroprotection (Lesage, 2003). Furthermore, Miller et al. (2003) reported that acute hypoxia occluded human TREK-1 expressed in the HEK293 cells under ischemic and/or acidic conditions. On the contrary, other reports demonstrated that TREK-1 was not oxygen sensitive (Buckler and Honore, 2005; Caley et al., 2005). To discuss the function of the tandem pore domain K$^+$ channels in the chemosensory organ, we reported that the immunoreactivities for TASK and TREK subfamilies in the carotid body (Yamamoto et al., 2002; Yamamoto and Taniguchi, 2004). Furthermore, no immunoreactivity for TRAAK was found in the paraganglion cells in the sympathetic ganglia (Yamamoto and Taniguchi, 2003). In the present study, therefore, we summarize the immunohistochemical localization of tandem pore domain K$^+$ channels in the rat paraganglion cells.

2. TASK AND TREK IN THE RAT CAROTID BODY

According to our immunohistochemical results, channels belonging to TASK and TREK family were distributed in the rat glomus cells in the carotid body (Fig. 1, Table 1). We found almost all glomus cells were immunoreactive for TASK-1, TASK-2, TASK-3, TREK-1, TREK-2 and TRAAK. These results may indicate that various tandem pore domain K^+ channels regulate K^+ current of the glomus cells. Because TASK-1 and TASK-3 are suppressed by lower oxygen levels (Kemp *et al.*, 2003), TASK-1 and TASK-3 in the glomus cells may participate in the sensory mechanism to hypoxia. Furthermore, since it has been reported that TASK channels

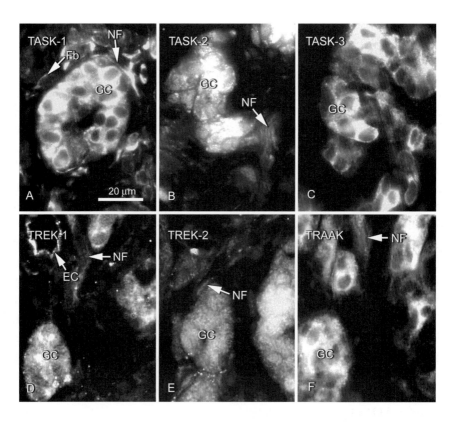

Figure 1. Immunoreactivity for TASK-1 (A), TASK-2 (B), TASK-3 (C), TREK-1 (D), TREK-2 (E) and TRAAK (F) in the rat carotid body. Clusters of the glomus cells (GC) are immunoreactive for these six tandem P domain K^+ channels. NF, nerve fibers; Fb, fibroblast; EC, vascular endothelial cells.

Table 1. Immunoreactivity for two pore K⁺ channels in the rat carotid body

	TASK-1	TASK-2	TASK-3	TREK-1	TREK-2	TRAAK
Type I cells	++	++	++	++	++	++
Type II cells	++	-	-	±	±	-
Nerve fibers	+	++	-	++	++	+
Vascular smooth muscle cells	++	+	++	-	-	-
Vascular endothelial cells	++	+	-	+	+	+

++, Intensely positive; +, weakly positive; ±, faintly positive; -, negative.

are closed by extracellular H^+ (Lesage, 2003), it is suggested that such channels also play a role in the acid sensing of glomus cells. On the other hand, Miller et al. (2004) reported that human TREK-1 is inhibited during alkalosis but activated during acidosis by hypoxia. The existence of TREK family in the glomus cells may enhance depolarization by hypoxia during respiratory and/or metabolic alkalosis. However, this supposition may be contradictory to that TREK-1 was not oxygen sensitive (Buckler and Honore, 2005; Caley et al., 2005). Further studies are needed to clarify the functions of TREK family in the glomus cells.

The nerve fibers in the carotid body were also immunoreactive for TASK-1, TASK-3, TREK-1, TREK-2 and TRAAK according to our results. It is reported that a part of discharges form carotid sinus nerve was independent of elevation of Ca^{2+} in the glomus cells (Roy et al., 2000). Thus, some nerve endings in the carotid body seem to be sensitive to hypoxia, although it is unknown whether they are also sensitive to CO_2/H^+ or not. TASK and TREK channels may contribute to chemosensory function of the nerve endings in the carotid body.

Furthermore, TASK and TREK channels were also observed in the vascular smooth muscle and endothelial cells in the carotid body as summarized in Table 1. These channels may modulate vascular constriction to regulate blood flow in the carotid body.

3. TANDEM PORE DOMAIN K⁺ CHANNELS IN THE PARAGANGLION CELLS IN THE AORTIC BODY, SUPERIOR LARYNGEAL NERVE AND SYMPATHETIC GANGLIA

We previously reported that TASK-3 immunoreactivity was found in the glomus or paraganglion cells in the carotid and aortic bodies and sympathetic ganglia (Yamamoto et al., 2003; Fig. 2A, B, D). Furthermore, TASK-3 immunoreactivity was also observed in the paraganglion cells within the superior laryngeal nerve (Fig. 2C) which are also excited by hypoxia (O'Leary et al., 2004). On the other hand, the paraganglion cells in the sympathetic ganglia may have oxygen sensing property, because paraganglion cells in the superior cervical ganglion respond to hypoxia with an increase in dopamine turnover (Dalmaz et al., 1990). However, immunoreactivities for TREK family are different in the sympathetic paraganglion cells from the glomus cells in the carotid and aortic bodies. In the sympathetic ganglia, TRAAK immunoreactivity was not observed in the paraganglion cells (Yamamoto et al., 2003; Fig 2E), while immunoreactivities for TREK-1 and TREK-2 were observed in the paraganglion cells in addition to the nerve cell bodies and nerve fibers (Fig. 2F). According to these findings, absence of TRAAK expression in paraganglion cells in the sympathetic ganglia may indicate that membrane

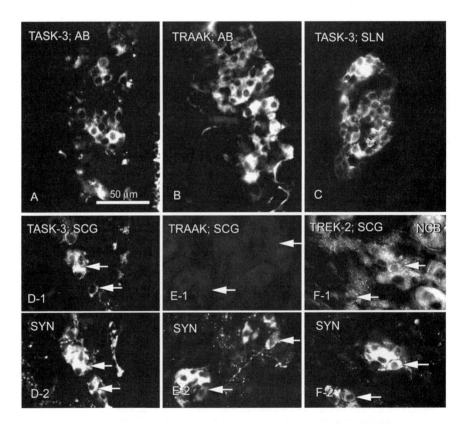

Figure 2. A, B, Immunoreactivity for TASK-3 (A) and TRAAK (B) in the rat aortic body. C, Immunoreactivity for TASK-3 in the cluster of paraganglion cells within the superior laryngeal nerve. D-F, Immunoreactivity for TASK-3 (D-1), TRAAK (E-1) and TREK-2 (F-1) in the paraganglion cells in the superior cervical ganglia. D-2, E-2 and F-2 shows the synaptophysin immunoreactivity in the same position of D-1, E-1 and F-1, respectively. Arrows indicate the paraganglion cells.

properties of sympathetic paraganglion cells are different from those of the glossopharyngeal/vagal paraganglia, namely, carotid and aortic bodies. Although the exact function of TRAAK channels in paraganglion cells is still unknown, it is possible that chemosensing cells might be classified into two types on the basis of TRAAK expression.

4. CONCLUSION

According to the immunohistochemical localization of TASK and TREK channels, tandem pore domain K^+ channels seem to be important for K^+ current modulation of glomus cells and/or afferent nerve endings in the rat carotid body. Especially, they play important roles on chemoreception for both hypoxia and hypercapnia. However, several problems remain to be unsolved at present as follows: exact function of TREK on chemosensing, the relationship between tandem pore domain K^+ channels and another oxygen sensing potassium channels, e.g., voltage gated K^+ channels Kv 3.4, Kv 4.1 and Kv 4.3, which are also expressed in the glomus cells of carotid body (Sanchez et al., 2002), and functional significance of the heterogeneous expression of TRAAK in the paraganglion cells. Further experiments are needed to reveal functional significance of the tandem pore domain K^+ channels in the glomus/paraganglion cells.

REFERENCES

Buckler, K.J., and Honore, E. The lipid-activated two-pore domain K^+ channel TREK-1 is resistant to hypoxia: implication for ischcaemic neuroprotection. J Physiol 2005; 562:213-222

Buckler, K.J., Williams, B.A., and Honore, E. An oxygen-, acid- and anaesthetic-sensitive TASK-like background potassium channel in rat arterial chemoreceptor cells. J Physiol 2000; 525: 135-142

Caley, A.J., Gruss, M., and Franks, N.P. The effects of hypoxia on the modulation of human TREK-1 potassium channels. J Physiol 2005; 562: 205-212

Dalmaz, Y., Borghini, N., Pequignot, J.M., and Peyrin, L. "Presence of chemosensitive SIF cells in the rat sympathetic ganglia: a biochemical, immunocytochemical and pharmacological study." In: *Chemoreceptors and Chemoreceptor Reflexes*, H. Acker, A. Trzebski, R.G. O'Regan eds. New York; Plenum Press, 1990; pp 393–399

Hartness, M.E., Lewis, A., Searle, G.J., O'Kelly, I., Peers, C., and Kemp, P.J. Combined antisense and pharmacological approaches implicate hTASK as an airway O_2 sensing K^+ channel. J Biol Chem 2001; 276: 26499-26508

Kemp, P.J., Searle, G.J., Hartness, M.E., Lewis, A., Miller P., Williams, S., Wooton, P., Adriaensen, D., and Peers, C. Acute oxygen sensing in cellular models: relevance to the physiology of pulmonary neuroepithelial and carotid bodies. Anat Rec 2003; 270A: 41-50.

Kim, D. Fatty acid-sensitive two-pore domain K^+ channels. Trend. Pharmacol Sci 2003; 24: 648-654

Lesage, F. Pharmacology of neuronal background potassium channels. Neuropharmacology 2003; 44: 1-7

Lewis, A., Hartness, M.E., Chapman, C.G., Fearon, I.M., Meadows, H.J., Peers, C., and Kemp, P.J. Recombinant hTASK1 is an O_2-sensitive K^+ channel. Biochem Biophys Res Commun 2001; 285: 1290-1294

Miller, P., Kemp, P.J., Lewis, A., Chapman, C.G., Meadows, H.J., and Peers, C. Acute hypoxia occludes hTREK-1 modulation: re-evaluation of the potential role of tamdem P domain K^+ channels in central neuroprotection. J Physiol 2003; 548: 31-37

Miller, P, and Kemp, P.J. Polymodal regulation of hTREK1 by pH, arachidonic acid, and hypoxia: physiological impact in acidosis and alkalosis. Am J Physiol 2004; 286: C272-C282

O'Leary, D.M., Murphy, A., Pickering, M., and Jones, J.F.X. Arterial chemoreceptors in the superior laryngeal nerve of the rat. Respir Physiol Neurobiol 2004; 141: 137-144

Patel, A.J., and Lazdunski, M. The 2P-domain K^+ channels: role in apoptosis and tumorigenesis. Pflügers Arch 2004; 448: 261-273

Roy, A., Razanov, C., Mokashi, A., and Lahiri, S. PO_2-PCO_2 stimulus interaction in $[Ca^{2+}]_i$ and CSN activity in the adult rat carotid body, Respir Physiol 2000; 122: 15-26

Sanchez, D., Lopez-Lopez, J.R., Perez-Garcia, M.T., Sanz-Alfayate, G., Obeso, A., Ganfornina, M.D., and Gonzalez, C. Molecular identification of $KV\alpha$ subunits that contribute to the oxygen sensitive K^+ current of chemoreceptor cells of the rabbit carotid body. J Physiol 2002; 542: 369-382

Yamamoto, Y., Kummer, W., Atoji, Y., and Suzuki, Y. TASK-1, TASK-2, TASK-3 and TRAAK immunoreactivities in the rat carotid body. Brain Res 2002; 950:304-307

Yamamoto, Y., and Taniguchi, K. Heterogeneous expression of TASK-3 and TRAAK in rat paraganglionic cells. Histochem Cell Biol 2003; 120: 335-339

Yamamoto, Y., and Taniguchi, K. Expression of tandem P domain K^+ channels, TREK-1, TREK-2 and TRAAK, in the rat carotid body. Anat Sci Int 2004; 79 Suppl: 372

Neuroglobin, a New Oxygen Binding Protein is Present in the Carotid Body and Increases after Chronic Intermittent Hypoxia

C. DI GIULIO, G. BIANCHI, M. CACCHIO, [1]L. ARTESE, [1]M. PICCIRILLI, V. VERRATTI, R. VALERIO, AND [2]R. ITURRIAGA

Department of Biomedical Sciences, [1]Department. of Odontostomastology. "G. d'Annunzio" University, Chieti, Italy and [2]Laboratory of Neurobiology, P. Catholic University of Chile, Santiago, Chile

1. INTRODUCTION

Neuroglobin (Ngb), a 151-amino-acid protein with a predicted molecular mass of 17 kD was recently identified as a member of the vertebrate globin family (Burmester and Hankeln, 2004; Mammen et al., 2002). Ngb, is predominantly expressed in nerve cells, particularly in the brain and in the retina (Burmester et al., 2000; Zhu et al., 2002), but is also expressed in other tissues (Burmester and Hankeln, 2004). The protein has three-on-three α-helical globin fold and are endowed with a hexa-coordinate heme-Fe atoms, which displays O_2 affinities and binds CO (Burmester & Hankeln, 2004).The physiological role of Ngb is not well understood, but it has been proposed that Ngb participates in several processes such as oxygen transport, oxygen storage, and NO detoxification (Burmester and Hankeln, 2004). Ngb as well as hemoglobin is a respiratory protein that reversibly binds gaseous ligands (NO and O_2) by means of the Fe-containing porphyrin ring. Ngb is concentrated in neuronal cellular regions that contain mitochondria, and its distribution is correlated with oxygen consumption rates (Pesce et al., 2003).

It has been proposed that Ngb enhances oxygen supply to neural components and may contribute to neuronal survival, because the level of Ngb is augmented during ischemia and hypoxia (Sun et al., 2001). Since the carotid body (CB) is the main oxygen chemoreceptor in the arterial blood and its high oxygen consumption, we tested if Ngb is present in the CB of normoxic rats and if chronic intermittent hypoxic may increase its level.

2. METHODS

Two groups of six male Wistar rats weighting 200-250 g were used. One group was maintained at room air (FIO_2 = 21 %) and served as controls. The other group was kept for 12 days in a Plexiglas chamber in chronic intermittent hypoxia: FIO_2 10-11% for 12 hs, followed by 12 hs of normoxia. The chamber temperature and the PCO_2 level were kept between physiological ranges. The rats were anaesthetized with Nembutal (40 mg/kg, ip) and the CBs were dissected. The CB tissue was immersed overnight in ice cold 4 % paraformaldehyde in 0.1 M phosphate buffered saline (PBS). Tissues were then rinsed in 15 % sucrose PBS for 1 hr, and stored at 4 °C in 30 % sucrose PBS for 2 hrs. Histological section of 10 µm (n = 6 for each sample) were serially cut with a cryomicrotome (Reichert-Jung Frigocut 2800), thaw-mounted in microscope slides, fixed by immersion in acetone at 4°C for 5 min, and air-dried. Slides were stored at 4°C until use.

The presence of Ngb in the CB tissue was detected by immunohistochemistry using a polyclonal antibody (E-16, Santa Cruz Biotech Antibodies). Slides were preincubated in PBS for 5 min and then with an antibody for Ngb of goat origin (diluted 1:100 in PBS) for 30 min. at 37 °C. Slices were then washed twice in PBS for 5 min, and in Tris-HCl buffer, pH 7.6 for 10 min. A second antibody, rabbit antigoat IgG, was added for 10 min and slides were washed with PBS. The Ngb immunoreactivity (NGB-ir) staining was analyzed using the software package Image Pro Plus 4.5 for the densitometric analysis. Five random fields were chosen in each CB. The Ngb immunoreactivity was measured in optic integrated units.

3. RESULTS

NGB is present in the normoxic CB, presumable in the glomus cells. The intensity of the Ngb immunoreactivity increased significantly in the chronic intermittent hypoxic CBs. Fig. 1 shows the Ngb immunohistochemistry distribution in a control CB (Fig. 1 A, B) and in a hypoxic CB (Fig. 1 C, D).

The analysis of the Ngb immunoreactivity in the CB shows that the integrated optical intensity of Ngb increased significantly from $0.29 \pm 0.067\%$ in the control to $0.62 \pm 0.086\%$ in the hypoxic CBs ($P < 0.01$, Fig. 2).

Neuroglobin, a New Oxygen Binding Protein 17

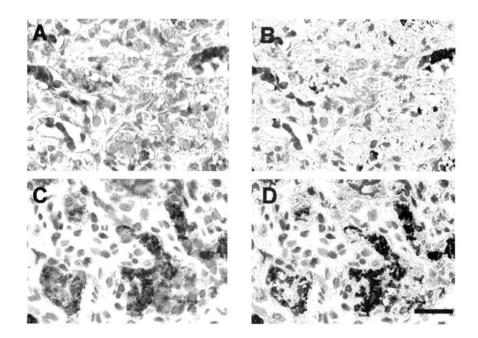

Figure 1. Immunohistochemical analysis of NGB in the rat CB, A and B, control CB. C and D, hypoxic CB. B and D, after having distinguished the NGB expression by the image processing (Magnification: 400 x).

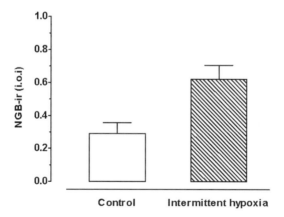

Figure 2. Integrated optical intensity of Ngb immunoreactivity in 6 control and 6 hypoxic CBs. The difference between both groups was statistically significant ($P < 0.01$).

4. DISCUSSION

To our knowledge, this is the first study showing that Ngb is expressed in the CB tissue under normoxic condition, and its expression increased during hypoxia. Regardless the small size, the CB has the highest blood flow reported for any organ and high oxygen consumption (Gonzalez et al., 1994). Therefore, it is likely that Ngb may work in the CB as a "neuronal globin protein", providing oxygen to the respiratory chain of CB cells. In addition, Ngb also may act as a sensor to detect cellular oxygen concentration, with an affinity comparable to myoglobin (P_{50} of 1-2 Torr; Burmester & Hankeln, 2004).

Although, Ngb may storage and transport oxygen, it has been proposed that Ngb participates in several processes. Hypoxia upregulates the expression of Ngb in neurons, suggesting that Ngb protects neurons against hypoxic damage. In fact, Sun et al. (2001) found that hypoxia stimulates Ngb transcription in cultured neurons, and antisense inhibition of Ngb expression increases hypoxic neuronal injury, whereas overexpression of Ngb provides resistance to hypoxia. These findings are consistent with a role for Ngb in promoting neuronal survival after hypoxic insults. Ngb may also be involved in the detoxification of nitric oxide and other reactive oxygen species (ROS) that probably are generated under hypoxic conditions. Thus, the presence of Ngb as "a respiratory protein" in CB may have important physiological effects such as NO and ROS detoxication.

A prominent feature of cell adaptation to hypoxia is the increased expression of hypoxic-inducible proteins (Bunn & Poyton, 1996; Lahiri et al., 2002). In the CB, it is well known that hypoxia stimulates the production of hypoxia-inducible factor-1, nitric oxide synthase, and tyrosine hydroxylase (Di Giulio et al., 2003). All of these proteins may exert protective effects through diverse mechanisms, but their hypoxic-responsiveness depends on O_2-binding proteins such as Ngb (Dewilde et al., 2001), which can sense hypoxia and trigger appropriate cell adaptative responses (Burmester & Hankeln, 2004).

In summary, in addition to an oxygen-storage role for Ngb in the CB, we cannot excluded that Ngb may act as an oxygen sensor to detect tissue oxygen levels, or has a protective role for neuronal and glomus cell survive. Further studies are needed to establish the role of Ngb in the CB.

ACKNOWLEDGEMENTS

Work partially supported by FONDECYT 1030330 (RI)

REFERENCES

Bunn HF & Poyton RO. (1996). Oxygen sensing and molecular adaptation to hypoxia. Physiol Rev 76: 839-885.
Burmester T & Hankeln T. (2004). Neuroglobin: a respiratory protein of the nervous system. News Physiol Sci 19: 110-113.

Burmester B, Weich S, Reinhardt T, & Hankeln T. (2000) A vertebrate globin expressed in the brain. Nature 407, 520–523.

Dewilde S, Kiger L, Burmester T, Hankeln T, Baudin-Creuza V, Aerts T, Marden MC, Caubergs R, & Moens L. (2001). Biochemical characterization and ligand-binding properties of neuroglobin, a novel member of the globin family. J Biol Chem 276: 38949–38955.

Di Giulio C, Bianchi G, Cacchio M, Macri MA, Ferrero G, Rapino C, Verratti V, Piccirilli M, & Artese L. (2003) Carotid body HIF-1α, VEGF and NOS expression during aging and hypoxia. Adv Exp Med Bio. 536: 603-610.

Gonzalez C, Almaraz,L, Obeso,A, & Rigual R. (1994) Carotid body chemoreceptors: from natural stimuli to sensory discharges. Physiol Rev 74, 829-898.

Lahiri S, Di Giulio C, & Roy A. (2002). Lesson from chronic intermittent and sustained hypoxia. Respir Physiol Neurobiol.130: 223-233.

Mammen PPA., Shelton JM., Goetsch SC, Williams SC, Richardson JA, Garry MG, & Garry DJ. (2002). Neuroglobin, a novel member of the globin family, is expressed in focal regions of brain. J Histochem Cytochem 50: 1591-1598.

Pesce A., Dewilde S., Nardini M., Moens L., Ascenzi P., Hankeln T., Burmester T., Bolognesi M. (2002) Human brain neuroglobin structure reveals a distint mode of controlling oxygen affinity. Structure 11: 1087-1095.

Prabhakar NR, Fields RD, Baker T, & Fletcher EC. (2001). Intermittent hypoxia: cell to system. Am Physiol Lung Cell Mol Physiol 281: 524-528.

Sun Y, Jin K, Mao XO, Zhu Y, & Greenberg DA. (2001). Neuroglobin is up-regulated by a protects neurons from hypoxic-ischemic injury. Proc Natl Acad Sci. 98: 15306-15311.

Zhu Y, Sun Y, Jin K, & Greenberg DA. (2002) Hemin induces neuroglobin expression in neural cells. Blood. 100: 2494–2498.

Hypoxia-Inducible Factor (HIF)-1α and Endothelin-1 Expression in the Rat Carotid Body during Intermittent Hypoxia

SIU-YIN LAM, GEORGE L. TIPOE[1], EMILY C. LIONG[1], AND MAN-LUNG FUNG

Departments of Physiology and [1]Anatomy, The University of Hong Kong, Faculty of Medicine Building, 21 Sassoon Road, Pokfulam, Hong Kong, China

1. INTRODUCTION

Physiological responses to hypoxia involve changes in gene expression that are mediated by the transcriptional activator HIF-1. HIF-1 is a heterodimeric transcription factor consisting of two subunits, HIF-1α and HIF-1β (Semenza, 2000; Wang et al., 1995). The expression of HIF-1α protein is closely regulated by oxygen tension in the cell, whereas HIF-1β expression is constitutive and independent of oxygen levels (Kallio et al., 1999; Semenza, 2000; Wang et al., 1995). It has been shown that HIF-1α plays a physiological role in chronic hypoxia (CH). HIF-1α serves as a key controller for the transcriptional regulation of the expression of a spectrum of oxygen-regulated genes, such as erythropoietin, vascular endothelial growth factor (VEGF) and VEGF receptors, for the cellular response to hypoxia in tissues including the carotid body (CB) (Fung, 2003; Glaus et al., 2004; Semenza, 2000; Tipoe and Fung, 2003).

The production of endothelin-1 (ET-1) by pulmonary endothelium is stimulated by hypoxia (Yoshimoto et al., 1991) and the expression of ET-1 is regulated by HIF-1 (Semenza, 2000). In the CB, ET-1 stimulates the excitability of the chemoreceptors (Chen et al., 2000; Chen et al., 2002) and induces mitosis of the chemosensitive type-I cells (Paciga et al., 1999). Furthermore, studies have shown that plasma ET-1 is increased at high altitude and the level is inversely proportional to arterial oxygen saturation (Goerre et al., 1995; Cruden et al., 1999).

Recently, it has been shown that intermittent hypoxia (IH) facilitates hypoxic chemosensitivity in rat CB and augments hypoxic ventilatory chemoreflex (Peng et al., 2004). IH also facilitates the hypoxia-evoked neurotransmitter release from the CB as well as PC12 cells (Kim et al., 2004). The CB plays important roles in pathophysiological changes in hypoxia (Prabhakar and Peng, 2004) and the expression of HIF-1 and ET-1 may be involved in the molecular mechanism for the changes of the organ in IH stress. The aim of this study was to examine the time-course of CB expression of HIF-1α and ET-1 in CH and IH rats in order to understand the molecular regulation that was significant to the ventilatory acclimatization of the hypoxic response at high altitude as well as the development of pathophysiological events during chronic hypoxemia in disease.

2. METHODS

The experimental protocol for this study was approved by the Committee on the Use of Live Animals in Teaching and Research of The University of Hong Kong. Sprague-Dawley rats aged 28 days were randomly divided into normoxic, CH and IH group. While the control was maintained in room air, both CH and IH rats were kept in acrylic chambers for normobaric hypoxia and had free access to water and chow. The oxygen fraction inside the chamber was kept at $10 \pm 0.5\%$, 24 hours per day for the CH group and was cyclic from 21 to $5 \pm 0.5\%$ per minute, 8 hours per day for the IH group. The desired oxygen level was established by a mixture of room air and nitrogen that was regulated and monitored by an oxygen analyzer (Vacumetrics Inc., CA, USA). Carbon dioxide was absorbed by soda lime granules and excess humidity was removed by a desiccator. The chamber was opened twice a week for an hour to clean the cages and replenish food and water. The rats were exposed to hypoxia for 3, 7, 14 and 28 days and were immediately used in experiments after taken out of the chamber.

Following deep anesthesia with halothane, the rat was decapitated and the carotid bifurcation was excised rapidly. The CB was carefully dissected free from the bifurcation and was fixed in neutral buffered formalin for 72 hour. Tissues were processed routinely for histology and embedded in paraffin blocks. Serial sections of 5 μm thickness were cut and mounted on silanized slides (DAKO, Denmark). Sections were kept in the oven overnight at 56°C. Consequently, sections were dewaxed with xylene and rehydrated with a series of decreasing grade of ethanol solutions. Sections were immunostained with antiserum to the following proteins: HIF-1α (mouse monoclonal IgG antibody, 1:25 dilution, Cat # 400080, Calbiochem, CA, USA); endothelin-1 (mouse monoclonal IgG antibody, 1:100 dilution, Cat # CP44, Oncogene, CA, USA), using LSAB kit (K0690, DAKO) and were deparaffinized and rehydrated. Sections for HIF-1α were immersed in antigen retrieval solution (0.1 M citric acid buffer, pH 6.0) for 10 min at 98°C. To block endogenous peroxidase activity, the sections were immersed in 3% hydrogen peroxide for 5 min at room temperature. All sections were immersed in a solution containing 0.01% trypsin and $CaCl_2$ for 5 min at room temperature. Sections were pre-incubated with 20% normal serum for 2 hour to reduced non-specific binding for the antiserum. Then sections were incubated with the corresponding primary antibodies in 0.05 M Tris-HCl buffer, respectively, containing 2% bovine serum albumin overnight at 4°C. Sections were washed three times in PBS, and then incubated with biotinylated link agent and streptavidin peroxidase for 30 min at room temperature. Finally, sections were washed and the peroxidase was visualized by immersing in 0.05% diaminobenzidine (DAB) containing 0.03% hydrogen peroxide in Tris-HCl buffer (pH 7.5) for 3-5 min. Sections were rinsed in distilled water and counterstained mildly with hematoxylin. Positive staining was indicated by a brown color. Control sections were incubated with either normal mouse or rabbit IgG and stained uniformly negative (not shown in the figures).

The immunoreactivities of HIF-1α and ET-1 were measured using the Leica QWIN Imager Analyzer (Cambridge, UK). Immunostained sections were captured with a CCD JVC camera using a Zeiss Axiophot microscope at 40X objective. The luminance incident light passing through each section was calibrated using the set-up menu where the grey pixel values were set to 0 and 1.00. Once the setup was done, five fields per section from one CB of one animal were measured. The percent area of positive stain for ET-1 protein was measured by detecting the positive brown cytoplasmic stain divided by the sum areas of the reference field. The percentage of HIF-1α positive was calculated by the number of positive HIF-1α nuclei divided by the total number of nuclei in the reference field. A total of 20 fields for 4 CBs from four different animals at each time-point namely day 3, 7, 14 and 28 were determined. The mean value of the 20 fields was calculated to represent each time-point.

GraphPad Prism® software (GraphPad Software, Inc., San Diego, USA) was used to analyze the data. A non-parametric Mann-Whitney U-test was used to compare differences between time-points. A p-value < 0.05 was considered statistically significant.

3. RESULTS

Immunohistochemical studies revealed that both the HIF-1α and ET-1 proteins were positively stained in most of cells throughout the CBs of rats exposed to CH and IH treatment (n=4). The HIF-1α immunoreactivity (IR) was mainly found in the nuclei, whereas positive staining of ET-1 was generally found in the cytoplasm.

In the IH group, the IR of HIF-1α increased in 3-day IH group (Fig. 1). The percent of cells with positive nuclear staining reached a peak level at day 7 but then returned to normoxic levels by day 28 (Figs. 1A and B). In the CH group, the HIF-1α expression elevated initially at day 3 and gradually reached a plateau in 2-4 weeks (Figs. 1A and B).

The positive cytoplasmic IR staining of ET-1 expression was markedly increased in the 3-day groups of both CH and IH rats, although the elevation was less in the IH groups than that in the CH group (Fig. 2). By day 14, the ET-1 expression reduced to a sustained level above the control in the CH group but was completely recovered to the normoxic levels in the IH group (Fig. 2B). This indicates that the ET-1 expression play an important role in the enhancement of CB excitability during an early time-course of hypoxia.

Furthermore, the immunostaining was not observed in CB of the corresponding control sections incubated with normal serum instead of the primary antibodies (Data not shown).

4. DISCUSSION

Results of the HIF-1α expression in both CH and IH are in agreement with the hypothesis that HIF-1α plays an active role in the physiological response to hypoxia. CH increased the nuclear expression of HIF-1α protein in the CB cells. This finding is consistent with previous studies showing that protein expression of HIF-1α and DNA-binding activity of HIF-1 in the nucleus of cultured cells are increased by hypoxia (Jiang et al., 1996). Also, mRNA and protein

A.

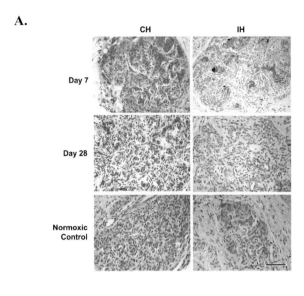

B.

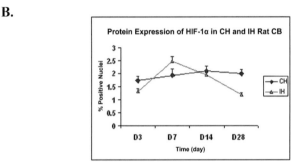

Figure 1. A: Immunohistochemical localization of HIF-1α in 7- and 28-day treatment of CH and IH and their corresponding normoxic group in rat CB. Bar = 40 μm. B: Protein expression of HIF-1α in CH and IH rat CB. Data are normalized with their corresponding normoxic group.

expression of HIF-1α but not HIF-1β is elevated in the brain (Chavez et al., 2000) and lung (Palmer et al., 1998) of rats exposed to CH. In a similar manner, IH increased the nuclear expression of HIF-1α protein in the CB glomus cells as well. Previous report showed that IH activates tyrosine hydroxylase (TH), the rate-limiting step in catecholamine biosynthesis, activity in PC12 cells due to increased serine phosphorylation, especially that of Ser-40 mediated by protein kinase A and Ca^{2+}/calmodulin-dependent protein kinase and the subsequent removal of endogenous product inhibition of TH via disruption of catecholamine binding (Kumar et al., 2003). It has also been shown in humans that IH associated with recurrent apneas leads to the development of pathophysiological conditions such as hypertension (Fletcher, 2001), whereas CH, as it occurs at

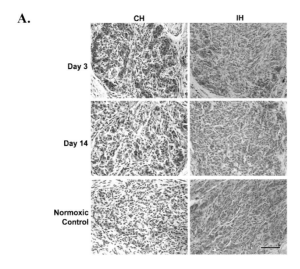

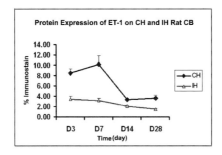

Figure 2. A: Immunohistochemical localization of ET-1 in 3- and 14-day treatment of CH and IH and their corresponding normoxic group in rat CB. Bar = 40 μm. B: Protein expression of ET-1 in CH and IH rat CB. Data are normalized with their corresponding normoxic group.

high-altitude dwelling, does not result in such adverse effects. The difference in the pathological consequences of CH and IH suggests distinct mechanisms underlying these two patterns of hypoxia. The differential regulation of HIF-1 in CH and IH needs further elucidation.

It is known that CH induces hypertrophy and hyperplasia in the CB. ET-1 is expressed in the CB (He et al., 1996; Chen et al., 2000; Chen et al., 2002) and it stimulates the proliferation of cultured glomus cells (Paciga et al., 1999). The intracellular calcium response to ET-1 in the glomus cells may play a role in the proliferation and structural changes of the chemoreceptors during CH. Accordingly, the fact that a lower amount of ET-1 expression in the IH than that of the CH group may account for the insignificant enlargement of CB in IH, due to less ET-1 mitogenic effect on the glomus cells. In addition, the elevation of intracellular calcium is an essential step for vesicular release of neurotransmitters for the chemoreception. Our results show that ET-1 expression is elevated in both CH and IH treatment, suggesting ET-1 can increase the excitability of the chemoreceptor to hypoxia. Also, the increase in the ET-1 expression is more significant in an early time-course of hypoxia indicating an important role of ET-1 in the enhancement of CB excitability in both CH and IH groups. Thus, results are in agreement with the hypothesis that ET-1 plays an active role in the physiological adaptation of the CB in response to hypoxia although ET-1 may have a prominent role in the hypertrophy and hyperplasia in the CB of CH, but not of the IH.

In conclusion, the increased HIF-1α expression could elevate HIF-1 activity that increases the ET-1 expression in the CBs, despite significant differences in the temporal patterns of the HIF-1α and ET-1 expression between CH and IH, which may account for some of the morphological and functional discrepancies.

ACKNOWLEDGEMENTS

We thank Mr. W.B. Wong and Ms. K.M. Leung for their technical assistance. This work was supported by research grant (7223/02M) from the Research Grants Council, Hong Kong.

REFERENCES

Chavez JC, Agani F, Pichiule P, and LaManna JC., 2000, Expression of hypoxia-inducible factor-1alpha in the brain of rats during chronic hypoxia. J Appl Physiol. 89(5):1937-42.
Chen J, He L, Dinger B, and Fidone S., 2000, Cellular mechanisms involved in rabbit carotid body excitation elicited by endothelin peptides. Respir Physiol. 121(1):13-23.
Chen Y, Tipoe GL, Liong E, Leung S, Lam SY, Iwase R, Tjong YW, and Fung ML., 2002, Chronic hypoxia enhances endothelin-1-induced intracellular calcium elevation in rat carotid body chemoreceptors and up-regulates ETA receptor expression. Pflugers Arch. 443(4):565-73.
Cruden NL, Newby DE, Ross JA, Johnston NR, and Webb DJ., 1999, Effect of cold exposure, exercise and high altitude on plasma endothelin-1 and endothelial cell markers in man. Scott Med J. 44(5):143-6.
Fletcher EC., 2001, Invited review: Physiological consequences of intermittent hypoxia: systemic blood pressure. J Appl Physiol. 90(4):1600-5.
Fung ML., 2003, Hypoxia-inducible factor-1: a molecular hint of physiological changes in the carotid body during long-term hypoxemia? Curr Drug Targets Cardiovasc Haematol Disord. 3(3):254-9.
Glaus TM, Grenacher B, Koch D, Reiner B, and Gassmann M., 2004, High altitude training of dogs results in elevated erythropoietin and endothelin-1 serum levels. Comp Biochem Physiol A Mol Integr Physiol. 138(3):355-61.
Goerre S, Wenk M, Bartsch P, Luscher TF, Niroomand F, Hohenhaus E, Oelz O, and Reinhart WH., 1995, Endothelin-1 in pulmonary hypertension associated with high-altitude exposure. Circulation. 15;91(2):359-64.
He L, Chen J, Dinger B, Stensaas L, and Fidone S., 1996, Endothelin modulates chemoreceptor cell function in mammalian carotid body. Adv Exp Med Biol. 410:305-11.
Jiang BH, Semenza GL, Bauer C, and Marti HH., 1996, Hypoxia-inducible factor 1 levels vary exponentially over a physiologically relevant range of O2 tension. Am J Physiol. 271(4 Pt 1):C1172-80.
Kallio PJ, Wilson WJ, O'Brien S, Makino Y, Poellinger L., 1999, Regulation of the hypoxia-inducible transcription factor 1alpha by the ubiquitin-proteasome pathway. J Biol Chem. 274(10):6519-25.

Kim DK, Natarajan N, Prabhakar NR, and Kumar GK., 2004, Facilitation of dopamine and acetylcholine release by intermittent hypoxia in PC12 cells: involvement of calcium and reactive oxygen species. J Appl Physiol. 96(3):1206-15.

Kumar GK, Kim DK, Lee MS, Ramachandran R, and Prabhakar NR., 2003, Activation of tyrosine hydroxylase by intermittent hypoxia: involvement of serine phosphorylation. J Appl Physiol. 95(2):536-44.

Paciga M, Vollmer C, and Nurse C., 1999, Role of ET-1 in hypoxia-induced mitosis of cultured rat carotid body chemoreceptors. Neuroreport. 10(18):3739-44.

Palmer LA, Semenza GL, Stoler MH, and Johns RA., 1998, Hypoxia induces type II NOS gene expression in pulmonary artery endothelial cells via HIF-1. Am J Physiol. 274(2 Pt 1):L212-9.

Peng YJ, Rennison J, and Prabhakar NR., 2004, Intermittent hypoxia augments carotid body and ventilatory response to hypoxia in neonatal rat pups. J Appl Physiol. 97(5):2020-5.

Prabhakar NR, and Peng YJ., 2004, Peripheral chemoreceptors in health and disease. J Appl Physiol. 96(1):359-66.

Semenza GL., 2000, HIF-1: mediator of physiological and pathophysiological responses to hypoxia. J Appl Physiol. 88(4):1474-80.

Tipoe GL, and Fung ML., 2003, Expression of HIF-1alpha, VEGF and VEGF receptors in the carotid body of chronically hypoxic rat. Respir Physiol Neurobiol. 138(2-3):143-54.

Wang GL, Jiang BH, Rue EA, and Semenza GL., 1995, Hypoxia-inducible factor 1 is a basic helix-loop-helix-PAS heterodimer regulated by cellular O_2 tension. Proc Natl Acad Sci USA. 92(12):5510-4.

Yoshimoto S, Ishizaki Y, Sasaki T, and Murota S., 1991, Effect of carbon dioxide and oxygen on endothelin production by cultured porcine cerebral endothelial cells. Stroke. 22(3):378-83.

Expression of HIF-2α and HIF-3α in the Rat Carotid Body in Chronic Hypoxia

SIU-YIN LAM, EMILY C. LIONG[1], GEORGE L. TIPOE[1], AND MAN-LUNG FUNG

Departments of Physiology and [1]Anatomy, The University of Hong Kong, Faculty of Medicine Building, 21 Sassoon Road, Pokfulam, Hong Kong, China

1. INTRODUCTION

Hypoxia is a crucial physiological stimulus in development and plays a key role in the pathophysiology of cancer, stroke, pulmonary disease, and other major causes of mortality (Iyer et al., 1998). Responses to changes in oxygen concentrations are primarily regulated by hypoxia inducible factors (HIFs). HIFs are heterodimeric transcription factors that regulate a number of adaptive responses to low oxygen tension. They are composed of oxygen-regulated α- and a constitutive non oxygen-regulated β- subunits and are belonged to the basic helix-loop-helix-PAS (bHLH-PAS) superfamily (Bruick, 2003). In mammals, three genes have been shown to encode HIF-α subunits namely HIF-1α, -2α and -3α. The HIF-1α protein is more widely expressed, while its homologs, HIF-2α/Endothelial PAS domain protein (EPAS-1) (Tian et al., 1997) and HIF-3α (Gu et al., 1998) are tissue and developmental specific in their expression. HIF-1α is expressed in the brain, heart, lung (Jain et al., 1998) and also in the carotid body (CB) (Baby et al., 2003; Tipoe and Fung, 2003). Whereas HIF-2α is expressed in the endothelial cells of various tissues, such as brain, heart, and liver, and the mRNA is also observed in alveolar epithelial cells in the lung (Ema et al., 1997). The EPAS-1 expression in mice embryo was induced by hypoxia for proper cardiac function (Tian et al., 1998). Furthermore, all the HIF-α subunits have been found in the kidney where diverse functions of the three had been shown. HIF-1α and -2α activate the expression of the HIF-mediated gene such as erythropoietin (EPO), whereas HIF-3α is likely an inhibitor of EPO gene transcription (Hara et al., 2001; Jelkmann, 2004).

HIF activates transcription of genes that increase systemic oxygen delivery or provide cellular metabolic adaptation under conditions of hypoxia. Abnormal HIF expression is related to numerous diseases of the vascular system, including heart disease, cancer and chronic obstructive pulmonary disease (Covello and Simon, 2004). The expression of HIF-1α subunit in the rat CB plays an essential role in the transcriptional regulation of the structural remodeling and functional modulation of the organ in chronic hypoxia (CH) (Fung, 2003). However, less is known about the expression and function of HIF-2α and HIF-3α in the rat CB. The aim of the present study was to examine the mRNA and protein expression of HIF-2α and -3α in the CB of rats in normoxia and in CH breathing 10% O_2 in isobaric chamber for up to 4 weeks.

2. METHODS

The experimental protocol for this study was approved by the Committee on the Use of Live Animals in Teaching and Research of The University of Hong Kong. Sprague-Dawley rats aged 28 days were randomly divided into normoxic and CH groups (n=4 for each group in each set of experiment). While the control was maintained in room air, CH rats were kept in a 300-litre acrylic chamber for normobaric hypoxia in the same room and had free access to water and chow. The oxygen fraction inside the chamber was kept at $10 \pm 0.5\%$, 24 hour per day. The desired oxygen content was established by a mixture of room air and nitrogen that was regulated and monitored by an oxygen analyzer (Vacumetrics Inc., CA, USA). Carbon dioxide was absorbed by soda lime granules and excess humidity was removed by a desiccator. The chamber was opened twice a week for an hour to clean the cages and replenish food and water. The rats were exposed to hypoxia for 7 days and were immediately used in experiments after taken out of the chamber.

2.1 Immunohistochemistry

Following deep anesthesia with halothane, the rat was decapitated and the carotid bifurcation was excised rapidly. The CB was carefully dissected free from the bifurcation and was fixed in neutral buffered formalin for 72 hour. Tissues were processed routinely for histology and embedded in paraffin blocks. Serial sections of 5 μm thickness were cut and mounted on silanized slides (DAKO, Denmark). Sections were kept in the oven overnight at 56°C. Consequently, sections were dewaxed with xylene and rehydrated with a series of decreasing grade of ethanol solutions. Sections were immunostained with antiserum to the following proteins: EPAS-1 (goat polyclonal antibody, 1:200 dilution, Cat # sc-8712, Santa Cruz, CA, USA); HIF-3α (goat polyclonal antibody, 1:250 dilution, Cat # sc-8718, Santa Cruz, CA, USA), using LSAB kit (K0690, DAKO) and were deparaffinized and rehydrated. Sections were immersed in antigen retrieval solution (0.1 M citric acid buffer, pH 6.0) for 10 min at 98°C. To block endogenous peroxidase activity, the sections were immersed in 3% hydrogen peroxide for 5 min at room temperature. Sections were pre-incubated with 20% normal serum for 2 hour to reduced non-specific binding for the antiserum. Then sections were incubated with the corresponding primary antibodies in 0.05 M Tris-HCl buffer, respectively, containing 2% bovine serum albumin overnight at 4°C. Sections were washed three times in PBS, and then incubated with biotinylated link agent and streptavidin peroxidase for 30 min at room temperature. Finally, sections were washed and the peroxidase was visualized by immersing in 0.05% diaminobenzidine (DAB) containing 0.03% hydrogen peroxide in Tris-HCl buffer (pH 7.5) for 3-5 min. Sections were rinsed in distilled water and counterstained mildly with hematoxylin. Positive staining was indicated by a brown color. Control sections were incubated with either normal mouse or rabbit IgG and stained uniformly negative.

2.2 Double Staining Procedures

DAKO Envision® Doublestain System (K-1395) and two sets of primary antibody were used. The first set was directed against EPAS-1 (goat polyclonal antibody, 1:100 dilution, Cat # sc-8712, Santa Cruz, CA, USA) and tyrosine hydroxylase (TH) (rabbit IgG antibody, 1:100 dilution, Cat # AB151, Chemicon International Inc., CA, USA). The second set of primary antibody was directed against HIF-3α (goat polyclonal antibody, 1:250 dilution, Cat # sc-8718, Santa Cruz, CA, USA) and TH (rabbit IgG antibody, 1:100 dilution, Cat # AB151, Chemicon International Inc., CA, USA). After incubating with the corresponding first primary antibody (HIF-2α and HIF-3α respectively), sections were incubated with peroxidase labeled polymer for 30 min at 37°C. The peroxidase was visualized with substrate DAB chromogen for 5-10 min. Sections were then incubated in a Doublestain Block solution (from DAKO Doublestain kit) which served to remove any potential cross-reactivity between reactions along with blocking endogenous alkaline phosphatase (AP) that may be present. The second primary antibody (TH) was then incubated for 1 h at 37°C. The AP was visualized by Fast Red solution for 5 min. Control sections were incubated with either normal goat or rabbit IgG and stained uniformly negative.

2.3 Reverse Transcriptase Polymerase Chain Reaction (RT-PCR)

Four CBs were dissected out from the bifurcation of carotid arteries. The CBs were then pooled together for the isolation of total RNA and RT-PCR was performed on both HIF-2α and HIF-3α genes. Isolated RNA (5 μg) was subjected to first strand cDNA synthesis using random hexamer primers and Superscript II transcriptase (GIBCO, USA) in a final volume of 20 μl. After incubation at 42°C for 1 hour, the reaction mixture was treated with RNase H before proceeding PCR analysis. The final mixture (2 μl) was directly used for PCR amplification. Messenger RNAs (mRNAs) of HIF-2α, HIF-3α and β-actin were detected with primers with the following sequences: HIF-2α (sense: CCC-CAG-GGG-ATG-CTA-TTA-TT; antisense: GGC-GAA-GAG-CTT-ATA-GAT-TA); HIF-3α (sense: AGA-GAA-CGG-AGT-GGT-GCT-GT; antisense: ATC-AGC-CGG-AAG-AGG-ACT-TT). All RNA was shown to be free of DNA contamination by RT-PCR without addition of reverse transcriptase. The PCR conditions for: (1) HIF-2α was 35 cycles of denaturing, 95°C, 30 s; annealing, 60°C, 1 min; elongating, 72°C, 30 s; (2) HIF-3α was 35 cycles of denaturing, 95°C, 30 s; annealing, 60°C, 1 min; elongating, 72°C, 1 min. The amplified mixture was finally separated on 2% agarose gel electrophoresis and the amplified DNA bands were detected using ethidium bromide staining.

GraphPad Prism® software (GraphPad Software, Inc., San Diego, USA) was used to analyze the data. A non-parametric Mann-Whitney U-test was used to compare differences between time-points. A p-value < 0.05 was considered statistically significant.

3. RESULTS

To determine the localization of protein expression of HIF-2α and HIF-3α in the rat CB, immunohistochemical studies using primary monoclonal antibody specific to HIF-2α (EPAS-1) and HIF-3α were performed. In general, a strongly positive immunostaining was observed in majority of cells throughout the CBs of rats exposed to CH and the staining was found in both the nuclei and cytoplasm (Figs. 1A and C). Double-labeling studies showed that the staining was colocalized with TH (Figs. 2A and C) in the type-I glomus cells. These suggest that both the HIF-2α and HIF-3α proteins are expressed in the type-I chemosensitive glomus cells of the CB and the stainings were in high intensity in the CH group.

In contrast, mild expression levels of HIF-2α (Figs. 1B and 2B) and HIF-3α (Figs. 1D and 2D) protein expression were found in the normoxic group. Positive immunostaining was sparsely scattered and weakly expressed in the nuclei and cytoplasm of the type I cells in the CB sections when compared with those in the CH group. Hence, the staining was more intense in the CH group than that of the normoxic control, indicating an upregulation of the expression. In addition, the immunostaining was not observed in the corresponding control sections incubated with normal serum instead of the primary antibodies (Data not shown).

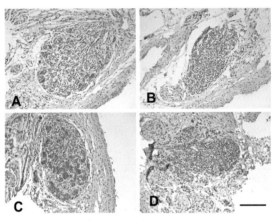

Figure 1. Upper row: Immunohistochemical localization of HIF-2α in (A) CH and (B) normoxic rat CB. Lower row: Immunohistochemical localization of HIF-3α in (C) CH and (D) normoxic rat CB. Bar = 40 μm.

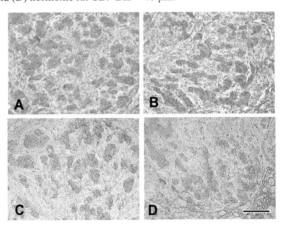

Figure 2. Immunohistochemical localization of HIF-2α and TH in the CB of (A) CH and (B) normoxic rat, and HIF-3α and TH in the CB of (C) CH and (D) normoxic rat. Immunoreactivity for HIF-2α and HIF-3α (brown stain) was localized in the nuclei with perinuclear TH-staining (red stain) in glomus cells clustering in glomeruli. Bar = 40 μm.

The mRNA expression of HIF-2α and HIF-3α in the CB was investigated using RT-PCR with specific primers. The mRNA transcripts of HIF-2α (Fig. 3) and HIF-3α (Fig. 4) were observed in the normoxic group. The mRNA levels were significantly higher in the CH than those of the normoxic group, whereas the β-actin expression remained unchanged (Figs. 3 and 4).

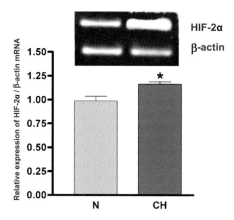

Figure 3. RT-PCR analysis of the mRNA expression of HIF-2α in normoxia (N) and CH rat CB. Expected size for HIF-2α is 298 bp and for β-actin is 436 bp. The data are expressed as means ±S.E.M. (n=5 for each group). (*) $p < 0.01$.

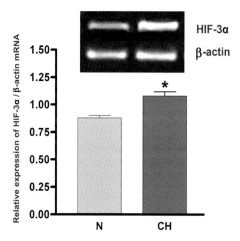

Figure 4. RT-PCR analysis of the mRNA expression of HIF-3α in normoxia (N) and CH rat CB. Expected size for HIF-3α is 301 bp and for β-actin is 436 bp. The data are expressed as means ±S.E.M. (n=5 for each group). (*) $p < 0.01$.

4. DISCUSSION

In the present study, the expression and localization of the HIF-2α and -3α subunits were unequivocally demonstrated in the rat CB. The results of RT-PCR showed that rat CB expresses mRNA of HIF-2α and -3α. Immunohistochemistry demonstrated the expression and localization of HIF-2α and -3α proteins in the CB. In addition, HIF-2α and -3α immunoreactivities were specifically localized in the type-I glomus cells of the CB, as shown by the colocalization with TH-immunostaining. Moreover, there were increases in the expression of HIF-2α and -3α, in both mRNA and protein levels, induced by the stress of CH. The current findings are the first to present solid evidence that HIF-2α and -3α are constitutively expressed in the rat CB and the expression of which are activated by CH.

HIF-1α plays a central role in the transcriptional regulation of a number of oxygen-regulated genes for the cellular response of tissues under low oxygen tensions. It was shown that HIF-1α mediates the structural and physiological changes induced by hypoxia in the CB (Fung, 2003; Roy et al., 2004). Our results suggest that in addition to HIF-1α, two other known HIF-α subunits (i.e. HIF-2α and -3α) are constitutively expressed in the rat CB. Although HIF-2α and HIF-1α share structural similarity, the two proteins expressed in different tissues and cell types. For example, HIF-2α is expressed in endothelial cells that line the walls of blood vessels in umbilical cord, whilst HIF-1α is expressed in smooth muscle cells that surround blood vessels (Tian et al., 1997). The difference in the expression sites might implicate functional discrepancy of the proteins. It was shown that the homozygous EPAS-1-deficient embryos die from circulatory failure during mid-gestational embryonic development, suggesting the elementary role of HIF-2α in the control of cardiac development. Also, the embryos had substantially reduced catecholamine levels as EPAS-1 is also expressed intensively in the organ of Zuckerkandl (OZ), the principle source of catecholamine production organ in mammalian embryos (Tian et al., 1998). The colocalization of HIF-2α and TH in the rat CB demonstrated that HIF-2α was specifically expressed in the chemosensitive type-I glomus cells. Hence, HIF-2α might play a role in the control of the synthesis and release of catecholamines in the CB.

Increased expression of HIF-1α in hypoxia is significant in the regulation of hypoxia-inducible genes in CB glomus cells (Fung, 2003). Besides, our results suggest an activation of mRNA and protein expression of the HIF-2α and -3α by CH in the rat CB. Previous study has shown that EPAS-1 expression in the OZ is hypoxia inducible and, at the mid-gestation of development, acting as a sensor of hypoxia in the embryo. In response to hypoxia, EPAS-1 can provoke the expression of genes required for either the synthesis or release of catecholamines, such as the TH gene (Tian et al., 1998). Similarly, HIF-2α may have functional roles in sensing hypoxia and in the upregulation of catecholamine synthesis in the CB type-I cells during CH.

Less is known about the expression and function of HIF-3α. Although HIF-3α has been found in human kidney (Hara et al., 2001) and rat liver (Kietzmann et al., 2001), the function of which is still uncertain. Nevertheless, transient transfection experiments demonstrated that the HIF-3α-ARNT interaction can occur *in vivo*, and that the activity of HIF-3α is upregulated in response to cobalt chloride or low oxygen tension (Gu et al., 1998). Our results undoubtedly showed the expression of HIF-3α in the rat CB and the expression of which is activated by CH, although the functional significance is still unclear and further elucidation is needed.

In conclusion, HIF-2α and HIF-3α are constitutively expressed in the rat CB and the transcriptional upregulation of the expression of the α-subunits may play a complementary role to HIF-1α in the structural and functional changes of the CB in CH.

ACKNOWLEDGEMENTS

We thank Mr. W.B. Wong and Ms. K.M. Leung for their technical assistance. This work was supported by research grant (7223/02M) from the Research Grants Council, Hong Kong.

REFERENCES

Baby SM, Roy A, Mokashi AM, and Lahiri S., 2003, Effects of hypoxia and intracellular iron chelation on hypoxia-inducible factor-1alpha and -1beta in the rat carotid body and glomus cells. Histochem Cell Biol. 120(5):343-52.

Bruick RK., 2003, Oxygen sensing in the hypoxic response pathway: regulation of the hypoxia-inducible transcription factor. Genes Dev. 17(21):2614-23.

Covello KL, and Simon MC., 2004, HIFs, hypoxia, and vascular development. Curr Top Dev Biol. 62:37-54.

Ema M, Taya S, Yokotani N, Sogawa K, Matsuda Y, and Fujii-Kuriyama Y., 1997, A novel bHLH-PAS factor with close sequence similarity to hypoxia-inducible factor 1 alpha regulates the VEGF expression and is potentially involved in lung and vascular development. Proc Natl Acad Sci U S A. 94(9):4273-8.

Fung ML., 2003, Hypoxia-inducible factor-1: a molecular hint of physiological changes in the carotid body during long-term hypoxemia? Curr Drug Targets Cardiovasc Haematol Disord. 3(3):254-9.

Gu YZ, Moran SM, Hogenesch JB, Wartman L, and Bradfield CA., 1998, Molecular characterization and chromosomal localization of a third alpha-class hypoxia inducible factor subunit, HIF3alpha. Gene Expr. 7(3):205-13.

Hara S, Hamada J, Kobayashi C, Kondo Y, and Imura N., 2001, Expression and characterization of hypoxia-inducible factor (HIF)-3alpha in human kidney: suppression of HIF-mediated gene expression by HIF-3alpha. Biochem Biophys Res Commun. 287(4):808-13.

Iyer NV, Kotch LE, Agani F, Leung SW, Laughner E, Wenger RH, Gassmann M, Gearhart JD, Lawler AM, Yu AY, and Semenza GL., 1998, Cellular and developmental control of O_2 homeostasis by hypoxia-inducible factor 1 alpha. Genes Dev. 12(2):149-62.

Jain S, Maltepe E, Lu MM, Simon C, and Bradfield CA., 1998, Expression of ARNT, ARNT2, HIF1 alpha, HIF2 alpha and Ah receptor mRNAs in the developing mouse. Mech Dev. 73(1):117-23.

Jelkmann W., 2004, Molecular biology of erythropoietin. Intern Med. 43(8):649-59.

Kietzmann T, Cornesse Y, Brechtel K, Modaressi S, and Jungermann K., 2001, Perivenous expression of the mRNA of the three hypoxia-inducible factor alpha-subunits, HIF1alpha, HIF2alpha and HIF3alpha, in rat liver. Biochem J. 354(Pt 3):531-7.

Roy A, Volgin DV, Baby SM, Mokashi A, Kubin L, and Lahiri S., 2004, Activation of HIF-1alpha mRNA by hypoxia and iron chelator in isolated rat carotid body. Neurosci Lett. 363(3):229-32.

Tian H, McKnight SL, and Russell DW., 1997, Endothelial PAS domain protein 1 (EPAS1), a transcription factor selectively expressed in endothelial cells. Genes Dev. 11(1):72-82.

Tian H, Hammer RE, Matsumoto AM, Russell DW, and McKnight SL., 1998, The hypoxia-responsive transcription factor EPAS1 is essential for catecholamine homeostasis and protection against heart failure during embryonic development. Genes Dev. 12(21): 3320-4.

Tipoe GL, and Fung ML., 2003, Expression of HIF-1alpha, VEGF and VEGF receptors in the carotid body of chronically hypoxic rat. Respir Physiol Neurobiol. 138(2-3):143-54.

Modulation of Gene Expression in Subfamilies of TASK K$^+$ Channels by Chronic Hyperoxia Exposure in Rat Carotid Body

INSOOK KIM[a], DAVID F. DONNELLY[b], AND JOHN L. CARROLL[a]

[a]*Dept of Pediatrics, University of Arkansas Medical Sciences, Little Rock, USA,* [b]*Dept of Pediatrics, Yale University, New Haven, USA,*

1. INTRODUCTION

The carotid body (CB) is a chemosensory organ which detects a decrease in PaO$_2$ or pHa and increases spiking levels on the carotid sinus nerve. Chemosensitivity normally increases after birth but this maturation is impaired by post-natal exposure to hyperoxia, resulting in a large reduction in spiking rates during normoxia and hypoxia (Donnelly, 2005) and reduction in carotid type I cell depolarization in response to anoxia (Kim, 2003). Previous studies have indicated that detection of hypoxia or acidity is mediated by modulation of a leak potassium conductance of which TASK-1 and TASK-3 are likely candidates. Accordingly, we hypothesized that post-natal hyperoxia exposure will alter the developmental profile of TASK channels within the carotid body cells.

Carotid body RNA was extracted from control (normoxia-exposed) rats at P14-P16 (N14) and rats exposed to post-natal hyperoxia from birth until P14-P16 (H14). All 5 known TASK K+ channels, TASK-1 (Duprat, 1997; Kim, 1999), -2 (Reyes, 1998), -3 (Kim, 2000, Rajan, 2000), -4 (Decher, 2001), and -5 (Ashmole, 2001; Kim, 2001), were tested using quantitative real time RT(reverse transcriptase)-PCR with SYBR green (iCycler IQ system) reporting system. Expression levels of 18s rRNA as reference gene for relative quantitative RNA analyses.

Relative quantification was applied to compare gene expression levels of the TASK K+ channels between N14 and H14. The expression ratio was calculated using a mathematical model based on difference between the threshold detection cycle (C$_T$) of the target gene, TASK K+ channels, and the C$_T$ of the reference gene, 18s rRNA (Pfaffl, 2001). The present results demonstrate that post-natal hyperoxia causes a downregulation of TASK-1, TASK-2 and TASK-5 channels.

2. METHODS

2.1 Chronic Hyperoxia Treatment

For chronic hyperoxia treatment, timed-pregnant Sprague-Dawley rats were placed in hyperoxia (60 % O_2) chamber (Oxycycler Model A84XOV, Reming Bioinstruments Co, Redfield, NY) 1-2 days prior to expected delivery and were allowed to give birth. Pups and dams were maintained in the chamber until use at P14-P16.

2.2 Isolation of Rat CB and tRNA Extraction

Carotid body RNA was extracted from control (normoxia-exposed) rats at P14-P16 (N14) and rats exposed to post-natal hyperoxia from birth until P14-P16 (H14). Isolated CBs were placed into 1.5ml centrifuge tube and spun down, washed one more time with ice-cold PBS, and processed to extract the Total RNA(tRNA) using AquaPure RNA isolation kit (Bio-Rad). To exclude genomic DNA contamination, RNA was treated with RQ1 RNase-free DNase (Promega). cDNA was synthesized by using iScript cDNA synthesis kit (Bio-Rad). To get the maximum concentration of tRNA, the pelleted tRNA was reconstituted in the minimum volume of reconstitution solution. RNA similarly extracted from heart and brain served as positive controls. No-template-control (NTC) and cDNA without reverse transcriptase (-RT) served as negative controls.

2.3 Real Time RT-PCR

Primers for TASK channels and reference genes were designed by using Beacon Designer 2.0; rTASK-1 (accession # NM_033376, rat), mTASK-2 (accession # AF319542, mouse), rTASK-3 (accession # NM_053405, rat), hTASK-4 (accession # AF339912, human), rTASK-5 (accession # NM_130813, rat), ß-actin (accession # NM_031144, rat), and 18s rRNA (accession # X01117, rat).

The efficiency of each TASK channel primers was tested with a series of cDNA dilution. To compare their gene expression levels between control rats and hyperoxia treated rats, relative expression ratio of each channel was utilized. In each PCR run, cDNAs synthesized from CB of control rats (N14) and CB of hyperoxia treated rats (H14) were tested simultaneously with primers of the target genes and the house keeping gene. The anticipated product was confirmed based on the melt curve.

The relative expression ratios for the candidate gene vs the housekeeping genes were calculated by the following equation (Pfaffl, 2001). The ratio of a target gene is expressed in a sample versus a control in comparison to a reference gene.

$$\text{ratio} = \frac{(E_{\text{target}})^{\Delta C_{T\text{target}}^{(\text{control-sample})}}}{(E_{\text{ref}})^{\Delta C_{T\text{ref}}^{(\text{control-sample})}}}$$

E_{target} is the real-time PCR efficiency of target gene transcript; E_{ref} is the real-time PCR efficiency of a reference gene transcript; $\Delta C_T^{\text{target}}$ is the C_T of control – sample of the target gene transcript; $\Delta C_T^{\text{ref}} = C_T$ control – sample of reference gene transcript. The relative ratio >1 means that the gene expression in N14 CB cells is up-regulated from N1 CB cells. The relative ratio close to 1, means that there are not significant changes on gene expression during CB maturation.

3. RESULTS

The quality of each primer, except TASK-4 channel, was evaluated with series of dilution of cDNA of N14 CB. For relative quantification of TASK channels expression in N14 (normoxia exposure, 14 days old) and H14 (hyperoxia exposure, 14 days old) carotid bodies, the arithmetic formula $2^{-\Delta CT}$ was used and took into account the amount of target, normalized to an endogenous reference (18s rRNA). The relative expression of target gene was determined according to the following relation:

$$\Delta C_T = C_T^{TASK} - C_T^{18srRNA} \quad (1)$$
$$\text{Relative Expression} = 1/(2^{-\Delta CT}) \quad (2)$$

where, C_T is the cycle threshold for TASK channels or 18s rRNA determined empirically. The mean relative expression values and standard errors were calculated from the three individual sets of pooled CB.

Preliminary results indicate that TASK-3 in carotid bodies from N14 and H14 was the most highly expressed. In N14 CB cells, TASK-3 was many-fold more highly expressed than TASK-1. TASK-5 was also expressed more than TASK-1, but TASK-2 was expressed slightly less than TASK-1 (TASK-3> TASK-5> TASK-1> TASK-2). In H14 CB, TASK-3 was also many-fold more highly expressed than TASK-1, and both TASK-5 and TASK-2 were expressed more than TASK-1 (TASK-3> TASK-5> TASK-2> TASK-1).

Comparing N14 vs. H14, results to-date show a profound reduction in TASK-1, -2, and -5 channel gene expression in carotid bodies from the H14 group. In the hyperoxia exposed group, TASK-1, -3, and TASK-5 expression were down-regulated by >70%, while TASK-2 expression was not significantly affected.

These data suggest that reduced expression of TASK K^+ channels, TASK-1, TASK-3, and TASK-5 channels, might play a role in the impairment of CB O_2 sensitivity due to 2 weeks postpartum hyperoxia exposure, resulting in a large reduction in spiking rates during normoxia and hypoxia and reduction in carotid type I cell depolarization in response to anoxia.

4. DISCUSSION

These preliminary results suggest that post-natal hyperoxia exposure causes a major reduction in expression of several TASK K^+ channels in the carotid body. Previously, TASK channels have been implicated as serving an important role in hypoxia sensing/transduction in carotid body glomus cells, and the present results are consistent with this role.

We used real-time PCR on whole carotid body to the study the effects of perinatal hyperoxia on TASK channel expression. Based on total mRNA extracted from carotid bodies, the message for TASK1 and TASK3 are significantly downregulated by post-natal hyperoxia. Some caution is necessary in interpreting the results since the reference gene expression (18s rRNA) would be expressed in all cell types within the carotid body. Thus, a reduced expression of TASK channels could be due to an alteration in the proportion of glomus cells to total glomic tissue. However, Erickson and colleagues found that post-natal hyperoxia reduced carotid volume but the proportion of glomus cells to total tissue was unchanged (Erickson, 1998). Thus, perinatal hyperoxia in the rat decreases glomic and non-glomic compartments proportionally, strongly suggesting that our findings reflect changes in TASK channel expression rather than hyperoxia-induced changes in cell type proportion. Therefore, the use of whole CB as an initial approach to examining the effects of hyperoxia on TASK K^+ channel expression seems justified. Additional studies using single-cell or few-cell real-time PCR to study glomus cell ion channel expression are currently underway in our laboratory.

Previous results from our laboratory (Donnelly, 2005), demonstrated that immediately following a period of post-natal hyperoxia, spiking activity is profoundly reduced on single, chemoreceptor receptor afferent nerves. This compliments previous work demonstrating a reduced ventilatory response to acute hypoxia and reduced chemoreceptor nerve spiking activity following a period of post-natal hyperoxia followed by a recovery period of 4-5 months (Bisgard, 2003). The mechanistic basis for the reduced chemoreceptor response to hypoxia is a focus of our present laboratory work, and the present results suggest that part of the impairment may be due to reductions in TASK channel expression.

TASK channels have been previously implicated as important in glomus cell hypoxia sensing/transduction. Previous work by Buckler and colleagues demonstrate that a TASK-like current mediates glomus cell depolarization during hypoxia and work in our laboratory demonstrated that this current is upregulated during normal development. Thus, the most direct interpretation of the present result is that post-natal hyperoxia impairs the normal increase in expression of TASK channels and this accounts for the reduced chemoreceptor response to acute hypoxia.

ACKNOWLEDGEMENTS

Dr. Kim is supported by the University of Arkansas for Medical Sciences Dean's/CUMG Research Development Fund and Dr. Carroll is supported by NIH RO1-HL054621-08.

REFERENCES

Ashmole I, Goodwin PA, and Stanfield PR (2001) TASK-5, a novel member of the tandem pore K+ channel family. Pflugers Arch. 442 (6):828-833

Bisgard GE, Olson EB, Wang ZY, Bavis RW, Fuller DD, and Mitchell GS (2003) Adult carotid chemoafferent responses to hypoxia after 1, 2, and 4 wk of postnatal hyperoxia. *J.Appl.Physiol* 95 (3):946-952

Decher N, Maier M, Dittrich W, Gassenhuber J, Bruggemann A, Busch AE, and Steinmeyer K (2001) Characterization of TASK-4, a novel member of the pH-sensitive, two-pore domain potassium channel family. *FEBS Lett.* 492 (1-2):84-89

Donnelly DF, Kim I, Carle C, Carroll JL (2005) Perinatal hyperoxia for 14 days increases nerve conduction time and the acute unitary response to hypoxia of rat carotid body chemoreceptors. J Appl Physiol. 2005 Feb 24; [Epub ahead of print]

Duprat F, Lesage F, Fink M, Reyes R, Heurteaux C, and Lazdunski M (1997) TASK, a human background K+ channel to sense external pH variations near physiological pH. EMBO J. 16(17):5464-71.

Erickson JT, Mayer C, Jawa A, Ling L, Olson EB, Vidruk EH, Mitchell GS, and Katz DM (1998) Chemoafferent degeneration and carotid body hypoplasia following chronic hyperoxia in newborn rats. *J.Physiol* 509 (Pt 2):519-526

Kim, I, Boyle KM, Carle C, Donnelly DF, and Carroll JL (2003) Perinatal hyperoxia reduces the depolarization and calcium increase in response to an acute hypoxia challenges in rat carotid chemoreceptor cells. Experimental Biology meeting abstracts [on CD-ROM]. *The FASEB Journal*, 17, Abstract #LB128

Kim D and Gnatenco C (2001) TASK-5, a new member of the tandem-pore K(+) channel family. Biochem.Biophys.Res.Commun. 284 (4):923-930

Kim Y, Bang H, and Kim D (1999) TBAK-1 and TASK-1, two-pore K(+) channel subunits: kinetic properties and expression in rat heart. Am J Physiol. 277(5 Pt 2):H1669-78

Kim Y, Bang H, and Kim D (2000) TASK-3, a new member of the tandem pore K(+) channel family. J Biol Chem. 275(13):9340-7.

Pfaffl MW (2001) A new mathematical model for relative quantification in real-time RT-PCR, Nucleic Acids Research 29(9): 2002-07

Rajan S, Wischmeyer E, Xin G, Preisig-Muller R, Daut J, Karschin A, and Derst C (2000) TASK-3, a novel tandem pore domain acid-sensitive K+ channel. An extracellular histiding as pH sensor. *J.Biol.Chem.* 275 (22):16650-16657

Reyes R, Duprat F, Lesage F, Fink M, Salinas M, Farman N, and Lazdunski M (1998) Cloning and expression of a novel pH-sensitive two pore domain K+ channel from human kidney. *J.Biol.Chem.* 273 (47):30863-30869

Postnatal Changes in Gene Expression of Subfamilies of TASK K$^+$ Channels in Rat Carotid Body

INSOOK KIM[a], JUNG H. KIM[b], AND JOHN L. CARROLL[a]

[a]Dept of Pediatrics, University of Arkansas for Medical Sciences, Little Rock, USA, [b]Dept of Systems Engineering, University of Arkansas at Little Rock, Little Rock, USA

1. INTRODUCTION

The carotid body (CB) is a chemosensory organ monitoring blood O_2 level, CO_2 level, and pH. It is known that CB O_2 sensitivity is minimal after birth and increases with age (Bamford, 1999), but the mechanisms of CB development are poorly understood. Previous studies have shown that CB glomus cell background K$^+$ current is inhibited by acute hypoxia and glomus cell O_2-sensitive background currents increase with age (Kim, 2003). It has been proposed these currents are carried by TASK-like K$^+$ channels. Therefore, we hypothesized that expression of one or several TASK K$^+$ channels might change during CB development and contribute to the development of glomus cell oxygen sensitivity.

The TASK (TWIK-related acid-sensitive K$^+$) channel family includes five subfamilies; TASK-1 (Duprat, 1997; Kim, 1999), TASK-2 (Reyes, 1998), TASK-3 (Kim, 2000, Rajan, 2000), TASK-4 (Decher, 2001), and TASK-5 (Ashmole, 2001; Kim, 2001). Among the TASK five subfamilies, TASK-1 and TASK-3 have been suggested to be possible O_2-sensitive channels (Buckler, 2000; Hartness, 2001). Therefore, TASK-1 and TASK-3 channels would be likely candidates for differential expression during carotid body development. However, possible differential gene expression of other TASK subfamilies during CB development cannot be excluded. To detect the developmental differences of TASK channel gene expression, real time quantitative PCR (qPCR) was utilized. All five TASK channels were tested on total RNA (tRNA) extracted from CB of normoxic P0-P1day old (N1) and P14-P16 days old (N14) rats. 18s rRNA gene was used as the reference gene. In neonatal rats, reared in normoxia, relative quantification was applied to compare gene expression levels of the CB TASK K$^+$ channels on day 1 (N1) vs. day 14 (N14). The ratio was calculated using a mathematical model based on the difference between C_T of target gene, TASK K$^+$ channels, and C_T of reference gene, 18s rRNA (Pfaffl, 2001).

2. METHODS

2.1 Isolation of Rat CB and *t*RNA Extraction

Carotid body (CB) was isolated from 2 age groups, P0-P1 (N1) and P14-P16 (N14) day old rats, as described previously (Wasikco et al., 1999). Timed pregnant Sprague-Dawley rats were reared in the ACHRI animal facilities. For CB isolation, rats were anesthetized with isoflurane and decapitated. The carotid bifurcations were dissected and placed in ice-cold phosphate buffered saline, removed from the bifurcations and placed in cold sterile PBS. Isolated CBs were placed into 1.5ml centrifuge tube and spun down, washed one more time with ice-cold PBS, and processed to extract the total RNA(*t*RNA) using AquaPure RNA isolation kit (Bio-Rad). To get the maximum concentration of *t*RNA, the pelleted *t*RNA was reconstituted in the minimum volume of reconstitution solution. Positive controls for the gene expression of each channel, were determined with *t*RNA of rat heart or brain tissue.

2.2 Real Time RT-PCR

Primers for TASK channels and reference gene were designed by using Beacon Designer 2.0; rTASK-1 (against rat gene), mTASK-2 (against mouse gene), rTASK-3 (against rat gene), hTASK-4 (against human gene), rTASK-5 (against rat gene), and 18s rRNA (against rat gene). Designed primers were ordered from Integrated DNA Technologies (IDT) and tested on cDNA from positive control tissues prior to testing on cDNA of CB. As the reference gene, 18S rRNA was used. All primers were tested on cDNA of brain tissues (cerebellum) or heart (ventricle) as positive control.

To exclude genomic DNA contamination, tRNA was treated with RQ1 RNase-free DNase (Promega). cDNA was synthesized by using iScript cDNA synthesis kit (Bio-Rad). To check the DNA contamination, cDNA without reverse transcriptase (-RT) was synthesized. The efficiency of each TASK channel primer was tested with series of cDNA dilution. To compare their gene expression levels between immature and matured CB, a relative expression ratio of each channel was utilized. In each PCR run, cDNAs synthesized from immature (N1) and mature (N14) CB were tested with primers of the target gene and the house keeping gene simultaneously. To prevent false detection with the real time PCR, no-template-control (NTC) was tested on every run for each testing primer set. For each running, each sample was tested as triplet. The real time PCR run on iCycler iQ real-time Detection system (Bio-Rad) by using SYBR Green Supermix (Bio-Rad).

3. RESULTS

To detect genomic contamination during tRNA extraction, cDNAs without reverse transcriptase (-RT cDNA) were synthesized for every extracted *t*RNA and tested on real time PCR. None of PCR products was detected from -RT cDNA. To test all designed primers for the five TASK channels, TASK-1, -2, -3, -4, and -5, and 18s rRNA, were tested with positive control tissues. TASK-1 and TASK-5 primer sets could produce PCR product from cDNA of rat heart. TASK-2 and TASK-3 primer sets could produce the PCR products from cDNA of rat cerebellum. But, TASK-4 primer set could not detect any PCR product from neither rat heart nor rat cerebellum cDNA. The quality of each primer, except TASK-4 channel, was evaluated with series of dilution of cDNA of rat N14 CB. The corresponding real-time PCR efficiency rate (E) in an exponential phase was calculated according to the equation:

$$E = 10^{[-1/\text{slope}]} \text{ (Pfaffl, 2001)}.$$

The cDNAs synthesized from N1 and N14 rat CB were run at the same time on real time PCR. All five TASK primer sets as target genes as well as 18s rRNA primer set as the reference gene were tested at the same time. The hTASK-4 primer set won't produce the PCR product from neither N1 nor N14 CB cDNA, as in the case of control tissues cDNA, heart or brain. The rest of TASK channels and 18s rRNA were detected from both N1 and N14 CB cDNA. The relative expression of TASK channels were calculated using the following equation.

For relative expression of TASK channels in each N1 or N14 CB, the arithmetic formula $2^{-\Delta C_T}$ was used and took into account the amount of target, normalized to an endogenous reference (18sr RNA). The relative expression of a target gene was determined according to the following relation:

$$\Delta C_T = C_T^{TASK} - C_T^{18srRNA} \quad (1)$$

$$\text{Relative Expression} = 1/(2^{-\Delta C_T}), \quad (2)$$

where C_T is the cycle threshold for TASK channels or 18s rRNA determined empirically. The mean relative expression values and standard errors were calculated from the three individual sets of pooled CB.

Preliminary results indicate that TASK-3, in both N1 and N14 CB, was the most highly expressed. In N1 (immature) CB cells, TASK-3 was expressed ~10 times higher than TASK-1, and both TASK-5 and TASK-2 were expressed more than TASK-1 (TASK-3> TASK-5> TASK-2> TASK-1). In N14 (mature) CB cells, TASK-3 was also expressed ~10 times higher than TASK-1. TASK-5 was expressed more than TASK-1, but TASK-2 was expressed slightly less than TASK-1 (TASK-3> TASK-5> TASK-1> TASK-2).

The relative ratios of TASK channel gene expression between N1 (immature) vs. N14 (mature) CB cells, were calculated by the following equation which is a mathematical model of a relative expression ratio in real-time PCR (Pfaffl 2001). The ratio of a target gene is expressed in a sample versus a control in comparison to a reference gene.

$$\text{ratio} = \frac{(E_{\text{target}})^{\Delta C_{T\text{target}}(\text{control-sample})}}{(E_{\text{ref}})^{\Delta C_{T\text{ref}}(\text{control-sample})}}$$

E_{target} is the real-time PCR efficiency of a target gene transcript; E_{ref} is the real-time PCR efficiency of a reference gene transcript; ΔC_T^{target} is the C_T of control (N1) – sample (N14) of the target gene transcript; $\Delta C_T^{ref} = C_T$ control (N1) – sample (N14) of a reference gene transcript. The relative ratio >1 means that the gene expression in N14 CB cells is up-regulated from N1 CB cells. The relative ratio of 1 means that there are no significant changes in gene expression during CB maturation.

Results to date show that TASK-1 and TASK-3 were up regulated by ~50% during CB maturation, while TASK-2 and TASK-5 expression did not change with development. These data suggest that up-regulation of TASK-1 and TASK-3 genes may play a role in postnatal development of O_2 sensitivity during CB maturation.

4. DISCUSSION

The small size of the rat carotid body presents numerous technical challenges for the study of glomus cell K^+ channel expression during postnatal development. We chose, as a first approach to this question, to apply real-time PCR to the study of TASK channel expression levels using whole carotid bodies. Although results were referenced to 18s rRNA and provide a composite view of changes in TASK gene expression in the whole CB, our findings do not provide information on TASK channel expression at the individual glomus cell level. Using the whole CB approach, for a TASK channel subtype expressed only in glomus cells, a postnatal change in the relative proportion of glomus cells to other cell types may give the appearance of changing expression during development. Unfortunately, the relative proportion of cell types in rat CB between birth and 14 days of age has not been previously studied. However, Kariya et al. reported that the rabbit carotid body is ultrastructurally mature at birth, although synapses may continue to develop after birth (Kariya 1990). Therefore, the use of whole CB as an initial approach to examining development of TASK K^+ channel expression seems justified. Additional studies using single-cell or few-cell real-time PCR to study glomus cell ion channel expression are currently underway in our laboratory.

It has been shown that glomus cell K^+ channels with the properties of TASK-1 and TASK-3 are inhibited by hypoxia and therefore are O_2-sensitive (Buckler 2000; Kemp 2001). In rats, glomus cell O_2 sensitivity, measured as the $[Ca^{++}]_i$ response to acute hypoxia challenge, develops after birth and achieves apparent maturity in 2-3 weeks postpartum (Bamford 1999). However, the mechanism of glomus cell O_2 sensitivity maturation is not yet known. Our findings are consistent with the hypothesis that the developmental increase of O_2 sensitivity in rat CB glomus cells is due, at least in part, increased expression of TASK-1 and/or TASK-3 K^+ channels. Although the relative quantification analysis showed that TASK-5 and TASK-2 channels are relatively highly expressed, their expression level did not change with age.

ACKNOWLEDGEMENTS

Dr. Kim is supported by the University of Arkansas for Medical Sciences Dean's/CUMG Research Development Fund and Dr. Carroll is supported by NIH RO1-HL054621-08.

REFERENCES

Ashmole I, Goodwin PA, and Stanfield PR (2001) TASK-5, a novel member of the tandem pore K+ channel family. Pflugers Arch. 442 (6):828-833

Bamford OS, Sterni LM, Wasicko MJ, Montrose MH and Carroll JL (1999) Postnatal maturation of carotid body and type I cell chemoreception in the rat. Am J Physiol 276: L875-L884

Buckler KJ, Williams BA and Honore E (2000) An oxygen-, acid- and anaesthetic-sensitive TASK-like background potassium channel in rat arterial chemoreceptor cells. J Physiol 525 Pt 1: 135-142

Decher N, Maier M, Dittrich W, Gassenhuber J, Bruggemann A, Busch AE, and Steinmeyer K (2001) Characterization of TASK-4, a novel member of the pH-sensitive, two-pore domain potassium channel family. FEBS Lett. 492 (1-2):84-89

Duprat F, Lesage F, Fink M, Reyes R, Heurteaux C, and Lazdunski M (1997) TASK, a human background K+ channel to sense external pH variations near physiological pH. EMBO J. 16(17):5464-71

Hartness ME, Lewis A, Searle GJ, O'Kelly L, Peers C, and Kemp PJ (2001) Combined antisense and pharmacological approaches implicate hTASK as an airway O(2) sensing K(+) channel. *J.Biol.Chem.* 276 (28):26499-26508

Kariya I, Nakjima T, Ozawa H (1990) Ultrastructural study on cell differentiation of the rabbit carotid body. Arch Histol Cytol. 53(3):245-58

Kim D and Gnatenco C (2001) TASK-5, a new member of the tandem-pore K(+) channel family. Biochem.Biophys.Res.Commun. 284 (4):923-930

Kim, I, Boyle KM, Carle C, Donnelly DF, and Carroll JL (2003) Perinatal hyperoxia reduces the depolarization and calcium increase in response to an acute hypoxia challenges in rat carotid chemoreceptor cells. Experimental Biology meeting abstracts [on CD-ROM]. *The FASEB Journal*, 17, Abstract #LB128

Kim Y, Bang H, and Kim D (1999) TBAK-1 and TASK-1, two-pore K(+) channel subunits: kinetic properties and expression in rat heart. Am J Physiol. 277(5 Pt 2):H1669-78

Kim Y, Bang H, and Kim D (2000) TASK-3, a new member of the tandem pore K(+) channel family. J Biol Chem. 275(13):9340-7

Pfaffl MW (2001) A new mathematical model for relative quantification in real-time RT-PCR, Nucleic Acids Research 29(9): 2002-07

Rajan S, Wischmeyer E, Xin G, Preisig-Muller R, Daut J, Karschin A, and Derst C (2000) TASK-3, a novel tandem pore domain acid-sensitive K+ channel. An extracellular histiding as pH sensor. *J.Biol.Chem.* 275 (22):16650-16657

Reyes R, Duprat F, Lesage F, Fink M, Salinas M, Farman N, and Lazdunski M (1998) Cloning and expression of a novel pH-sensitive two pore domain K+ channel from human kidney. *J.Biol.Chem.* 273 (47):30863-30869

Wasicko MJ, Sterni LM, Bamford OS, Montrose MH, and Carroll JL (1999) Resetting and postnatal maturation of oxygen chemosensitivity in rat carotid chemoreceptor cells. *J.Physiol* 514 (Pt 2):493-503

Morphological Changes in the Rat Carotid Body in Acclimatization and Deacclimatization to Hypoxia

[1]HIDEKI MATSUDA, [2]HARUHISA HIRAKAWA, [3]SHIGERU OIKAWA, [4]YOSHIAKI HAYASHIDA, [5]TATSUMI KUSAKABE

[1]*Department of Otorhinolaryngology, Yokohama City University School of Medicine, 3-9, Fukuura, Kanazawa-ku, Yokohama,* [2]*Department of Physiology, National Defense Medical College, Tokorozawa,* [3]*Department of Medicine, Labour Welfare Corporation Ehime Rosai Hospital, Ehime,* [4]*International Buddhist University, Osaka, and* [5]*Laboratory for Anatomy and Physiology, Department of Sport and Medical Science, Kokushikan University, Tokyo, Japan*

1. INTRODUCTION

The carotid bodies are enlarged in the rats exposed to long term hypoxia. In some studies the animals were exposed to hypoxia for relatively short periods, and in other studies for relatively long periods. However, most authors use the term "chronic hypoxia" in their publications. This terminology can cause much confusion. On the other hand, there are no morphological studies of the carotid bodies after the termination of chronic hypoxia except in a few instances (Heath et al., 1973). Recently high altitude training has been used to try to improve some physical conditions. High altitude exercise can help to make clear morphological changes in chemoreceptor organs during acclimatization to hypoxia and during deacclimatization after chronic hypoxia is terminated.

We summarize the morphological changes and changes in the peptidergic innervation in rat carotid bodies during the course of hypoxic adaptation, and during the course of recovery to evaluate the different levels of acclimatization and deacclimatization. The original findings have been detailed in the recent publications (Kusakabe et al., 2003, 2004).

2. MORPHOLOGICAL CHANGES IN THE CAROTID BODIES AFTER HYPOXIC EXPOSURE AND AFTER THE TERMINATION OF CHRONIC HYPOXIA

The carotid bodies of rats exposed to hypoxia for 2, 4, and 8 weeks were found to be enlarged several fold in comparison with those of normoxic control rats. The rate of enlargement was different for the carotid bodies exposed for three different periods (Fig. 1). The mean short axis of the carotid bodies of the rats exposed to hypoxia for 2, 4, and 8 weeks was 1.2, 1.3, and 1.5 times longer than in normoxic controls, respectively (Fig. 2). The mean long axis was 1.3, 1.6, and 1.7 times longer than in normoxic controls, respectively (Fig. 2). With a prolonged hypoxic exposure, the percentage of blood vessels with relatively wide lumens, more than 21 µm, increased, and the percentage of vessels with relatively narrow lumens, less than 5-10 µm, decreased (Fig. 3). Thus, the enlargement of the hypoxic carotid bodies was mainly due to vascular dilation as suggested by Blessing and Wolff (1973), and Laidler and Kay (1975). Pequignot and Hellström (1984) reported that vascular dilation is already evident in the rat carotid bodies after 1 week of exposure to hypoxia. It seems likely that the enlargement of the carotid bodies with vascular expansion begins soon after the start of hypoxic exposure. As far as enlargement of the carotid bodies is concerned, the use of the term "chronic hypoxia" has little meaning as a general expression regardless of the duration of hypoxic exposure.

The carotid bodies 1 week after the termination of chronic hypoxia were significantly diminished in size in comparison with the carotid bodies of rats exposed to hypoxia for 8 weeks, and the carotid bodies 8 weeks after the termination of hypoxia were similar to the normoxic controls in size (Figs.1 and 2). This indicates that recovery in the carotid bodies had already started relatively early, i.e., 1 week after the termination of chronic hypoxia, and complete recovery occurred by 4-8 weeks after the termination of hypoxia. In carotid bodies 1 week after the termination of the chronic hypoxia, the percentage of blood vessels with relatively wide lumens greater than 26 µm decreased, and the percentage of vessels with relatively narrow lumens of less than 5 µm increased (Fig. 3). These percentages of vasculature are similar to those in normoxic control carotid bodies. As stated above, the enlargement of chronically hypoxic carotid bodies is mainly due to vascular dilation (Blessing and Wolff, 1973; Laidler and Kay, 1975; Pequignot and Hellström, 1984). Naturally, shrinking of the carotid bodies after the termination of hypoxia may

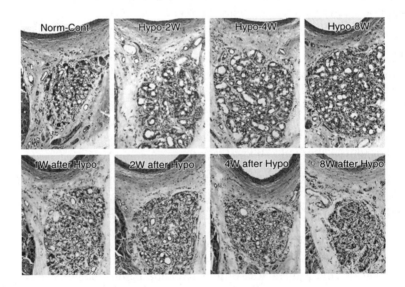

Figure 1. Hematoxylin-eosin stained sections from the center of a control normoxic carotid body (Norm-Cont), a carotid body after 2, 4, and 8 weeks of hypocapnic hypoxic exposure (Hypo-2W, -4W, and -8W), and a carotid body 1, 2, 4, and 8 weeks after the termination of chronically hypocapnic hypoxia (1W, 2W, 4W, and 8w after Hypo). The hypoxic carotid bodies are enlarged with vascular expansion in comparison with normoxic controls. The carotid bodies after the termination of chronic hypoxia were diminished in size (T. Kusakabe et al., 2005).

be due to vascular contraction. In the course of recovery, vascular contraction is also evident in the carotid bodies 1 week after the termination of hypoxia. Thus, it seems likely that shrinking of the carotid bodies with vascular contraction begins soon after the termination of hypoxic exposure.

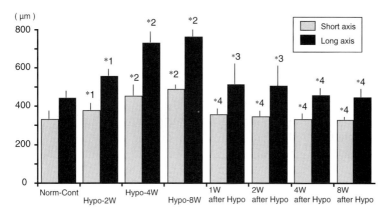

Figure 2. Histograms comparing the short and long axes of normoxic control carotid bodies, after 2, 4, and 8 weeks of hypoxic exposure, and 1, 2, 4, and 8 weeks after the termination of hypoxia. *1 p<0.01, and *2 p<0.005 in comparison with the normoxic control column. *3 p<0.01, and *4 p<0.005 in comparison with the Hypo-8W column (T. Kusakabe et al., 2005).

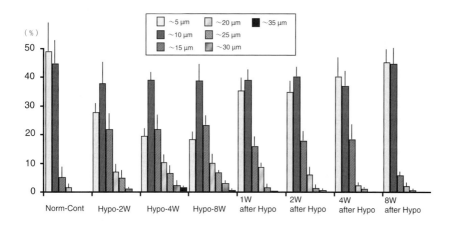

Figure 3. Histograms representing the percentage of blood vessels of seven ranges of diameter in normoxic control carotid bodies, after 2, 4, and 8 weeks of hypoxic exposure, and 1, 2, 4, and 8 weeks after the termination of hypoxia.

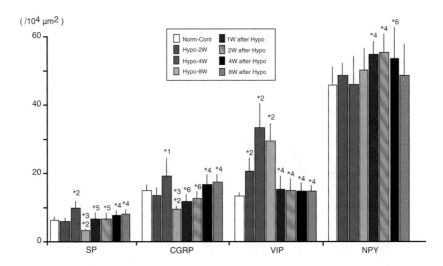

Figure 4. Histograms comparing the density of varicosities per unit area in normoxic control carotid bodies (Norm-Cont), after 2, 4, and 8 weeks of hypoxic exposure, and 1, 2, 4, and 8 weeks after the termination of hypoxia. *1 $p<0.01$, and *2 $p<0.005$ in comparison with the normoxic control column, and *3 $p<0.005$ in comparison with the 4-week column. *4 $p<0.005$, *5 $p<0.01$, and *6 $p<0.05$ in comparison with the Hypo-8W column.

3. PEPTIDERGIC INNERVATION IN THE CAROTID BODIES AFTER HYPOXIC EXPOSURE AND AFTER TERMINATION OF CHRONIC HYPOXIA

In the case of changes in peptidergic innervation, differing durations of hypoxic exposure become an important subject for discussion. Mean density per unit area of substance P (SP) and calcitonin gene-related peptide (CGRP) immunoreactive fibers was transiently high in the carotid bodies after 4 weeks of hypoxic exposure, and decreased significantly to nearly or under 50% after 8 weeks of hypoxic exposure (Fig. 4). Density of vasoactive intestinal polypeptide (VIP) immunoreactive fibers increased significantly in all periods of hypoxic exposure, and was especially high after 4 weeks of hypoxic exposure (Fig. 4). Density of neuropeptide Y (NPY) immunoreactive fibers was unchanged in the carotid bodies during hypoxic exposure (Fig. 4). SP and CGRP fibers in the carotid bodies after 4 weeks of hypoxic exposure may be involved in both chemosensory and vascular dilatory systems, although we previously reported that SP and CGRP fibers in carotid bodies after 3 months of hypoxic exposure may not be involved in chemosensory mechanisms because the density of immunoreactive fibers at this stage decreases to under 50% (Kusakabe et al., 1998, 2000). It seems likely that VIP is more effective in the carotid bodies after 4 weeks of exposure than in carotid bodies after longer exposure, and that physiological involvement of NPY fibers is invariable from short to prolonged hypoxic exposure. Considered together with our recent findings (Kusakabe et al., 2000), the peptidergic innervation after 4-8 weeks of hypoxic exposure have acclimatizing stages, and the innervation after 3 months of exposure indicates a completely acclimatized stage.

In the period of 1-8 weeks after the termination of chronic hypoxia, the most striking features of the peptidergic innervations are the maintenance, or even the increase, in the density of NPY fibers, and an immediate decrease in the density of VIP fibers (Fig. 4). We previously speculated that at least part of the vascular dilation in the chronically hypoxic rat carotid bodies may depend on the vasodilatory effect of VIP, and concluded that VIP fibers are indirectly involved in chemosensory mechanisms by controlling local carotid body circulation (Kusakabe et al., 1998). In various mammalian vasculatures, NPY is thought to have a vasoconstrictory effect (Brain et al., 1985; Edvinsson et al., 1983; Lundberg et al., 1982). The percentage of blood vessels with relatively narrow lumens decreased in the carotid bodies 1 week after the termination of chronic hypoxia. It seems likely that shrinking of the carotid body with vascular contraction is caused by the increased density of vasoconstrictive NPY. At least part of the vascular constriction in the carotid bodies in recovery stages may depend on the vasoconstrictory effect of NPY. The morphological changes during recovery stages may be under the control of peptidergic innervation.

ACKNOWLEDGEMENTS

We are grateful to Prof. R.C. Goris of the Department of Anatomy, Yokohama City University School of Medicine, for his help in editing the manuscript. This work was supported by grants-in aid 13680048 and 14570073 from the Ministry of Education, Science, Sports, and Culture, Japan, and in part by grants in the project research at Kokushikan University.

REFERENCES

Blessing M.H. and Wolff H., 1973, Befunde an glomus caroticum der ratte nach aufenhalt im einer simulierten höhe von 7500m. *Virchow Arch. Pathol. Anat. Physiol. 360, 78-97.*
Brain S.D., Williams, T.J., Tippins, J.R., Moris, H.R., and MacIntyre, I., 1985, Calcitonin gene-related peptide is a potent vasodilator. *Nature (Lond) 313, 54-56.*
Edvinsson L., Emson P., McCulloch J., Teramoto K., and Uddman R., 1983, Neuropeptide Y: cerebrovascular innervation and vasomotor effects in the cat. *Neurosci. Lett. 43, 79-84.*
Heath D., Edwarda C., Winson M. and Smith P., 1973, Effects on the right ventricle, pulmonary vasculature, and carotid bodies of the rat of exposure to, and recovery from, simulated high altitude. *Thorax 28, 24-28.*
Kusakabe T., Hayashida Y., Matsuda H., Gono Y., Powell F.L., Ellisman M.H., Kawakami T. and Takenaka T., 1998, Hypoxic adaptation of the peptidergic innervation in the rat carotid body. *Brain Res. 806, 165-174.*
Kusakabe T., Hayashida Y., Matsuda H., Kawakami T. and Takenaka T., 2000, Changes in the peptidergic innervation of the carotid body a month after the termination of chronic hypoxia. *Adv. Exp. Med. Biol. 475, 793-799.*
Kusakabe T., Hirakawa H., Matsuda H., Kawakami K., Takenaka T., and Hayashida Y., 2003, Peptidergic innervation in the rat carotid body after 2, 4, and 8 weeks of hypocapnic hypoxic exposure. *Histol. Histopathol. 18, 409-418.*
Kusakabe T., Hirakawa H., Oikawa S., Matsuda H. Kawakami T., Takenaka T. and Hayashida Y., 2004, Morphological changes in the rat carotid body 1, 2, 4, and 8 weeks after the termination of chronically hypocapnic hypoxia. *Histol. Histopathol. 19, 1133-1140.*

Kusakabe T., Matsuda H. and Hayashida Y., 2005 in press, Hypoxic adaptation of the rat carotid body. *Histol. Histopathol.*
Laider P. and Kay J.M ., 1975, A quantitative morphological study of the carotid bodies of rats living at a simulated altitude of 4300 meters. *J. Pathol. 117, 183-191.*
Lundberg J.M., Terenis L., Hökfelt T., Martling C.R., Tatemoto K., Mutt V., Polak J., Bloom S., Goldstei, M., 1982, Neuropeptide Y (NPY)-like immunoreactivity in peripheral noradrenergic neurons and effects of NPY on sympathetic function. *Acta Physiol. Scand. 116, 477-480.*
Pequignot J.M., Hellström S. and Johansson C., 1984, Intact and sympathectomized study of the carotid bodies of long term hypoxic rats: A morphometric ultrastructural study. *J. Neurocytol. 13, 481-493.*

Effect of Carbon Dioxide on the Structure of the Carotid Body

A comparison between normoxic and hypoxic conditions

[1]TATSUMI KUSAKABE, [2]HARUHISA HIRAKAWA, [3]SHIGERU OIKAWA, [4]HIDEKI MATSUDA, AND [5]YOSHIAKI HAYASHIDA

[1]*Laboratory for Anatomy and Physiology, Department of Sport and Medical Science, Kokushikan University, Tokyo,* [2]*Department of Physiology, National Defense Medical College, Tokorozawa,* [3]*Department of Medicine, Labour Welfare Corporation Ehime Rosai Hospital, Ehime,* [4]*Department of Otorhinolaryngology, Yokohama City University School of Medicine, 3-9, Fukuura, Kanazawa-ku, Yokohama, and* [5]*International Buddhist University, Osaka, Japan*

1. INTRODUCTION

Three types of hypoxia with different levels of carbon dioxide (hypocapnic, isocapnic, and hypercapnic hypoxia) have been called systemic hypoxia (Hirakawa et al., 1997). Recently, the changes in general morphology and in peptidergic innervation in the carotid bodies of rats exposed to systemic hypoxia were examined to evaluate the effect of arterial CO_2 tension (Kusakabe et al., 1998, 2000, 2002). The carotid bodies of the systemic hypoxic rats were found to be enlarged several fold, but the degree of enlargement was different for each (Kusakabe et al., 2003). The mean diameter of the hypercapnic hypoxic carotid bodies were smaller than the hypocapnic and isocapnic hypoxic carotid bodies. The vasculature in the carotid bodies of chronically hypercapnic hypoxic rats was found to be enlarged in comparison with that of normoxic control rats, but the rate of vascular enlargement was smaller than that in hypocapnic and isocapnic hypoxic carotid bodies. This indicates that the morphological changes in the hypoxic carotid bodies may depend on the arterial CO_2 tension. However this hypothesis may be restricted to the carotid bodies in hypoxic conditions. To clarify this we compared the morphological changes and those in the peptidergic innervation between the carotid bodies of the rats exposed to hypercapnic hypoxia and those exposed to normoxic hypercapnia.

2. MATERIALS AND METHODS

2.1 Hypercapnic Exposure

Eight-week-old rats were placed in an air-tight acrylic chamber with two holes. One hole, located at the top of a side wall of the chamber, was connected to a multi-flowmeter (MODEL-1203, KOFLOC, Japan), and was used to deliver a gas mixture (hypercapnic hypoxia: 10% O_2 in N_2 and 6% CO_2:

total 10 L/min; normoxic hypercapnia: 16% O_2 in N_2 and 6% CO_2, total 20 L/min) into the chamber. The second hole was located at the bottom of the opposite wall of the chamber, and was used to flush out the gas mixture. In normoxic conditions, O_2 concentration in the chamber was decreased to 16% to avoid hyperoxia due to tachypnea. The CO_2 was added to the hypoxic gas mixture at the concentration of 6% because this much was necessary to maintain an arterial partial pressure of CO_2 close to hypercapnic measured during exposure. The flow of air, N_2, and CO_2 was regulated by a multi-flowmeter, and O_2 and CO_2 concentration within the chamber were monitored with a gas analyzer (Respina IH 26, San-ei, Japan). The temperature within the chamber was maintained at 25°C. This hypoxic condition has been confirmed to be hypercapnic to rats in a previous study (Hayashida et al., 1996). Animals were exposed in this chamber for 8 weeks with food and water available *ad libitum*. Control rats were housed for the same period in the same chamber. The chamber was opened for 10 min every 3 days for husbandry.

All experiments with animals were performed in accordance with "Principles of laboratory animal care" (NIH publ. no. 86-23, revised 1985) and with "Guiding Principles for the Care and Use of Animals in the Fields of Physiological Sciences" published by the Physiological Society of Japan.

2.2 Tissue Preparation

The animals were intraperitoneally anesthetized with sodium pentobarbital (0.05 mg/g), and perfused through a thin nylon tube inserted into the ventricle with 0.1M heparinized phosphate buffer saline (PBS), followed by freshly prepared Zamboni's fixative solution (4% paraformaldehyde and 0.2% picric acid in 0.1M PBS) at a constant flow rate. The pair of carotid bodies was then removed under a dissecting microscope, and immersed in the same fixative for an additional 6-8 h at 4°C. After a brief washing in PBS, the specimens were transferred to 30% sucrose in PBS at 4°C for 24 h. The specimens were cut serially at 16 µm on a cryostat, and mounted in four series on poly-L-lysine coated slides.

2.3 Immunohistochemistry

The sections were processed for immunohistochemistry according to the peroxidase-antiperoxidase (PAP) method. The immunostaining procedure has been detailed in a previous report (Kusakabe, et al., 1991). In brief, the sections were incubated at 4°C overnight with the primary antisera against the following neuropeptides: SP (1:1500; Cambridge Research Biochemicals, Northwich, UK), CGRP (1:1500; Cambridge Research Biochemicals, Northwich, UK), VIP (1:2000; Incstar, Stillwater, USA), and NPY (1:2000; Incstar, Stillwater, USA). The peroxidase activity was demonstrated with 3,3´-diaminobenzidine. The reaction for neuropeptides was verified by treating sections with primary antibody which had been inactivated by overnight incubation with 50-100 µM of its peptide. Some sections were also stained with hematoxylin eosin for general histology.

2.4 Data Analysis

In hematoxylin and eosin stained sections through the center of the carotid bodies, their short and long axes and the diameter of blood vessels were measured with an ARGUS 100 computer and image processor (Hamamatsu-Photonics, Japan). The measurement was performed on 6 sections taken from 6 carotid bodies of 3 rats exposed to hypercapnia and exposed to air for each of three periods. The values taken from hypercapnic carotid bodies were expressed as means ± S.D. (n=6), and those from normoxic control carotid bodies were also expressed as means ± S.D. (n=6). The number of blood vessels of seven different ranges of diameter, less than 5 µm (~5), 6-10 µm (~10), 11-15 µm, (~15) 16-20 µm (~20), 21-25 µm (~25), 26-30 µm (~30), and 31-35 µm (~35), in normoxic control and hypercapnic carotid bodies was expressed as percentage of total number of blood vessels.

The density of immunoreactive fibers in the normoxic and hypoxic carotid bodies was represented as the number of varicosities per unit area (104 µm2) of parenchyma. The manner of measurement is detailed in recent reports (Kusakabe et al., 1998, 2000, 2002). The number of varicosities was counted on 6 sections of hypercapnic carotid bodies, and on 6 sections from normoxic control carotid bodies.

The values were expressed as means ± S.D., and statistical comparisons between the control and experimental values were determined using Student's t-test.

3. RESULTS

3.1 General Histology of Normoxic Carotid Bodies and of Carotid Bodies After 8 Weeks of Hypercapnic Exposure

In hematoxylin and eosin stained sections through the center of the normoxic rat carotid bodies, the bodies were oval in shape and were mainly composed of clusters of glomus cells and blood vessels with narrow lumens (Fig. 1A). The mean short and long axes of the normoxic carotid bodies were 329.0±35.3 µm and 439.7±28.5 µm, respectively (Fig. 2).

The carotid bodies of hypercapnic hypoxic rats were found to be enlarged several-fold in comparison with the carotid bodies of normoxic controls (Fig. 1B). The enlarged hypoxic carotid bodies contained many blood vessels whose diameter was larger than vessels in normoxic carotid bodies, but the rate of vascular enlargement in the hypercapnic hypoxic carotid bodies was smaller than in previously reported isocapnic and hypocapnic hypoxic carotid bodies (Kusakabe et al. 1998, 2000). The mean short and long axes of the hypercapnic hypoxic carotid bodies were 390.6±37.9 µm and 664.5±59.6 µm, respectively (Fig. 2). These values were significantly ($p<0.005$) larger than those in the normoxic controls.

The carotid bodies of normoxic hypercapnic rats were similar in size to those of normoxic controls (Fig. 1C). In the sections through the center of the carotid bodies, the mean short and long axes were 308.0±12.6 µm and

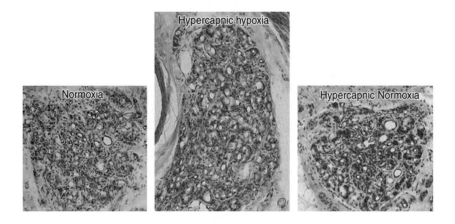

Figure 1. A comparison of hematoxylin-eosin stained sections of the control normoxic (A), hypercapnic hypoxic (B), and normoxic hypercapnic (C) carotid bodies.

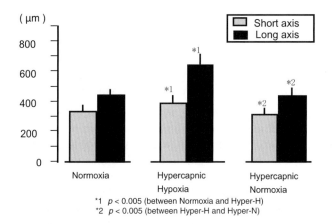

Figure 2. A comparison of the diameter of the carotid body in normoxia, hypercapnic hypoxia, and normoxic hypercapnia.

436.9±39.7 μm, respectively (Fig. 2). There was no significant difference between the diameter of the hypercapnic carotid bodies and those of the normoxic controls.

In the normoxic control carotid bodies, about 90% of the blood vessels were small with diameters less than 10 μm, and blood vessels of greater than 15 μm in diameter were under 10% (Fig. 3A). In the hypercapnic hypoxic carotid bodies, the percentage of relatively small vessels and the percentage of relatively large vessels were similar to the percentage in normoxic control carotid bodies, although the carotid bodies themselves were significantly larger than in

Effect of Carbon Dioxide on CB Structure 59

normoxic controls (Fig. 3B). The percentage of blood vessels of seven different ranges of diameter in the normoxic hypercapnic carotid bodies was similar to the percentage in normoxic control bodies (Fig. 3C).

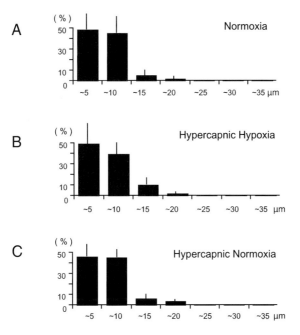

Figure 3. A comparison of the diameter (short axis) of blood vessels within the carotid body in normoxia, hypercapnic hypoxia, and normoxic hypercapnia.

3.2 Peptidergic Nerve Fibers in Control and Hypercapnic Carotid Bodies

Immunoreactivity of different four neuropeptides, SP, CGRP, VIP, and NPY, was recognized in the nerve fibers distributed throughout the parenchyma of the carotid body (Kusakabe et al., 1998b, 2000, 2002). These immunoreactive fibers appeared as thin processes with a number of varicosities. NPY-immunoreactive varicose fibers were more numerous than SP-, CGRP-, and VIP-immunoreactive fibers. Most of them were associated with the vessels within the carotid body. The mean density of varicosities of SP, CGRP, VIP, and NPY fibers per unit area (104 µm2) was 6.1±0.8, 14.8±1.8, 13.1±2.2, and 45.7±4.8, respectively.

In the hypercapnic hypoxic carotid bodies, the density of NPY fibers per unit area (104 µm2) was significantly ($p<0.005$) increased from 45.7±4.8 to 64.8±7.1, although the density of SP and CGRP fibers per unit area was significantly decreased, and the density of VIP fibers was unchanged.

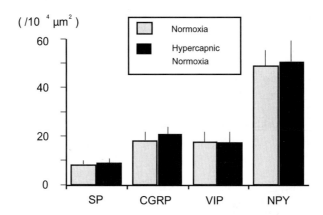

Figure 4. Histogram comparing the density of varicosities of SP, CGRP, VIP, and NPY immunoreactive fibers per unit area in normoxic control conditions and normoxic hypercapnia.

In the normoxic hypercapnic carotid bodies, the distribution pattern and the density of SP, CGRP, VIP, and NPY immunoreactive fibers was similar to that in the control carotid bodies (Fig. 4).

No glomus cells with the immunoreactivity of these four neuropeptides were observed in the normoxic and hypercapnic carotid bodies.

4. DISCUSSION

It has been suggested that hypercapnia increases the cerebral and gastrointestinal blood flow (Fensternacher and Rapoport, 1984). Thus, CO_2 tension causes vasodilation in both central and peripheral vascular systems. In this study, the percentage of vascular enlargement in hypercapnic hypoxic carotid bodies is smaller than in previously reported chronically hypocapnic and isocapnic hypoxic carotid bodies (Kusakabe et al., 1998a, 2000), and the percentage of blood vessels of seven different ranges of diameter in normoxic hypercapnic carotid bodies was similar to that in normoxic controls. This may indicate that high CO_2 tension in hypoxic conditions causes the carotid body vasculature to constrict, and high CO_2 tension in normoxic conditions does not cause vascular dilation in the carotid body.

According to Fukuda et al., (1987), increases in the carotid sinus nerve discharge produced by hypercapnia are smaller than increases caused by hypoxia. The responses to CO_2 may be relatively low in the rat carotid body.

In conclusion, high CO_2 tension in normoxic conditions does not cause morphological changes in the rat carotid body or changes in the peptidergic innervation within the carotid body. Considered together with our recent findings on the carotid bodies in systemic hypoxia, CO_2 may have some additive effects on the chemoreceptor organs in hypoxic conditions.

ACKNOWLEDGEMENTS

We are grateful to Prof. R.C. Goris of the Department of Anatomy, Yokohama City University School of Medicine, for his help in editing the manuscript. This work was supported by grants-in aid 13680048 and 14570073 from the Ministry of Education, Science, Sports, and Culture, Japan, and in part by grants in the project research at Kokushikan University.

REFERENCES

Fenstermacher, J.D. and Rapoport, S.I. 1984, Blood-brain barrier. In: Handbook of Physiology. Volume IV, Microcirculation, Part 2, Waverly Press, Baltimore   Pp 969-1000.

Fukuda Y., Sato A., and Trzebski A., 1987,   Carotid vhemoreceptor discharge   responses to hypoxia and hypercapnia in normotensive and spontaneously hypertensive rats.   *J. Auton. Nerv. Syst., 19, 1-11.*

Hayashida, Y., Hirakawa, H., Nakamura, T. and Maeda, M., 1996,   Chemoreceptors in autonomic responses to hypoxia in conscious rats.   *Adv. Exp. Med. Biol.*   410:439-442.

Hirakawa H., Nakamura T. and Hayashida Y., 1997,   Effect of carbon dioxide on autonomic cardiovascular responses to systemic hypoxia in conscious rats.   *Am. J. Physiol. 273, R747-R754.*

Kusakabe T., Anglade P. and Tsuji S., 1991,   Localization of substance P, CGRP, VIP, neuropeptide Y, and somatostatin immunoreactive nerve fibers in the carotid labyrinths of some amphibian species.   *Histochemistry 96, 255-260.*

Kusakabe T., Powell F.L. and Ellisman M.H., 1993,   Ultrastructure of the glomus cells in the carotid body of chronically hypoxic rats: with special reference to the similarity of amphibian glomus cells.   *Anat. Rec. 237,   220-227.*

Kusakabe T., Hayashida Y., Matsuda H., Gono Y., Powell F.L., Ellisman M.H.,   Kawakami T. and Takenaka T., 1998,   Hypoxic adaptation of the peptidergic innervation in the rat carotid body.   *Brain Res.   806, 165-174.*

Kusakabe T., Hayashida Y., Matsuda H., Kawakami T. and Takenaka T., 2000, Changes in the peptidergic innervation of the carotid body a month after the termination of chronic hypoxia. *Adv. Exp. Med. Biol. 475, 793-799.*

Kusakabe T., Hirakawa H., Matsuda H., Yamamoto Y., Nagai T., Kawakami T., Takenaka T. and Hayashida Y., 2002, Changes in the peptidergic innervation in the carotid body of rats exposed to hypercapnic hypoxia: an effect of arterial CO_2 tension. *Histol. Histopathol. 17, 21-29.*

Kusakabe et al., Matsuda H. and Hayashida Y., 2003, Rat carotid bodies in systemic hypoxia: Involvement of arterial CO_2 tension in morphological changes. *Adv. Exp. Med. Biol. 536, 611-617.*

S-Nitrosoglutathione (SNOG) Accumulates Hypoxia Inducible Factor-1α in Main Pulmonary Artery Endothelial Cells but not in Micro Pulmonary Vessel Endothelial Cells

S. FUJIUCHI,[1] Y. YAMAZAKI,[1] Y. FUJITA,[1] Y. NISHIGAKI,[1] A. TAKED,[1] Y. YAMAMOTO,[1] T. FIJIKANE,[1] T. SHIMIZU,[1] S. OSANAI,[2] T. TAKAHASHI[2] AND K. KIKUCHI[2]

[1] *Department of Clinical Research, Dohoku National Hospital, National Hospital Organization, 7 chome Hanasaki, Asahikawa 070-8644, Japan and* [2] *The first department of internal medicine, Asahikawa Medical College.*

1. INTRODUCTION

Adequate cellular oxygen tension is essential for maintaining a variety of physiological process. Disorder of oxygen delivery eventually leads to the cell dysfunction. Therefore, sensing mechanism of cellular hypoxia is critical. Under hypoxic condition, a lot of protein is induced in mammalian cells for preventing hypoxic stress. Hypoxia inducible factor-1 (HIF-1) is a transcription factor protein that thought to be a one of the key molecule as gatekeeper of cellular hypoxia. HIF-1 regulates the expression of series of genes involved in angiogenesis, oxygen transport and glucose metabolism (1, 2). Most of these gene products utilize for the maintaining O2 homeostasis. HIF-1 is composed of two subunit called HIF-1α and HIF1β (3). In normoxic and hypoxic condition, HIF-1α and HIF1β mRNA are constitutively expressed (4). With regard to the protein level, HIF-1α is hydroxylated at Pro402 and Pro564 by the enzyme designated prolyl hydroxylase domain containing protein (PHD) under normoxia (5, 6). Hydroxylated HIF-1α binds to the von Hippel Lindau protein (pVHL), which is the substrate for ubiquitin ligase complex (7). Therefore, HIF-1α is rapidly degraded under normoxia by the ubiquitin-protease pathway (8, 9, 10). When cells are exposed to hypoxia, HIF-1α protein escapes from this degradation system. Subsequently, accumulated HIF-1α protein translocates to the nucleus, and dimerizes with HIF-1β 11). This heterodimeric protein binds to hypoxia-responsive element (HRE) and induces transcription of downstream genes (12). Thus, transcriptional activity of HIF-1 is primarily dependent on the HIF-1α expression.

It is accepted that hypoxia, transition metal such as CoCl2 directly inhibit PHD (6). Recently, several reports demonstrate that nitric oxide (NO) and S-nitrosoglutathione (SNOG), chemically diverse NO donor, also induced

HIF-1α (13, 14, 15). However, the effect of SNOG on pulmonary artery endothelial cell has not been understood well. In this study, we examined effect of SNOG on two different pulmonary artery endothelial cells with regard to HIF-1α expression.

2. MATERIALS AND METHODS

2.1 Materials

S-nitrosoglutathione (SNOG) was obtained from TOCRIS (Ellisville, MO, USA). Cycloheximide and cobalt chloride were from Sigma Chemical Co. (St. Louis, MO, USA). HIF-1α antibody and secondary antibody were purchased from BD Bioscience (Tokyo, Japan), New England Biolabs (Beverley, MA, USA), respectively.

2.2 Cell Culture

Human pulmonary artery endothelial cells (HPAEC) and human micro vessel endothelial cell-lung (HMVEC-L) were obtained from Cell Applications Inc. (San Diego, CA, USA), BioWhittaker (Walkersville, MD, USA), respectively. Cells were grown in the medium that provided by the manufactures and kept in a humidified chamber of 5% CO_2 in air at 37°C.

2.3 ELISA Based HIF-1 DNA Binding Assay

HIF-1 binding to the hypoxic responsive element (HRE) activity was performed using ELISA based binding assay kit (Active Motif) (Carlsbad, CA, USA) according to manufacture's instruction. Nuclear extracts were prepared using nuclear extract kit (Active Motif). Ten microgram of nuclear protein was subjected to the analysis. Relative absorbance is standardized with unstimulated control.

2.4 Western Blot Analysis

HIF-1α protein expression was evaluated by western blot analysis. The nuclear protein (10μg) was added to same volume of Laemmli buffer and boiled for 3 min. The proteins were separated on a 5-20% (w/v) gradient SDS polyacrylamide gel and transferred to PVDF membrane using wet transfer module. Blots were washed twice with TBST (150mM NaCl, 40mM Tris-HCl, pH7.4, 0.1% Tween20). Unspecific binding was blocked with TBST and 5% skim milk for 1h. The HIF-1α antibody (1:250 dilutions in TBST with 5% skim milk) was added and incubated overnight at 4°C. Afterward, the blot was washed for 30 min with TBST. Then blot was incubated with goat antimouse secondary antibody conjugated with peroxidase (1:2500 in TBST with 5% skim milk) (New England Biolabs) for 1h, followed by chemiluminescence detection system (Amersham, Buckinghamshire, UK).

2.5 Semi-quantitative RT-PCR Analysis for VEGF and HIF-1α mRNA

HPAEC and HMVEC-L (1×10^5) were grown in 6 well dishes. Total RNA was isolated using the RNeasy kit (Qiagen, Tokyo, Japan) and eluted to 30μl of DEPC water. Five out of 30 μl RNA solutions was applied to perform the reverse transcription with Omniscirpt RT kit (Qiagen). The real time quantification of VEGF, HIF-1α and 18s rRNA gene were carried out using specific primer and probe kit (Applied Biosystem, Tokyo, Japan) with ABI 7900 gene detection system (Applied Biosystem). Relative mRNA expression of target genes were standardized with 18s rRNA expression as internal control.

3. RESULT

3.1 HIF-1α Protein Expression and HIF-1 DNA Binding Activity were Induced by SNOG in HPAEC but not in HMVEC-L

HPAEC and HMVEC-L were cultured with SNOG (0.2, 0.4, 1mM) for 4h or with 1mM SNOG for 15 min to 4h. Nuclear extracts (5 μg) were subjected to western blot analyses. In HPAEC, HIF-1α expression was seen in time and SNOG dose dependent manner, however, only marginal expression was seen in HMVEC-L (Figure 1).

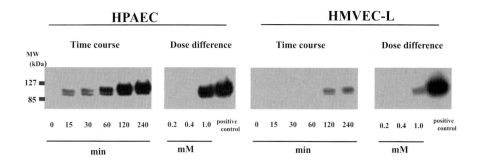

Figure 1. HIF-1α protein expression in HPAEC and HMVEC-L by SNOG. HPAEC and HMVEC-L were cultured with 1mM SNOG for 15 min to 4h or with SNOG (0.2, 0.4, 1mM) for 4h. HIF-1α was induced time- and SNOG dose dependently in HPAEC. By contrast, only marginal expression was seen in HMVEC-L. HIF-1α accumulation appeared 15 min after stimulation, lasting for 4 h.

HIF-1 DNA binding assay demonstrated that SNOG induced binding activity in HPAEC but not in HMVEC-L (Figure 2).

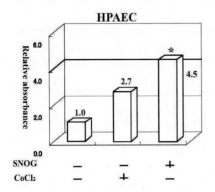 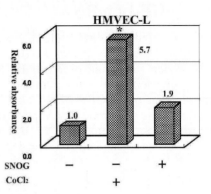

Figure 2. HIF-1 DNA binding activity analyzed by ELISA based assay. SNOG (1mM) or CoCl2 (250µM) was added to the medium for 4 h. In HPAEC, relative absorbance was increased to 4.5, 2.7 fold by the addition of SNOG, CoCl2, respectively. However, SNOG increased activity by 1.9 fold while CoCl2 induced activity by 5.7 fold compared with control in HMVEC-L. Experiment performed at least three times as duplicate. (*: p <0.05 *vs.* control)

On the other hand, CoCl2 induced either HIF-1α protein expression or HIF-1 DNA binding activity in both cells. These results indicated that SNOG sensing mechanism under normoxic condition is different between HPAEC and HMVEC-L.

3.2 VEGF mRNA Expression after SNOG Stimulation

HIF-1 induces several downstream genes that related to oxygen homeostasis. In order to elucidate whether induced HIF-1α protein by SNOG was biologically active, we analyzed VEGF mRNA expression. Vascular endothelial growth factor (VEGF) is one of the HIF-1 target genes. VEGF involved in development of vascular to the tissue in which low oxygen tension. SNOG (1mM) or CoCl2 (250µM) was added to the medium for 4h, followed by total RNA isolation. Semi-quantitative RT-PCR was performed employing TaqMan PCR method. Data were shown as relative mRNA expression standardized by unstimulated 18s rRNA gene as internal control. Although SNOG increased VEGF mRNA expression in HPAEC, there was no effect in HMVEC-L. These results suggested that HIF-1 activity induced by SNOG confirmed with DNA binding experiment is functional (Figure 3).

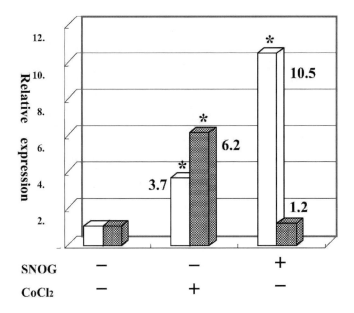

Figure 3. EGF mRNA expression after SNOG stimulation. SNOG (1mM) or CoCl2 (250μM) was added to the medium for 4 h. After RT reaction, VEGF mRNA expression was analyzed semi-quantitatively as described in materials and methods. In HPAEC (open bar), relative mRNA expression to 10.5, 3.7 fold by the addition of SNOG, CoCl2, respectively. However, SNOG increased activity by 1.2 fold while CoCl2 induced activity by 6.2 fold compared with control in HMVEC-L (hatched bar). These are the similar result seen in the HIF-1 DNA binding assay. Experiment performed triplicate. (*: $p < 0.05$ *vs.* control)

3.3 HIF-1α mRNA Expression

We next examined whether SNOG affects HIF-1α mRNA expression. Semi-quantitative RT-PCR analyses revealed that HIF-1α mRNA expression was unchanged regardless of presence or absence with SNOG (1mM) and CoCl2 (250μM) (Figure 4). To date, HIF-1α protein expression is regulated at post translational modification. Similar with the previous study, these results indicated that HIF-1α protein expression by SNOG is unlikely regulated transcriptional level.

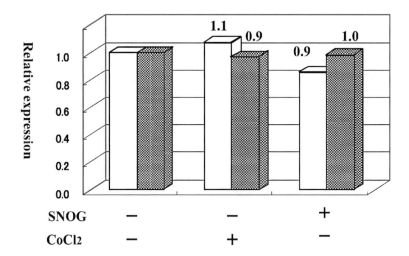

Figure 4. HIF-1α mRNA expression after SNOG stimulation. SNOG (1mM) or CoCl2 (250μM) was added to the medium for 4 h. After RT reaction, HIF-1α mRNA expression was analyzed semi-quantitatively as described in materials and methods. Either SNOG or CoCl2 did not change HIF-1α mRNA expression in HPAEC and HMVEC-L. HPAEC (open bar), HMVEC-L (hatched bar)

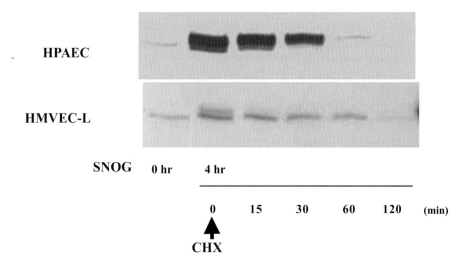

Figure 5. HIF-1α protein stabilization with SNOG. Cells were stimulated with SNOG (1mM) for 4 h followed by the addition of cycloheximide (100 μM). The addition of SNOG resulted in sustained protein expression over 60 min even if blocking of translation. This result suggested SNOG stabilizes already translated HIF-1α protein.

3.4 HIF-1α Protein is Stabilized by SNOG

To investigate whether HIF-1α protein synthesis is affected by SNOG, cells were stimulated with SNOG (1mM) for 4 h followed by the addition of cycloheximide (100 μM) to block *de novo* protein synthesis. Although the half-life of HIF-1α protein is estimated less than 10 min, the addition of SNOG resulted in sustained protein expression over 60 min even if blocking of translation (Figure 5). This result suggested SNOG stabilizes translated HIF-1α protein.

4. DISCUSSION

The effect of NO has been extensively studied in this decade. Nitric oxide involves a variety of cardiopulmonary, immunological, neuronal and apoptotic process as a mediator (16). In the human lung system, NO is produced from vascular endothelial cells, vascular smooth muscle cells, macrophages and bronchial epithelial cells upon stimulation such as inflammation, hypoxic stress (16). In addition to this endogenous NO production, inhaled NO is utilized for the therapy of persistent pulmonary hypertension of the neonate (PPHN) (17) respiratory distress syndrome (IRDS), primary pulmonary hypertension (PPH) (18). Since the pharmacological effects other than vasodilator has been described, it is important to explore the unexpected effects of NO in cultured human pulmonary vessel endothelial cells.

Hypoxia-inducible factor-1α is a transcriptional factor that regulates a series of genes which involves cellular and systemic O2 homeostasis (1, 2). HIF-1 is heterodimeric protein that composed of HIF-1α and HIF-1β (3). Under normoxia, HIF-1α is hydroxylated at Pro402 and Pro564 by the enzyme termed prolyl hydroxylase domain containing protein (PHD). The hydroxylated HIF-1α associates with pVHL. Since this complex is the target for ubiquitin protease system (7), HIF-1α protein is undetectable under normoxia. Once cells are exposed to hypoxia, dehydroxylation occurs at Pro402 and Pro564 by the inhibition of PHD enzymatic activity. The dehydroxylated HIF-1α no longer forms complex with pVHL. This results in HIF-1α protein accumulation in cytoplasm, translocation to the nucleus and dimerization with HIF-1β (11) which constitutively expressed regardless of oxygen tension. This heterodimeric protein binds to hypoxia-responsive element (HRE) and induces transcription of downstream genes (12). Thus, transcriptional activity of HIF-1 is primarily dependent on the HIF-1α protein expression.

There are several factors that modulate HIF-1 activity. Although earlier report demonstrated that NO inhibits HIF-1α expression under hypoxia, or after treatment with CoCl2 (19, 20), recent study indicated that co-culture with NO donor such as SNOG or NO evokes HIF-1α expression in cultured cells under normoxia (13, 21, 22). Since it has not been fully understood yet, we examined the effect of SNOG with regard to HIF-1α expression on two different human pulmonary artery endothelial cells which derived from main (HPAEC) and micro (HMVEC-L) pulmonary artery. As the result, we found that the different profile of HIF-1α protein expression by SNOG under normoxic condition in HPAEC and HMVEC-L. Our data suggested that HIF-1α activation upon NO might have different mechanism between main and peripheral pulmonary artery endothelial cells.

In addition to endogenous NO that is produced by vascular endothelial cells, micro pulmonary vascular endothelial cells are likely to be exposed to exogenous NO from activated alveolar macrophages. Another possible clinical situation, the patients under inhaled NO therapy in which micro pulmonary artery endothelial cells are likely to be exposed to high concentration NO.

Epstein et al. first reported egl-9 gene product in *C.elegans* is responsible for HIF-1α hydroxylase at specific amino acid residue (6). At present, four EGL-9 homologues identified in mammalian cells and termed prolylhydroxylase domain containing protein (PHD) 1-4. Hypoxia or transition metal such as CoCl2 directly inhibits PHD (6). In our experiment CoCl2 induced HIF-1α activity in both cells. These results suggested that PHD functions as sensors of transition metal in both cell lines. Interestingly, NO activates HIF-1α by inhibiting PHD (15). Isaacs et al reported that pVHL and oxygen-independent HIF-1α degradation pathway (23). There might have different protein degradation pathway between HPAEC and HMVEC-L.

There are some other possibilities that could explain this difference. In order to obtain sufficient HIF-1 transcriptional activity, it is necessary to recruit cofactor complex with HIF-1α transactivation domain (24). We could not deny this issue.

In conclusion, our data demonstrated that even if pulmonary vascular endothelial cell, HIF-1α induction by SNOG is different according to the anatomical region. Further investigation will be required for clarify molecular mechanism with regard to HIF-1α activity in pulmonary endothelial cells.

REFERENCES

1) Guillemin K, Krasnow MA. The hypoxic response: huffing and HIFing. Cell 1997;89:9-12.
2) Semenza GL. Hypoxia-inducible factor 1: master regulator of O2 homeostasis. Curr. Opin. Genet. Dev. 1998;8:588-594.
3) Wang GL, Jiang BH, Rue EA and Semenza GL Hypoxia-inducible factor 1 is a basic-helix-loop-helix-PAS heterodimer regulated by cellular O2 tension. Proc. Natl. Acad. Sci. U S A. 1995;92:5510-5514.
4) Wiener.CM, Booth G, Semenza GL. In vivo expression of mRNAs encoding hypoxia-inducible factor 1. Biochem. Biophys. Res. Commun. 1996;225:485-488.
5) Bruick RK, McKnight SL. A conserved family of prolyl-4-hydroxylases that modify HIF. Science 2001;294:1337-1340.
6) Epstein AC, Gleadle JM, McNeill LA, Hewitson KS, O'Rourke J, Mole DR, Mukherji M, Metzen E, Wilson MI, Dhanda A, Tian YM, Masson N, Hamilton DL, Jaakkola P, Barstead R, Hodgkin J, Maxwell PH, Pugh CW, Schofield CJ, Ratcliffe PJ. C. Elegans EGL-9 and mammalian homologs define a family of dioxygenases that regulate HIF by prolyl hydroxylation. Cell 2001;107:43-54.
7) Bonicalzi ME, Groulx I, de Paulsen N, Lee S. Role of exon 2-encoded beta -domain of the von Hippel-Lindau tumor suppressor protein. J.Biol.Chem. 2001;276:1407-1416.
8) Salceda S, Caro J. Hypoxia-inducible factor 1alpha (HIF-1alpha) protein is rapidly degraded by the ubiquitin-proteasome system under normoxic conditions. Its stabilization by hypoxia depends on redox-induced changes. J.Biol.Chem. 1997;272:22642-22647.
9) Huang LE, Gu J, Schau M, Bunn HF. Regulation of hypoxia-inducible factor 1alpha is mediated by an O2-dependent degradation domain via the ubiquitin-proteasome pathway. Proc. Natl. Acad. Sci. USA 1998;95:7987-7992.
10) Kallio PJ, Wilson WJ, O'Brien S, Makino Y, Poellinger L. Regulation of the hypoxia-inducible transcription factor 1alpha by the ubiquitin-proteasome pathway. J.Biol.Chem. 1999;274:6519-6525.

11) Kallio PJ, Pongratz I, Gradin K, McGuire J, Poellinger L. Activation of hypoxia-inducible factor 1alpha: posttranscriptional regulation and conformational change by recruitment of the Arnt transcription factor. Proc. Natl. Acad. Sci. USA 1997;94:5667-5672.
12) Semenza GL, Jiang BH, Leung SW, Passantino R, Concordet JP, Maire P, Giallongo A. Hypoxia response elements in the aldolase A, enolase 1, and lactate dehydrogenase A gene promoters contain essential binding sites for hypoxia-inducible factor 1. J.Biol.Chem. 1996;271:32529-32537.
13) Sandau KB, Fandrey J, Brune B. Accumulation of HIF-1alpha under the influence of nitric oxide.Blood 2001; 97:1009-1015
14) Zhou J, Fandrey J, Schumann J, Tiegs G, Brune B. NO and TNF-alpha released from activated macrophages stabilize HIF-1alpha in resting tubular LLC-PK1 cells. Am. J. Physiol.Cell Physiol. 2003;284;C439-C446
15) Metzen E, Zhou J, Jelkmann W, Fandrey J, Brune B. Nitric oxide impairs normoxic degradation of HIF-1alpha by inhibition of prolyl hydroxylases. Mol. Biol. Cell 2003;14:3470-3481.
16) Gaston B, Kobzik L, Stamler JS "Distribution of Nitric Oxide Synthase in the Lung" In *Nitric Oxide and the Lung*, Zapol WM, Bloch KD ed. New York, NY: Marcel Dekker,1997
17) Roberts JD, Polaner DM, Lang P. Inhaled nitric oxide in persistent pulmonary hypertension of the newborn. Lancet 1992;340:818-819.
18) Pepke-Zaba J,HigenbottamTW. Inhaled nitric oxide as a cause of selective pulmonary vasodilatation in pulmonary hypertension. Lancet 1991;338:1173-1174.
19) Liu Y, Christou H, Morita T, Laughner E, Semenza GL, Kourembanas S. Carbon monoxide and nitric oxide suppress the hypoxic induction of vascular endothelial growth factor gene via the 5' enhancer. J Biol. Chem. 1998; 273:15257-15262.
20) Sogawa K, Numayama-Tsuruta K, Ema M, Abe M, Abe H, Fujii-Kuriyama Y. Inhibition of hypoxia-inducible factor 1 activity by nitric oxide donors in hypoxia. Proc. Natl. Acad. Sci. USA.1998;95:7368-7373.
21) Kimura H, Weisz A, Kurashima Y, Hashimoto K, Ogura T, D'Acquisto F, Addeo R, Makuuchi M, Esumi H. Hypoxia response element of the human vascular endothelial growth factor gene mediates transcriptional regulation by nitric oxide: control of hypoxia-inducible factor-1 activity by nitric oxide. Blood 2000; 95:189-197.
22) Palmer LA, Gaston B, Johns RA. Normoxic stabilization of hypoxia-inducible factor-1 expression and activity: redox-dependent effect of nitrogen oxides. Mol. Pharmacol. 2000;58:1197-1203.
23) Isaacs JS, Jung YJ, Mimnaugh EG, Martinez A, Cuttitta F, Neckers LM. Hsp90 regulates a von Hippel Lindau-independent hypoxia-inducible factor-1 alpha-degradative pathway. J. Biol. Chem. 2002; 277:29936-29944.
24) Gu J, Milligan J, Huang LE. Molecular mechanism of hypoxia-inducible factor 1alpha -p300 interaction. A leucine-rich interface regulated by a single cysteine. J. Biol. Chem. 2001;276:3550-3554.

Changes in Antioxidant Protein SP-22 of Chipmunk Carotid Bodies during the Hibernation Season

KOHKO FUKUHARA[1], YI WU[2], HIROKI NANRI[3], MASAHARU IKEDA[4], YOSHIAKI HAYASHIDA[5], KATSUAKI YOSHIZAKI[6], KAZUO OHTOMO[7]

[1]Division of Cell Biology and Histology, Department of Anatomy and Biochemistry, Akita University School of Medicine, Akita 010-8543, Japan, [2]Department of Biology, School of Biology and Chemistry Engineering, Guangzhou University, China, [3]Department of Nutritional Sciences, Faculty of Health and Welfare, Seinan-Jogakuin University, Kitakyusyu 803-0835, [4]Department of Health Development, University of Occupational and Environmental Health, Kitakyusyu 807-8555, [5]International Buddhist University, Osaka 583-8501, [6]Department of Physiology and [7]Department of Anatomy, Akita University School of Health Sciences, Akita 010-8543, Japan

1. INTRODUCTION

Hibernators survive repeated cycles of torpor and arousal during the hibernation season. During torpor, hibernating animals drastically reduce their heart rate, respiratory rate, body temperature, blood flow and oxygen consumption; however, during periodic arousal, this suppressed physiological state rapidly surges and returns to euthermy (Daan, 1991; Waßmer et al., 1997; Fukuhara et al., 2003; 2004).

The carotid body consists of two different cell types (type I and type II cells), vasculature and connective tissues. Type I cells may be chemoreceptor cells sensitive to hypoxia, low pH and hypercapnea, and have various neuroactive substances (Lundberg et al., 1979; Oomori et al., 1994). During hibernation, these substances such as tyrosine hydroxylase (TH), methionine-enkephaline and GABA increased in the chipmunk carotid body (Fukuhara et al., 2004). For TH in carotid bodies, the increased level during hibernation drastically fell during 2 h of arousal after hibernation in chipmunks (Fukuhara et al., 2004) and in bats (Fukuhara et al., 2003). Type I cells enlarged during hibernation, and in contrast, the diameter of blood vessels around type I cells decreased by 13% during hibernation (Fukuhara et al., 2004). These data suggest that the carotid body plays a role in the regulation mechanism of chemoreception, which occurs in response to not only the hibernating state but also to periodical arousal from hibernation.

Hibernating animals awaken periodically, and 60-80% of their winter energy resources are consumed during arousal (Wang, 1978). During these periods, oxygen consumption increases markedly, and then mitochondria may be exposed to various oxidative stresses. Mitochondria may be a major source of reactive oxygen species (ROS) production within cells as a by-product of aerobic energy metabolism. The mitochondrial protein SP-22 was originally isolated from the bovine adrenal cortex as a substrate protein for mitochondrial ATP-dependent protease (Watabe et al., 1994; 1995) and it is reported to play an important role in the antioxidant defense mechanism of mitochondria (Araki et al., 1999; Shibata et al., 2003).

In this study, we examined immunohistochemical changes in SP-22 in carotid bodies during the euthermic, hibernation and arousal phases. Moreover, we observed morphological changes in blood vessels within the carotid body parenchyma during hibernation, and 1 h and 2 h after the onset of arousal from hibernation.

2. MATERIALS AND METHODS

2.1 Materials

Twelve adult chipmunks (*Tamias sibiricus*) weighing 80-100g were used in this study. They were housed individually with free access to food and water, and were placed in a refrigerated incubator with ventilation in constant darkness. The temperature was initially set at 14°C, was gradually reduced by 8°C over a 2-month period, and was maintained at 8 ± 2°C during the hibernation season. The hibernation state was verified daily by checking the excrement. Hibernating chipmunks were those who had been followed up for a week after the first verification of hibernation and were still hibernating. Hibernation was tested by the response to mechanical stimulus. Hibernating animals were divided into three groups: hibernating group, 1 h arousal group and 2 h arousal group. Arousal was induced by removal from the refrigerated incubator to room temperature (20°C) because chipmunks are very susceptible to mechanical and thermal disturbance. Euthermic animals were those maintained under the same conditions conducive to hibernation but who failed to hibernate.

2.2 Tissue Preparation

Animals in euthermic and those aroused from hibernation were anesthetized with an intraperitoneal injection of sodium pentobarbital (50 mg/kg body weight), while hibernating animals were under low-temperature anesthesia. For fixation, these animals were perfused through the heart with ice-cold 4% (v/v) paraformaldehyde and 0.2% (v/v) picric acid in 0.1 M phosphate buffer (pH 7.4) at a constant flow rate. Carotid bodies containing the common carotid artery and its branches were dissected out and these tissues were post-fixed in the same fixative for 8–12 h at 4°C. After washing briefly in 10 mM phosphate-buffered saline (PBS, pH 7.4), the carotid bodies were removed from adjacent tissues and rinsed for 12–24 h in 10 mM PBS containing 30% (w/v) sucrose. They were then frozen and sectioned at 10 μm with a cryostat (Histostat Microtome 2200, Meiwa Shoji Co., Ltd., Japan). The sections were mounted on silane-coated slides (Matsunami, Japan).

2.3 Immunohistochemistry

The carotid bodies were subjected to immunohistochemical investigation using polyclonal rabbit antibodies against SP-22, which was originally isolated from the bovine adrenal cortex as a substrate protein for mitochondrial ATP-dependent protease (Watabe et al., 1994; 1995). The sections were processed for immunohistochemistry according to the peroxidase–antiperoxidase method, as described previously (Fukuhara et al., 2003). In brief, the sections were incubated with the antibody against SP-22 (1:100) in 10 mM PBS with 0.2% (v/v) Triton-X for 18–24 h at room temperature. After rinsing in PBS-T, the sections were incubated with goat anti-rabbit IgG (1:200; Jackson, USA) in 10 mM PBS-T (pH 7.4) for 2 h. Next, the sections were rinsed with PBS-T, and then reacted with rabbit peroxidase-antiperoxidase complex (1:200; Dako, Denmark) in 10 mM PBS-T for 2 h. Peroxide activity was visualized with 0.05% (w/v) 3,3'-diaminobenzidine tetrahydrochloride (Sigma, St. Louis, MO, USA) in 50 mM Tris-HCl buffer (pH 7.5) containing 0.01% (v/v) hydrogen peroxide for 10 min.

2.4 Measurements

Morphometric analysis of SP-22-immunostained sections was performed using NIH-Image public domain software. The diameters of small blood vessels were measured on micrographs at a final magnification of ×1,220. The results were expressed as the mean ± standard deviations.

3. RESULTS

3.1 Immunohistochemical Results

Immunoreactive SP-22 was widely distributed in the cytoplasm in type I cells of carotid bodies in euthermic, hibernating, and 1 h and 2 h after the onset of arousal from hibernation groups, respectively (Fig. 1). The immunoreactive area of SP-22 in type I cells under hibernation increased in comparison with that under euthermic animals. At 2 h after the onset of arousal from hibernation, not only the immunoreactive area but also the intensity of SP-22 increased in comparison with during torpor. The ratio of the SP-22 immunostained area to the unit area in carotid bodies increased by 13.5% during torpor, and then further increased by 30.8% 2 h after the onset of arousal from torpor compared with the level during euthermia (Fig. 2).

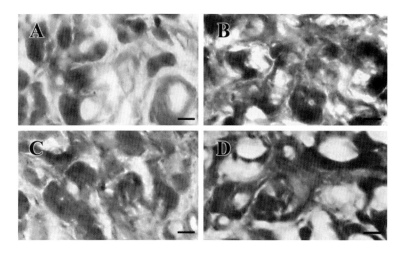

Figure 1. SP-22 immunoreactivity in chipmunk carotid bodies during euthermia (A), hibernation (B), 1 h (C) and 2 h (D) after the onset of arousal from hibernation. Scale bar = 30 μm.

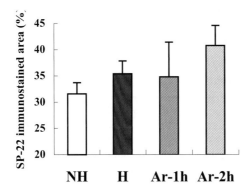

Figure 2. The percentage of the SP-22 immunostained area per section from euthermic (NH), hibernating (H), 1h (Ar-1h) and 2h (Ar-2h) after the onset of arousal from hibernation. Data were 31.2 ± 2.1% (n = 5, NH), 35.4 ± 2.4% (n = 5, H), 34.8 ± 6.6% (n = 7, Ar-1h) and 40.8 ± 3.8% (n = 11, Ar-2h), respectively.

3.2 Morphometric Changes of Blood Vessels

The diameter of small blood vessels in the carotid body parenchyma showed a 36% reduction during torpor compared with normothermia, but the diameter was enlarged approximately 2-fold 2 h after the onset of arousal from torpor, up to a 16% increase compared to that during euthermia.. Data are expressed as the mean ± SD (small blood vessel: euthermia n = 291, torpor n = 242, 1 h arousal n = 130, 2 h arousal n = 156) (Fig. 3).

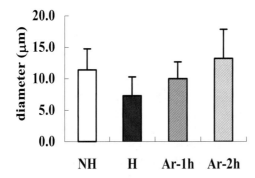

Figure 3. The mean diameter of small blood vessels in the carotid body parenchyma during euthermic (NH), hibernating (H), 1h (Ar-1h) and 2h (Ar-2h) after the onset of arousal from hibernation. Data of the mean diameter was 11.2 ± 3.3 (n = 291, NH), 7.0 ± 2.9 (n = 242, H) 10.0 ± 2.6 (n = 130, Ar-1h) and 13.2 ± 4.6 (n = 156, Ar-2h).

4. DISCUSSION

Hypothermia during hibernation is not continuous. Chipmunks wake at regular intervals, ranging from 1 to 7 days. During arousal, the body temperature increases to 37°C and they maintain this elevated body temperature for ~16 hours, subsequently reentering hibernation and maintaining a low body temperature until the next arousal. The progression from euthermy during arousal to torpor during hibernation takes about 24 hours but arousal from hibernation takes only 2 hours or less (Wang, 1988; Daan, 1991; Waβmer et al., 1997). Lee et al. (2002) observed about an 8-fold increase in oxygen consumption in the bat brain during 30 min of arousal from hibernation. In this study, the diameter of small blood vessels in the carotid body parenchyma showed 36% reduction during torpor compared with euthermia, but the diameter was enlarged approximately 2-fold 2 h after the onset of arousal from torpor. Vascular enlargement has been reported in rat carotid bodies after 8 weeks of hypoxic exposure (Kusakabe et al., 2004). It is considered that the hypoxic condition occurs during the course of periodical arousal from torpor. Urate levels as evidence of oxidative stress peaked at the time of maximal oxygen consumption during arousal and were lowest in torpid squirrels (Toien O et al., 2001). Drastic changes in blood flow and oxygen consumption during arousal from hibernation induced an increase in the risk of oxidative stress to sensitive tissues. Excess ROS as a hazardous substance is generated during arousal from hibernation (Toien et al., 2001; Drew et al., 2002; Hermes-Lima and Zenteno-Savin, 2002).

Hibernating animals may be considered to have an enhanced antioxidant defense system to protect their cells and organisms from oxidative damage during periodical arousal from hibernation. Previous studies indicate that several antioxidant substances participate in protection against oxidative damage during hibernation. Plasma ascorbate increased 3-4 fold in arctic and 13-lined squirrels during torpor (Drew KL et al., 1999), in contrast, tissue ascorbate concentrations increased significantly during arousal in the liver and spleen (Toien O et al., 2001). The activities of superoxide dismutase, ascorbate and glutathione peroxidase in brown adipose tissue are increased during hibernation (Buzadzic et al., 1990; Carey et al., 2003).

The carotid body is a highly vascularized organ. Stereological studies indicate that some 25% of the total volume of the carotid body is composed of blood vessels (Pallot DJ, 1987), and the carotid body has an enormous blood flow. It is suggested that oxygen consumption is some 70% greater within the carotid body than that in the cerebral cortex (Pallot DJ, 1987). Consequently, mitochondria may be exposed to several oxidative stresses during periodic arousal. Compared with summer-active squirrels, the levels of mitochondrial stress protein GRP75 were consistently higher in the intestinal mucosa of hibernators in each of five hibernation states (entrance, short-bout torpor, long-bout torpor, arousal and interbout euthermia) (Carey et al., 1999). Ultrastructural studies in dormice showed enlarged mitochondria with abundant vesicular cristae during hibernation in the adrenal cortex (Zancanaro et al., 1997) and in hepatocytes, pancreatic acinar cells and brown adipocytes (Malatesta M, 2001).

An antioxidant protein, SP-22, is specifically localized in mitochondria and is believed to play important roles in the regulation of cellular redox status by serving as a primary line of defense against H_2O_2 produced during respiration. The expression of SP-22 protein was enhanced about 1.5-4.6-fold when bovine aortic endothelial cells (BAEC) were exposed to various oxidative stresses (Araki et al., 1999). BAEC with an increased level of SP-22 protein caused by pretreatment with mild oxidative stress became tolerant to subsequent intense oxidative stress. In this study, the SP-22 immunostained area to the unit area in carotid bodies increased by 13.5% during torpor bouts, and then furthermore increased by 30.8% 2 h after the onset of arousal from torpor compared with the euthermic level. It is considered that SP-22 protein in carotid bodies may counter increased ROS production generated by the rapid increase in mitochondrial activity during periodic arousal from hibernation and may function as an antioxidant during the hibernation season.

These results indicated that the SP-22 of carotid bodies in hibernating animals may participate in maintaining the hibernation state and in marked physiological changes in periodic arousal during the hibernation season.

REFERENCES

Araki M, Nanri H, Ejima K, Murasato Y, Fujiwara T, Nakashima Y, Ikeda M (1999) Antioxidant function of the mitochondrial protein SP-22 in the cardiovascular system. J Biol Chem 274: 2271-8

Buzadzic B, Spasic M, Saicic ZS, Radojicic R, Petrovic VM, Halliwell B (1990) Antioxidant defenses in the ground squirrel (Citellus citellus). 2. The effect of hibernation. Free Radic Biol Med.9: 407-13.

Carey HV and Andrews MT and Martun SL (2003) Mammalian hibernation: cellular and molecular responses to depressed metabolism and low temperature. Physiol Rev 83: 1153-81

Carey HV, Sills NS, Gorham DA (1999) Stress proteins in mammalian hibernation. Am Zool 39: 825-835

Daan S, Barnes BM, Strijkstra AM (1991) Warming up for sleep? Ground squirrels sleep during arousals from hibernation. Neurosci Lett 128: 265-268

Drew KL, Osborne PG, Frerichs KU, Hu Y, Hallenbeck JM and Rice ME (1999) Ascorbate and glutathione regulation in hibernating ground squirrels. Brain Res 851: 1-8

Drew KL, Toien O, Rivera PM, Smith MA, Perry G, Rice ME (2002) Role of the antioxidant ascorbate in hibernation and warming from hibernation. Comp Biochem Physiol C Toxicol Pharmacol 133: 483-92

Fukuhara K, Senoo H, Yoshizaki K, Ohtomo K. (2003) Immunohistochemical study of the carotid body just after arousal from hibernation. Adv Exp Med Biol 536: 619-28

Fukuhara K, Yoshizaki K, Wu Y, Senoo H, Ohtomo K (2004) Immunohistochemical and morphological changes in chipmunk carotid body during hibernation. Akita J Med 31: 71-81

Hermes-Lima M, Zenteno-Savin T (2002) Animal response to drastic changes in oxygen availability and physiological oxidative stress. Comp Biochem Physiol C Toxicol Pharmacol.133: 537-56

Kusakabe T, Hirakawa H, Oikawa S, Matsuda H, Kawakami T, Takenaka T, Hayashida Y (2004) Morphological changes in the rat carotid body 1, 2, 4, and 8 weeks after the termination of chronically hypocapnic hypoxia. Histol Histopathol 19: 1133-40

Lee M, Choi I, Park K (2002) Activation of stress signaling molecules in bat brain during arousal from hibernation. J Neurochem 82: 867-73

Lundberg JM, Hokfelt T, Fahrenkrug J, Nilsson G, Terenius L (1979) Peptides in the cat carotid body (glomus caroticum): VIP-, enkephalin-, and substance P-like immunoreactivity. Acta Physiol Scand 107: 279-281

Malatesta M, Battistelli S, Rocchi MB, Zancanaro C, Fakan S, Gazzanelli G (2001) Fine structural modifications of liver, pancreas and brown adipose tissue mitochondria from hibernating, arousing and euthermic dormice. Cell Biol Int 25: 131-8

Oomori Y, Nakaya K, Tanaka H, Iuchi H, Ishikawa K, Satoh Y, Ono K (1994) Immunohistochemical and histochemical evidence for the presence of noradrenaline, serotonin and gamma-aminobutyric acid in chief cells of the mouse carotid body. Cell Tissue Res 278: 249-254

Pallot DJ (1987) The mammalian carotid body. Adv Anat Embryol Cell Biol 102: 1-91

Shibata E, Nanri H, Ejima K, Araki M, Fukuda J, Yoshimura K, Toki N, Ikeda M, Kashimura M (2003) Enhancement of mitochondrial oxidative stress and up-regulation of antioxidant protein peroxiredoxin III/SP-22 in the mitochondria of human pre-eclamptic placentae. Placenta 24: 698-705

Toien O, Drew KL, Chao ML, Rice ME (2001) Ascorbate dynamics and oxygen consumption during arousal from hibernation in arctic ground squirrels. Am J Physiol Regul Integr Comp Physiol 281: R572-83)

Wang LCH (1988) Mammalian hibernation: An escape from the cold. In: *Advances in Comparative and Environmental Physiology* (Gilles R, ed.), Berlin, Springer, Vol 2, pp 1-45

Wang LCH (1978) Time patterns and metabolic rates of natural torpor in the Richardson's ground squirrel. Can J Zool 57: 149-155)

Waβmer T, Wollnik F (1997) Timing of torpor bouts during hibernation in European hamsters (*Cricetus cricetus L*). J Comp Physiol B 167: 270-279

Watabe S, Kohno H, Kouyama H, Hiroi T, Hasegawa H, Yago N, Nakazawa T (1994) Purification and characterization of a substrate protein for mitochondrial ATP-dependent protease in bovine adrenal cortex. J Biochem (Tokyo) 115: 648-654

Watabe S, Hasegawa H, Takimoto K, Yamamoto Y, Takahashi SY (1995) Possible function of SP-22, a substrate of mitochondrial ATP-dependent protease, as a radical scavenger. Biochem Biophys Res Commun 213: 1010-1016

Zancanaro C, Malatesta M, Vogel P, Fakan S (1997) Ultrastructure of the adrenal cortex of hibernating, arousing, and euthermic dormouse, Muscardinus avellanarius. Anat Rec 249: 359-64

Potential Role of Mitochondria in Hypoxia Sensing by Adrenomedullary Chromaffin Cells

JOSEF BUTTIGIEG, MIN ZHANG, ROGER THOMPSON, COLIN NURSE

Department of Biology, McMaster University, Hamilton, Ontario, Canada L8S 4K1

1. INTRODUCTION

Exposure of the neonate to episodes of acute hypoxia during birth results in a variety of adaptive changes that include fluid re-absorption and secretion of surfactant in the lungs to promote air breathing (Slotkin and Seidler 1988). These physiological responses depend critically on catecholamine secretion from adrenomedullary chromaffin cells (AMC), which express a direct, developmentally-regulated hypoxia sensing mechanism, independent of the nervous system (Slotkin and Seidler 1988, 1986; Thompson et al., 1997). The hypoxic response in neonatal AMC, as well as their immortalized counterparts (i.e. MAH cells), appears to be mediated via inhibition of O_2-sensitive K^+ channels, though the signaling pathway is not completely understood (Fearon et al 2002; Thompson et al., 1997). These O_2-sensitive K^+ channels include large conductance Ca^{2+}-dependent K^+, i.e. BK or maxi-K^+, and delayed rectifier K^+ channels (Thompson and Nurse 1998; Thompson et al., 2002). Inhibition of these channels is thought to facilitate membrane depolarization, voltage-gated Ca^{2+} entry and catecholamine secretion (Thompson et al., 1997; Thompson and Nurse, 1998, 2000).

Though still controversial, a variety of mechanisms has been proposed to explain the O_2-sensitivity of specialized cells (Lopez-Barneo et al., 2001). A popular signaling pathway is based on hypoxia-induced changes in reactive oxygen species (ROS) derived from the protein complex NADPH oxidase (Fu et al. 2000) or the mitochondrial electron transport chain (ETC; Lopez-Barneo et al., 2001; Michelakis et al., 2004; Waypa et al., 2001). In neonatal AMC, reduction of ROS levels via NADPH oxidase does not appear to play a role in O_2- sensing since the hypoxic response in cells from NADPH oxidase-deficient mice (lacking the gp91^{phox} subunit) was indistinguishable from that in wild type cells (Thompson et al., 2002). However, this does not preclude the possibility that hypoxia may cause changes in ROS levels via other isoforms of NADPH oxidase, the mitochondrial ETC or Krebs cycle intermediates; e.g. α-ketoglutarate dehydrogenase (Michelakis et al., 2004; Starkov et al., 2004; Waypa et al., 2001).

There are conflicting reports on the critical sites of ROS changes in the mitochondria and whether the levels are increased or decreased during hypoxia (Michelakis et al., 2002; Starkov et al., 2004; Waypa et al., 2001). In this communication, we consider the hypothesis that hypoxia sensing in AMC involves a decrease in mitochondrial-derived ROS. To explore this possibility, we use primary cultures of neonatal AMC as well as cultures of immortalized MAH cells, with and without functioning mitochondria.

2. EXPERIMENTAL PROCEDURES

2.1 Cell Culture

Adrenomedullary Chromaffin Cells (**AMC**): Primary cultures enriched in adrenal chromaffin cells were prepared from neonatal (1 day- old) Wistar rats (Charles River, Quebec; Harlan, Madison, WI) as previously described (Thompson et al., 1997). All procedures were carried out according to the guidelines of the Canadian Council on Animal Care (CCAC).

MAH cell line: MAH cells (a generous gift from Dr. David Anderson and Dr. Laurie Doering) were grown in modified L-15/CO_2 as previously described (Fearon et al., 2002). Cultures were fed every 1–2 days and passaged every 3–4 days. Cells were then plated onto 35 mm culture dishes, coated with poly-D-lysine and laminin to promote cell adhesion.

Mitochondrial deficient MAH cells (ρ^0 **MAH cells**): MAH cells were grown in modified L-15/CO_2 medium supplemented with 0.6 % glucose, 1% penicillin/streptomycin, 10 % fetal bovine serum, 5 µM dexamethasone, 10 mM sodium pyruvate, 2 mM uridine, and 20 ng/ml of ethidium bromide, to inhibit mitochondrial function and division (Park et al., 2001). The medium was changed every 1-2 days. In these experiments, ρ^0 cells were plated at a density of 3×10^4 cells per ml. The ρ^0 status, i.e. the absence of mitochondrial DNA, was confirmed by testing for the lack of expression of mitochondrial DNA-encoded cytochrome oxidase I gene using PCR techniques. PCR amplification of the β-actin gene was used as control.

2.2 Electrophysiology

Voltage clamp data from neonatal AMC, wild type MAH, and ρ^0 MAH cells were obtained using the nystatin perforated-patch configuration of the whole-cell patch clamp technique as previously described (Thompson and Nurse, 1998; Thompson et al., 2002). Cells were held at -60 mV and step depolarized to the indicated test potential (between -70 and +70 mV in 10 mV increments) for 100 ms at a frequency of 0.1 Hz. In some experiments, ramp depolarizations from –100 to + 50 mV were used. The pipette solution contained in mM: K gluconate, 95; KCl, 35; NaCl, 5; CaCl2, 2; HEPES, 10; at pH 7.2, and nystatin (300-450 µg/ml). In these experiments, the bathing solution contained in mM: NaCl, 135; KCl, 5; $CaCl_2$, 2; $MgCl_2$, 2; glucose, 10; and HEPES, 10, at pH 7.4.

2.3 Luminol and Luciferase Chemiluminescence

Modified 24 well dishes were used in all chemiluminescence experiments designed to measure ROS levels using luminol, or ATP secretion using the luciferin-luciferase assay. Cells were plated on a layer of matrigel or laminin. Prior to obtaining chemiluminescence readings (relative light units or RLU), the growth medium was removed and replaced with 900 µl of Hepes-buffered bathing solution (see electrophysiology above). Following the addition of 200 µl of luciferin-luciferase solution (ATP determination kit; Molecular probes# A22066), or of bathing solution containing 120 µM luminol, the dish was placed in a Labsystem LuminoskanTM luminometer connected to a Pentium III computer as previously described (Buttigieg and Nurse, 2004). For stimulus application, 900 µl of normoxic (control) bathing solution was replaced with an equal volume of hypoxic solution (PO_2 = 15-20 mmHg). All luminescence records were obtained at 37°C.

3. RESULTS

3.1 Proximal but not Distal Mitochondrial Blockers Mimic Hypoxia

As previously reported (Thompson et al., 1997,2002; Thompson and Nurse 1998), acute hypoxia caused a reversible inhibition of outward K^+ current in neonatal AMC (Fig 1A). Application of the complex I blocker rotenone (1 µM) also caused a reversible inhibition of whole-cell outward K^+ current, similar to hypoxia (Fig 1B). Moreover, the presence of rotenone occluded the effect of hypoxia (Fig. 1B), suggesting both agents acted via a common pathway. In contrast, application of the complex IV blocker cyanide (2 mM) had negligible effect on outward K^+ current in several neonatal AMC (Fig 1C). In a few cases application of cyanide caused a reversible enhancement of outward K^+ current (not shown).

Since ATP is co-stored and co-released with catecholamines (Winkler and Westhead 1980), we monitored chromaffin cell secretion using the luciferin-luciferase assay for detection of extracellular ATP (Bumttigieg and Nurse, 2004). In neonatal AMC, hypoxia increased the luciferin-luciferase chemiluinescence signal (RLU) relative to that in normoxia (Fig. 1D), and this effect was mimicked by the complex I blocker rotenone (1 µM) (Fig. 1E; n =3). Consistent with the voltage clamp data, cyanide failed to stimulate ATP release from these cells (Fig 1F). Taken together, these data suggest that proximal, but not distal, ETC blockers mimic the effects of hypoxia in neonatal AMC.

3.2 Hypoxia and Rotenone Decrease ROS Levels in Neonatal Chromaffin Cells

To test directly whether ROS levels are altered during hypoxia we measured intracellular ROS production with the chemiluminescent probe, luminol (Michelakis et al., 2002). As exemplified in Fig. 2A, hypoxia caused a reversible decrease in ROS levels in neonatal AMC, and this effect was mimicked by rotenone (Fig. 2B). In contrast, cyanide (2-5 mM; n=3) produced an increase in ROS levels after 2 min exposure (not shown). These data support the hypothesis that O_2 sensing by neonatal AMC occurs via a decrease in intracellular ROS levels.

3.3 O$_2$-sensing in MAH Cells: Role of Mitochondria

We tested further for the importance of mitochondria in O$_2$-sensing in chromaffin cells by using the immortalized MAH cell line. Chromaffin-derived MAH cells deficient in functional mitochondria (i.e. ρ^0 MAH cells) were generated following exposure to low concentrations of ethidium bromide (Park et al., 2001). After multiple passages, ρ_0 MAH cells were generated and confirmation that they lacked functional mitochondria was obtained using PCR to probe for the mitochondrial marker, cytochrome oxidase I (COX I); this marker was absent in ρ^0 MAH cells (data not shown).

Preliminary data suggest that though normally-appearing K$^+$ currents could be recorded from ρ^0 MAH cells, there was no detectable hypoxic inhibition of outward K$^+$ currents as seen in wild type MAH cells (Fig 3). These data support a role for functioning mitochondria in O$_2$-sensing by MAH cells. Preliminary data also indicate that the secretory machinery in ρ^0 MAH cells is still intact since depolarization with 30 mM K$^+$ still evoked ATP release (unpublished observations).

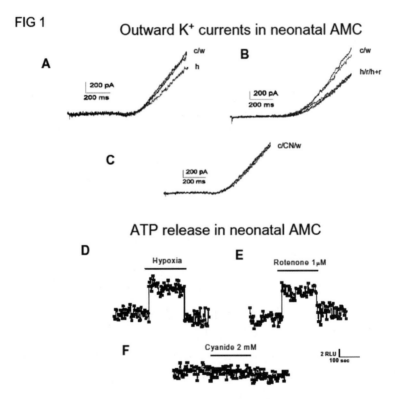

Figure 1. Effects of proximal and distal ETC inhibitors on K$^+$ current in neonatal AMC. Currents were evoked following ramp depolarizations from –100 to +50 mV; holding potential was –60 mV. Hypoxia (h) caused a reversible decrease in outward K$^+$ current (A), that was mimicked by the complex I blocker rotenone (r; 1 μM). The combined effect of hypoxia and rotenone (h + r) was non-additive (B). Application of the complex IV blocker cyanide (2 mM) had no effect in some cells (C), or increased K$^+$ current in others (not shown). Both hypoxia and rotenone increased secretion from neonatal AMC in D and E respectively, as monitored by ATP release using the luciferin-luciferase assay. In contrast, cyanide (2 mM) had no effect on secretion (F).

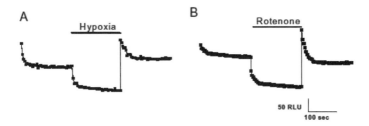

Figure 2. Effects of hypoxia and rotenone on ROS levels in neonatal AMC. Using luminol chemiluminescence (RLU) to measure ROS levels, both hypoxia and rotenone caused a reversible decrease in ROS in A and B respectively.

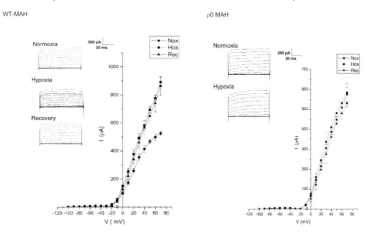

Figure 3. Hypoxia-evoked inhibition of K^+ current is dependent on functional mitochondria in immortalized chromaffin (MAH) cells. Left, hypoxia (PO_2 = 5-20 mmHg) caused a reversible decrease in outward K^+ current at more positive potentials; holding potential was -60 mV and the voltage was stepped from –100 to +60 mV in 10 mV increments. The corresponding I-V plot is shown to the right of sample traces for a typical example. Right, hypoxia had no detectable effect on K+ currents in mitochondrial-deficient (ρ^0) MAH cells, generated by the ethidium bromide technique (see Methods).

4. DISCUSSION

The mechanisms of O_2-sensing in the neonatal AMC are not well understood, though there is strong evidence that hypoxia acts via inhibition of outward K^+ current, which enhances cell depolarization and catecholamine secretion (Thompson et al., 1997, 2002; Thompson and Nurse 1998, 2000). In the present study we provide evidence supporting the hypothesis that hypoxia sensing in these cells involves the mitochondrial electron transport chain (ETC) and a decrease in ROS generation. The experimental approaches utilized blockers of the proximal and distal ETC, as well as the generation of mitochondrial-deficient ρ^0 cells, derived from an O_2-sensitive adrenal chromaffin cell line (MAH; Fearon et al., 2002).

Our data suggest that proximal but not distal ETC inhibitors mimic the hypoxic response in neonatal AMC. Thus, the complex I blocker rotenone mimicked hypoxia in causing inhibition of outward K^+ current and ATP secretion, whereas the complex IV blocker cyanide did not. Moreover, the combined effects of hypoxia and rotenone were non-additive, suggesting both acted via a common pathway. Additionally, both hypoxia and rotenone caused a reversible decrease in ROS levels, thereby supporting a role for changes in intracellular ROS as a key link between the O_2-sensor and K^+ channel inhibition. The hypothesis that functional mitochondria were required for O_2-sensing was supported in preliminary experiments on mitochondria-deficient ρ^0 MAH cells. These cells failed to respond to hypoxia with inhibition of K^+ current, as did their wild type counterparts (see Fearon et al., 2002). In our current working model for O_2-sensing in neonatal AMC, hypoxia causes a decrease in ROS levels (e.g. H_2O_2) possibly through interaction with some site in the proximal ETC. This decrease in ROS levels is proposed to lead, either directly or indirectly, to modulation of K^+ channel activity (Ortega-Saenz 2003; Michelakis et al., 2002; Ward et al., 2004), and catecholamine secretion. Further studies are required to test some of the predictions of this model.

ACKNOWLEDGEMENTS

This work was supported by the Heart and Stroke Foundation of Ontario and the Natural Science and Engineering Research Council of Canada (NSERC).

REFERENCES

Buttigieg J, Nurse C (2004). Detection of hypoxia-evoked ATP release from chemoreceptor cells of the rat carotid body. *Biochem. Biophys. Res. Commun.* **322**: 82-7.

Fearon IM, Thompson RJ, Samjoo I, Vollmer C, Doering LC, Nurse CA. (2002). O_2-sensitive K^+ channels in immortalized rat chromaffin-cell-derived MAH cells. *J Physiol.* **545**:807-18.

Fu XW, Wang D, Nurse CA, Dinauer MC, Cutz E. (2000). NADPH oxidase is an O_2 sensor in airway chemoreceptors: Evidence from K^+ current modulation in wild-type and oxidase-deficient mice. *Proc. Nat. Acad. Sci. USA.* **97**:4374-4379.

Lopez-Barneo J, Pardal R, Ortega-Saenz P. (2001). Cellular mechanism of oxygen sensing. *Annu Rev Physiol.* **63**:259-87.

Michelakis ED, Thebaud B, Weir EK, Archer SL (2004). Hypoxic pulmonary vasoconstriction: redox regulation of O_2-sensitive K^+ channels by a mitochondrial O_2-sensor in resistance artery smooth muscle cells. *J Mol Cell Cardiol.* **37**:1119-36.

Michelakis ED, Hampl V, Nsair A, Wu XC, Harry G, Haromy A, Gurtu R, Archer SL. (2002). Diversity in mitochondria function explains differences in vascular oxygen sensing. *Circ. Res.* **90**:1307-1215.

Ortega-Saenz P, Pardal R, Garcia-Fernandez M, Lopez-Barneo J. (2003). Rotenone selectively occludes sensitivity to hypoxia in rat carotid body glomus cells. *J Physiol.* **548**:789-800.

Park KS, Nam KJ, Kim JW, Lee YB, Han CY, Jeong JK, Lee HK, Pak YK (2001). Depletion of mitochondrial DNA alters glucose metabolism in SK-Hep1 cells. *Am J Physiol Endocrinol Metab.* **280**:E1007-14.

Seidler FJ, Slotkin TA. (1986). Ontogeny of adrenomedullary responses to hypoxia and hypoglycemia: role of splanchnic innervation. *Brain Res Bull.* **16**(1):11-4.

Slotkin TA, Seidler FJ (1988). Adrenomedullary catecholamine release in the fetus and newborn: secretory mechanisms and their role in stress and survival. *J. Dev. Physiol.* **10**:1-16.

Starkov AA, Fiskum G, Chinopoulos C, Lorenzo BJ, Browne SE, Patel MS, Beal MF. (2004). Mitochondrial alpha-ketoglutarate dehydrogenase complex generates reactive oxygen species. *J Neurosci.* **24**(36):7779-88.

Thompson, R.J., Jackson, A. and Nurse, C.A. (1997) Developmental loss of hypoxic chemosensitivity in rat adrenomedullary chromaffin cells. *J. Physiol. (Lond).* **498**:503-510.

Thompson RJ, Nurse CA (1998). Anoxia differentially modulates multiple K^+ currents in neonatal rat adrenal chromaffin cells. *J. Physiol. (Lond).* **512**:421-434.

Thompson, RJ, Nurse CA (2000). O_2-chemosensitivity in developing rat adrenal chromaffin cells. *Adv Exp Med Biol.* **475**:601-9.

Thompson, R.J., Farragher, S.M., Cutz, E. and Nurse, C.A. (2002) Developmental regulation of O_2 sensing in neonatal adrenal chromaffin cells from wild-type and NADPH-oxidase-deficient mice *Pflugers. Arch.* **444**: 539-48.

Ward JP, Snetkov VA, Aaronson PI (2004). Calcium, mitochondria and oxygen sensing in the pulmonary circulation. *Cell Calcium.* **36**:209-20.

Waypa GB, Chandel NS, Schumacker PT (2001). Model for hypoxic pulmonary vasoconstriction involving mitochondrial oxygen sensing. *Circ Res.* **88**:1259-1266.

Winkler H, Westhead E. (1980). The molecular organization of adrenal chromaffin granules. *Neuroscience.* **5**:1803-23.

Localization of Ca/Calmodulin-Dependent Protein Kinase I in the Carotid Body Chief Cells and the Ganglionic Small Intensely Fluorescent (SIF) Cells of Adult Rats

HISAE HOSHI, HIROYUKI SAKAGAMI, YUJI OWADA AND HISATAKE KONDO

Division of Histology, Department of Cell Biology, Graduate School of Medicine, Tohoku University, Sendai, JAPAN

1. INTRODUCTION

Control of cellular functions by extracellular signal-induced elevation of intracellular Ca is a common theme in excitable biosystems including the carotid body chemoreception. The Ca-signal is delivered to appropriate intracellular target proteins via phosphorylation catalyzed by multifunctional Ca/calmodulin-dependent protein kinases (CaM kinases), which are composed of types I, II and IV (Hanson and Schulman, 1992; Sakagami and Kondo, 1998). In order to see how CaM kinases are involved in the chemoreception, the present study examined the localization of CaM kinase I in immuno-light and electron microscopy in the rat carotid body as well as its adjacent superior cervical ganglion. The latter contains SIF (small intensely fluorescent) cells, at least some of which exhibit cytological features including the innervation similar to those of the carotid body chief cells (Kondo, 1977).

2. MATERIALS AND METHODS

2.1 Production and Characterization of the Antibody Against CaM Kinase I

The entire coding region of CaM kinase Iα was amplified by PCR with a rat brain first-strand cDNA library and a combination of primers (sense, 5'TGAATTC(EcoRI) ATGCCAGGGGCAGTGGAAGGCCCC3'; antisense, 5'AGAATTC (EcoRI) TCAGTCCATGGCCCTAGAGCTTGG3'). After digested with EcoRI restriction enzyme, the cDNA fragment was subcloned into the EcoRI site of the pGEX4T-1 expression vector (Pharmacia LKB Biotechnology, Piscataway, NJ, USA). The glutathione-S-transferase (GST)-CaM kinase Iαfusion protein was induced in bacteria by addition of isopropyl β D-thiogalactopyranoside and purified with glutathione-Sepharose 4B (Pharmacia LKB Biotechnology, Piscataway, NJ, USA) according to the manufacturer's protocols. Two New Zealand White rabbits were immunized by intradermal injection of GST-CaMKIα fusion protein in Freund's adjuvant (Difco Laboratories, Detroit, MI, USA) six times at a 3 weeks interval. To characterize the specificity of the antibody, HeLa cells were transfected with an empty vector or the expression vectors encoding either FLAG-tagged CaMKIα, CaMKIβ2, CaMKIγ1, or CaMKIδ. After harvested with boiling SDS-PAGE sample buffer, the lysates were subjected to immunoblot analysis. As a result, the obtained antibody reacted with synthetic rat CaM kinase Iα and Iδ, but not Iβ or Iγ in a single band respectively. More details on the procedure of the production of the polyclonal rabbit anti-rat CaM kinase I is described elsewhere (Sakagami et al., 2005)

2.2 Immunohistochemistry

Wistar-Imamichi rats on postnatal 7 weeks were perfused transcardially with 0.9% saline, followed by 4% paraformaldehyde in 0.1 M phosphate buffer (PB) under anesthesia of sodium pentobarbital (35 mg/kg body weight). Following perfusion, the carotid bifurcation including the carotid body and the superior cervical ganglion were immediately removed and postfixed in the same fixative overnight. After immersion in 30% sucrose/0.1 M PB overnight at 4C, the tissue blocks were sectioned on a cryostat and mounted onto gelatin-coated slide glasses. The sections were incubated with the present antiserum at a dilution of 1:1000 in PBS overnight at 4C. After washing three times with PBS, the sections were incubated in biotinylated goat anti-rabbit IgG for 1 h. The antigen-antibody reaction sites were detected by a method employing the avidin-biotin-HRP complex (Vector ABC kit, Vector Labs) and DAB (diamino benzidine). For immuno-electron microscopy, tissue sections after detection of the immune reaction sites were postfixed with 0.5 %OsO4 in PBS and embedded in Epon and sectioned with the ultramicrotome. In the control experiment, sections were incubated in the primary antiserum preabsorbed with 1 mM CaM kinase Iα or Iδ-MBP fusion protein, but all other steps were identical to those described above, and no significant immunoreaction was detected in any sections.

3. RESULTS

In immuno-light microscopy, numerous cells immunoreactive with the present antibody were found in the carotid body and they were closely aggregated in small groups to form lobules around capillaries (Fig 1a). In immuno-electron microscopy, the immunoreactive cells were characterized by numerous granular vesicles, 100-200 nm in diameter (Figs 2a, 2b), and they were identified as the chief cells (Kondo, 1971, 1976). The immunoreactive material was localized in the cytoplasm and it was not homogeneously distributed throughout the cells, but much denser in close association with the granular vesicles whose cores were electron-dense. No immunoreaction was detected in the nuclei. All the chief cells were not immunoreactive, but there were a substantial number of the chief cells which exhibited the immunoreactivity at negligible levels and located in direct apposition to the immunopositive chief cells (Fig 2b). No marked differences were discerned between the immunopositive and immunonegative chief cells in terms of the size of granular vesicles, their population density and other cytological features. No immunoreactivity was detected in nerve fibers innervating the chief cells, or sustentacular cells, or capillary endothelial cells.

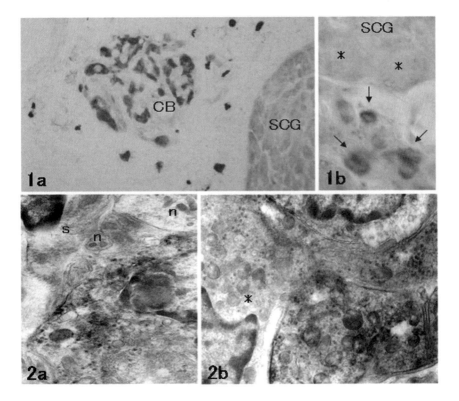

Figure 1a and 1b. Immuno-light micrographs of the carotid body (CB) and its adjacent superior cervical ganglion (SCG) of adult rat. Note numerous chief cells intensely immunoreactive for CaM kinase I in the carotid body (Fig 1a) and a few intensely immunoreactive SIF cells (arrows) within SCG, in contrast to the weak immunoreactivity in principal ganglion cells (asterisks) (Fig 1b). X150 (Fig 1a) X1,500(Fig 1b).

Figure 2a and 2b. Immuno-electron micrographs of chief cells of the carotid body. The immunoreactive material is deposited more densely in the cytoplasm in close association with the characteristic granular vesicles in immunopositive cell (Fig 2a). Note immunonegative chief cells (*) in apposition to the immunopositive chief cell (Fig 2b). Also note the absence of the immunoreactivity in nerve fibers (n) or sustentacular cells (s). X 16,000.

In the superior cervical ganglion, the immunoreactivity for CaM kinase I was weak in all the principal ganglion cells, In contrast, a few clusters of small cells were occasionally found to be intensely immunoreactive among weakly immunoreactive ganglion cells (Fig 1b). In immunoelectron microscopy the small and intensely immunoreactive cells were characterized by granular vesicles, 100-200 nm in diameter (Figs 3a, 3b), and they were identified as small intensely fluorescent (SIF) cells or small granule-containing cells (Matthews and Raisman, 1969). Similar to the chief cells, the immunoreactivity was localized in the cytoplasm and it was especially evident in close association with the granular vesicles whose cores were electron-dense. Immunonegative SIF cells were often seen next to immunoreactive SIF cells. No significant immunoreaction was seen in nerve fibers close to and within the SIF cell clusters, or ganglionic satellite cells or Schwann cells.

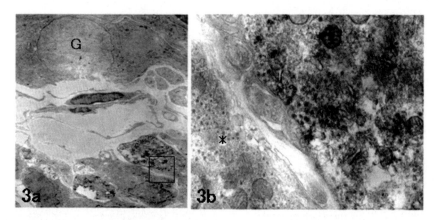

Figure 3a and 3b. Immuno-electron micrographs of ganglionic SIF cells. An area enclosed by a rectangle in Fig 3a is shown at higher magnification in Fig 3b. Note the immunoreactive material deposited more densely in the cytoplasm in close association with the characteristic granular vesicles and also note the occurrence of cell (*) immunonegative for CaM kinase I in close apposition to the immunopositive cell. G: principal ganglion cell. X 3,500 (Fig 3a), X16,000 (Fig 3b).

4. DISCUSSION

This is the first report demonstrating the intense immunoreactivity for CaM kinase I in the chief cells of the carotid body chemoreceptor. Because of the limited resolution in the ABC and DAB method for detection of the antigen-antibody reaction sites, the exact localization of the antigen CaM kinase I in the cytoplasm at ultrastructural levels requires for the re-examination using the immuno-gold method. However, the much denser localization of the DAB reaction deposits in portions of the cytoplasm close to the granular vesicles suggests some intimate functional relation of CaM kinase I with the secretion of catecholamines from the granular vesicles in the chief cells. However, since the present study disclosed the immunoreactivity for CaM kinase I in the adrenal chromaffin cells to be much weaker than that in the carotid body chief cells (data not shown), the intense immunoreactivity for CaM kinase I in the chief cells suggests that there may be some specific mechanisms in the secretion and/or postsecretion of catecholamines from the granular vesicles in these cells different from the chromaffin cells. It is further suggested that the specific mechanisms requiring CaM kinase I-involvement may be intimately related to the chemosensory transduction.

The occurrence of chief cells immunonegative for CaM kinase I in apposition to the immunopositive chief cells should also be noted when considering the significance of CaM kinase I in the carotid body. Although there have been reports suggesting the heterogenous population of the chief cells in terms of the granular vesicle size and shape, and the catecholamine species in the granular vesicles (McDonald and Mitchell, 1975), no clear differences were noted in the cytological features of the granular vesicles between the immunopositive chief cells and immunonegative ones in the present study. This suggests that the difference in the immunoreactivity for CaM kinase I represents the variety in some functional state of the chief cells. Therefore, it is necessary to examine whether or not the expression and/or localization of CaM kinase I is changeable in response to the chemoreceptive stimuli. In this regard, a previous finding should be noted that the expression of CaM kinase I in the retina is regulated by light stimulation (Tsumura et al., 1999).

The present study also disclosed for the first time the intense immunoreactivity for CaM kinase I in the ganglionic SIF cells. Because SIF cells immunonegative for CaM kinase I were found in close proximity to the positive SIF cells, and because the frequency of occurrence of the immunopositive SIF cells seems to be smaller than the frequency of occurrence of the cells by the conventional catecholamine-fluorescence (Matthews and Raisman, 1969), the CaM kinase I-immunopositive SIF cells represent a subpopulation, but not all, of the ganglionic SIF cells.

What is the functional significance of the occurrence of intense CaM kinase I-immunoreactivity in the ganglionic SIF cells in common with the carotid body chief cells? In this regard, the previous findings by Kondo (1976, 1977) should be noted: Some SIF cells receive an innervation similar to the carotid body chief cells in terms of the dominancy of afferent synapses in en-passant forms along the trajectory of innervating nerve fibers, suggesting that some SIF cells play chemoreceptive roles in the same way as the carotid body chief cells. Therefore, the intense immunoreactivity for CaM kinase I in the cytoplasm of these two cell types in common may suggest some specific involvement of CaMKI in the chemoreception played by the two cells in common.

It is necessary as the next step to obtain antibodies individually specific to each of CaM kinase Iα and Iδ, and to clarify exactly which of the two isoforms is responsible for the present immunoreactivity. It is further important to clarify the exact mechanism for the isoform(s) of CaM kinase I to be involved in the chemoreception.

REFERENCES

Hanson PL., Schulman H. Neuronal Ca/calmodulin-dependent protein kinase. Annu. Rev. Biochem. 1992; 61:559-601.

Kondo H. Innervation of the carotid body of the adult rat. A serial ultrathin section analysis. Cell Tiss Res. 1976; 173:1-15.

Kondo H. Innervation of SIF cells in the superior cervical and nodose ganglia: An ultrastructural study with serial sections. Biol Cell 1977; 30:253-264.

Matthews MR., Raisman G. The ultrastructure and somatic efferent synapses of small granule-containing cells in the superior cervical ganglion. J Anat. 1969; 105:255-282.

Mc Donald DM., Mitchell RA. The innervation of glomus cells, ganglion cells and blood vessels in the rat carotid body: a quantitative ultrastructural analysis. J Neurocytol 1975; 4: 177-230.

Sakagami H., Kondo H. Gene expression and regulation of Ca/calmodulin-dependent protein kinase type IV in the nervous tissues. In "Dynamic Cells: Cell Biology of the 21th Century", ed. By Yagihashi S, Kachi T. & Wakui M., Elsevier Sciences BV, 1998.

Sakagami H., Nishimura H., Takeuchi Y., Fukunaga K., Kondo H. Neuronal expression of Ca/calmodulin-dependent protein kinase Id and its nuclear translocation in response to Ca stimuli. 2005; in prep.

Tsumura T., Murata A., Yamaguchi F., Sugimoto K., Hasegawa H., Hatase O., Nairn AC., Tokuda M. The expression of Ca/calmodulin-dependent protein kinase in rat retina is regulated by light stimulation. Vision Res 1999; 39:3165-3173.

Dual Origins of the Mouse Carotid Body Revealed by Targeted Disruption of *Hoxa3* and *Mash1*

YOKO KAMEDA

Department of Anatomy, Kitasato University School of Medicine, Sagamihara, Kanagawa 228-8555, Japan

1. INTRODUCTION

The carotid body is the main arterial chemoreceptor that senses oxygen levels in the blood. In mammalian species, the carotid body is localized in the carotid bifurcation and innervated by the carotid sinus nerve consisting of sensory fibers from the glossopharyngeal nerve. The organ also receives the ganglioglomerular nerve issuing from the superior cervical ganglion of sympathetic trunk (Verna, 1979). In the mouse, especially, the carotid body joins with the superior cervical ganglion and is penetrated by nerve bundles derived from the ganglion (Kameda et al., 2002). In contrast to mammalian species, the carotid body of chickens is situated in the cervico-thoracic region together with the thyroid, parathyroid and ultimobranchial glands, which form a continuous series along the common carotid artery. The organ is located between the distal (nodose) ganglion of the vagus nerve and the recurrent laryngeal nerve and supplied richly with their branches (see Kameda, 2002 for references).

The carotid body is made up of two cell types, glomus cells (chief cells or type I cells) and sustentacular cells (type II cells). The glomus cells contain many dense-cored vesicles which store biogenic amines including serotonin (5-HT), dopamine and noradrenaline, catecholamine-synthesizing enzymes such as tyrosine hydroxylase (TH) and dopamine β-hydroxylase, and neuropeptides including enkephalin and neuropeptide Y (NPY)(Chiocchio et al, 1966; Oomori et al., 1994; Kameda et al., 2002). The glomus cell groups are enveloped by processes of sustentacular cells which are immunoreactive for glial markers, i.e., glial fibrillary acid protein (GFAP), S-100 protein and vimentin (Kameda, 1996; Nurse and Fearon, 2002; Kameda, 2005). The sustentacular cells contain aggregations of intermediate filaments in both cell bodies and processes and have supporting function. The glomus cells of the mouse carotid body share TuJ1, PGP9.5, TH and NPY immunoreactivities with the sympathetic ganglion. The cells also exhibit immunoreactivity for 5-HT, whereas the sympathetic neurons do not.

2. DEVELOPMENT OF THE CAROTID BODY

A specific condensation of mesenchymal cells in the wall of the third arch artery represents the first morphological sign of carotid body development in both mammals and birds (Kondo, 1975; Kameda et al., 1994). It is well known that the outer walls of the branchial arch arteries are formed by the neural crest cells that migrate from the hindbrain neural folds (Le Lièvre and Le Douarin, 1975; Waldo et al., 1999; Jiang et al., 2000). In the mouse embryos, the carotid body primordium is first recognized in the wall of the third arch artery at E 13.0. At this stage, the primordium is enclosed with the neuronal progenitors and nerve fibers immunoreactive for TuJ1 and PGP 9.5, which are continuous with the superior cervical sympathetic ganglion (Kameda et al., 2002). The neuronal progenitors also exhibit immunoreactivity for TH. From E 13.5, the cells immunoreactive for TuJ1, PGP9.5 and TH began to appear in the carotid body primordium. At E 15.5, the carotid body is filled with cells displaying neuronal markers, TuJ1 and PGP9.5, and there are also many TH- and NPY-immunoreactive cells in it. 5-HT-immunoreactive cells began to appear in the carotid body at E 16.5.

In the chick embryos, the carotid body rudiment is formed in the wall of the third arch artery at E 8.0. The rudiment is located close to the distal vagal ganglion which exhibits immunoreactivity for TuJ1, PGP9.5 and HNK-1. The cells exhibiting the neuronal markers, TuJ1, PGP9.5 and HNK-1, first surround and then invade the carotid body rudiment. The neuronal cells also represent glomus cell markers, TH, NPY, 5-HT and chromogranin A. They are distributed in the connective tissue surrounding the carotid body primordium at E 9 and appear within the primordium from 10 days onwards (Kameda et al., 1990). Furthermore, electron microscopic study has demonstrated that the epithelial cells containing secretory granules are first detected around the carotid body rudiment which still consists of mesenchymal cells and nerve fibers, and then appear within the rudiment (Kameda, 1994). The neuronal progenitors derived from the distal vagal ganglion invade not only the carotid body primordium but also the wall of the third arch artery. Therefore, the chicken glomus cells are widely distributed in the wall of the common carotid artery and its branches supplying the endocrine organs such as the carotid body, thyroid, parathyroid and ultimobranchial glands (Kameda, 2002). It has been shown by immunoelectron microscopy and in situ hybridization that glomus cells distributed in the wall of the common carotid artery contain many dense-cored vesicles immunoreactive for NPY and also express intense signals for NPY mRNA (Kameda et al., 1999).

3. THE DEFECT OF CAROTID BODY FORMATION IN *HOXA3* NULL MUTANT MOUSE

Hoxa3 is a member of the *Hox* family of transcription factors, which mediate the formation of the mammalian body plan along the antero-posterior axis (Trainor and Krumlauf, 2000). Hoxa3 is expressed in ectomesenchymal neural crest cells that contribute to the development and differentiation of the third pharyngeal arch (Watari et al., 2001). Mice with a targeted disruption of the *Hoxa3* gene fail to initiate formation of the thymus and parathyroid primordia, which arise from the third pharyngeal pouch endoderm through interactions with surrounding neural crest cells (Kameda et al, 2004). Furthermore, the *Hoxa3* homozygous null mutants show the bilateral defects of the common carotid artery, which is derived from the third branchial arch artery (Kameda et al., 2003). The tunica media of the great arteries derived from the arch arteries are formed by the ectomesenchymal neural crest cells. To assess the cause of the regression of the third arch artery, we have examined whether the neural crest cells of the *Hoxa3* null mutants are able to colonize the third pharyngeal arch. To visualize the neural crest cells, the *Hoxa3* heterozygous mutants were crossed with connexin43-lacZ transgenic mice in which β-galactosidase expression is restricted to the derivatives of neural crest (Lo et al., 1997). The migration of neural crest cells from the neural tube to the third branchial arch is not affected in the *Hoxa3* homozygous mutants (Chisaka and Kameda, 2005). The initial formation of the third arch artery is not disturbed.

The artery, however, regresses in the null mutants at E 11.5, when differentiation of the third pharyngeal arch begins. The pharyngeal arches are transient embryonic structure. In particular, the third arch rapidly grows and differentiates during E 11.5 – E 12.5. In the *Hoxa3* null mutant, the differentiation of the third pharyngeal arch is delayed and the hypoplastic third arch still remains at E 12.5. The number of proliferating cells in the third arch of null mutants is markedly decreased, compared with that in wild-type embryos. Taken together, these results indicate that Hoxa3 is required for the growth and differentiation of the third pharyngeal arch and its defective development may induce the regression of the third arch artery.

The *Hoxa3* homozygous null mutant mice have the absence of the carotid body (Kameda et al., 2002). The null mutant embryos are unable to initiate formation of the carotid body primordium at E 13.0, because the third arch artery disappears at E 11.5. The superior cervical sympathetic ganglion develops normally and its projections enclose the internal carotid artery in the *Hoxa3* null mutants. The neuronal precursors derived from the ganglion, however, fail to differentiate into glomus cells due to the absence of the carotid body primordium. The precursors may not acquire a glomus cell phenotype unless they colonize the carotid body primordium. In the mutant mice, the superior cervical ganglion exhibits hypetrophy, probably reflecting a defect in its target organ, the carotid body.

4. ABSENCE OF GLOMUS CELLS IN MASH1 NULL MUTANT MICE

Mash1, a mammalian homologue of *Drosophila achaete-scute (asc)* genes, is expressed in the three main divisions of the developing autonomic nervous system: the sympathetic, parasympathetic and enteric systems (Johnson et al, 1990; Guillemot et al., 1993). The sympathetic nervous system is derived from the neural crest cells that condense adjacent to the dorsal aorta (Le Douarin and Kalcheim, 1999). In *Mash1* homozygous null mutant mice, the neural crest cells arrive to the dorsal aorta and give rise to sympathetic progenitors, but the progenitors fail to undergo complete differentiation, resulting in the absence of the sympathetic ganglia (Hirsch et al., 1998). The *Mash1* null mutant embryos exhibit the normal formation of the carotid body rudiment at E 13.0 and the rudiment continues to develop. However, there are no cells expressing neuronal and glomus cell markers such as TuJ1, PGP 9.5, TH and NPY in the carotid body throughout development (Kameda, 2005). Furthermore, no 5-HT cells appear in the carotid body. Because the superior cervical ganglion is absent, neuronal progenitors derived from the ganglion lack in the mutant carotid body rudiment, resulting in the defect of the glomus cell formation. On the other hand, many cells immunoreactive for S-100 protein, a sustentacular cell marker, appear in the mutant carotid body at E 15.5, when the wild-type carotid body yet contains few S-100 immunoreactive cells. Electron microscopic study has demonstrated that in the *Mash1* null mutants at birth, most of the carotid body consists of the cells exhibiting mesenchymal-like morphology and of unmyelinated nerve fibers (Kameda, 2005). These ultrastructural features of the null mutant carotid body resemble those of the E 8.0 chick carotid body rudiment before it receives neuronal precursors from the distal vagal ganglion (Kameda, 1994). Thus, the carotid body of newborn *Mash1* null mutants is made up of sustentacular cells but lacks glomus cells.

In conclusion, the results indicate that the mouse carotid body is formed by at least two distinct lineages of precursors. One is derived from the superior cervical ganglion, Mash1-dependent and gives rise to glomus cells immunoreactive for TuJ1, PGP9.5, TH, NPY and 5-HT. The other is derived from mesenchymal cells of the third arch artery wall, Hoxa3-dependent and gives rise to sustentacular cells expressing S100, GFAP and vimentin.

REFERENCES

Chisaka O, Kameda Y (2005) *Hoxa3* regulates the proliferation and differentiation of the third pharyngeal arch mesenchyme in mice. Cell Tissue Res 320:77-89

Chiocchio SR, Biscardi AM, Tramezzani JH (1966) Catecholamines in the carotid body of the cat. Nature 212: 834-835

Guillemot F, Lo L-C, Johnson JE, Auerbach A, Anderson DJ, Joyner AL (1993) Mammalian *achaete-scute* homolog 1 is required for the early development of olfactory and autonomic neurons. Cell 75:463-476

Hirsch MR, Tiveron M-C, Guillemot F, Brunet JF, Goridis C (1998) Control of noradrenergic differentiation and Phox2a expression by MASH1 in the central and peripheral nervous system. Development 125:599-608

Jiang X, Rowitch DH, Soriano P, McMahon AP, Sucov HM (2000) Fate of the mammalian cardiac neural crest. Development 127: 1607-1616

Johnson JE, Birren SJ, Anderson DJ (1990) Two rat homologues of *Drosophila achaete-scute* specifically expressed in neuronal precursors. Nature 346:858-861

Kameda Y (1994) Electron microscopic study on the development of the carotid body and glomus cell groups distributed in the wall of the common carotid artery and its branches in the chicken. J Comp Neurol 348:544-555

Kameda Y (1996) Immunoelectron microscopic localization of vimentin in sustentacular cells of the carotid body and the adrenal medulla from guinea pigs. J Histochem Cytochem 44:1439-1449

Kameda Y (2002) Carotid body and glomus cells distributed in the wall of the common carotid artery in the bird. Micr Res Techn 59:196-206

Kameda Y (2005) *Mash1* is required for glomus cell formation in the mouse carotid body. Dev Biol, in press

Kameda Y, Amano T, Tagawa T (1990) Distribution and ontogeny of chromogranin A and tyrosine hydroxylase in the carotid body and glomus cells located in the wall of the common carotid artery and its branches in the chicken. Histochemistry 94:609-616

Kameda Y, Yamatsu Y, Kameya T, Frankfurter A (1994) Glomus cell differentiation in the carotid body region of chick embryos studied by neuron-specific class III β-tubulin isotype and Leu-7 monoclonal antibodies. J Comp Neurol 348:531-543

Kameda Y, Miura M, Ohno S (1999) Ultrastructural localization of neuropeptide Y and expression of its mRNA in the glomus cells distributed in the wall of the common carotid artery of the chicken. J Comp Neurol 413:232-240

Kameda Y, Nishimaki T, Takeichi MO, Chisaka O (2002) Homeobox gene *Hoxa3* is essential for the formation of the carotid body in the mouse embryos. Dev Biol 247:197-209

Kameda Y, Watari-Goshima W, Nishimaki T, Chisaka O (2003) Disruption of the *Hoxa3* homeobox gene results in anomalies of the carotid artery system and the arterial baroreceptors. Cell Tissue Res 311:343-352

Kameda Y, Arai Y, Nishimaki T, Chisaka O (2004) The role of *Hoxa3* gene in parathyroid gland organogenesis of the mouse. J Histochem Cytochem 52:641-651

Kondo H (1975) A light and electron microscopic study on the embryonic development of the rat carotid body. Am J Anat 144:275-294

Le Douarin NM, Kalcheim C (1999) The neural crest (Second Edition). Cambridge University Press, Cambridge.

Le Lièvre CS, Le Douarin NM (1975) Mesenchymal derivatives of the neural crest: analysis of chimaeric quail and chick embryos. J Embryol Exp Morph 34:125-154

Lo CW, Cohen MF, Huang GY, Lazatin BO, Patel N, Sullivan R, Pauken C, Park SMJ (1997) *Cx43* gap junction gene expression and gap junctional communication in mouse neural crest cells. Dev Genet 20:119-132

Nurse CA, Fearon IM (2002) Carotid body chemoreceptors in dissociated cell culture. Microsc Res Tech 59:249-255

Oomori Y, Nakaya K, Tanaka H, Iuchi H, Ishikawa K, Satoh Y, Ono K (1994) Immunohistochemical and histochemical evidence for the presence of noradrenaline, serotonin and gamma-aminobutyric acid in chief cells of the mouse carotid body. Cell Tissue Res 278:249-254

Trainor PA, Krumlauf R (2000) Patterning the cranial neural crest: hindbrain segmentation and *hox* gene plasticity. Nature Rev 1:116-124

Verna A (1979) Ultrastructure of the carotid body in the mammals. Int Rev Cytol 60:271-330

Waldo KL, Lo CW, Kirby ML (1999) Connexin 43 expression reflects neural crest patterns during cardiovascular development. Dev Biol 208:307-323

Watari N, Kameda Y, Takeichi M, Chisaka O (2001) *Hoxa3* regulates integration of glossopharyngeal nerve precursor cells. Dev Biol 240:15-31

Genetic Regulation of Chemoreceptor Development in DBA/2J and A/J Strains of Mice

ALEXANDER BALBIR[1], MARIKO OKUMURA[1], BRIAN SCHOFIELD[1], JUDITH CORAM[1], CLARKE G. TANKERSLEY[1], ROBERT S. FITZGERALD[1], CRISTOPHER P. O'DONNELL[2], MACHIKO SHIRAHATA[1]

Department of Environmental Health Sciences, Division of Physiology, Johns Hopkins University Bloomberg School of Public Health, Baltimore, MD, USA [1] *Department of Medicine, University of Pittsburgh Medical Center, Pittsburgh, PA, USA*[2]

1. INTRODUCTION

The role of the carotid body (CB) in response to hypoxia is very well defined (Fitzgerald and Shirahata, 1997). The hypoxic ventilatory response (HVR) is characterized by an increase in ventilation, but this response remains variable among individuals (Eisele et al., 1992; Vizek et al., 1987; Weil 1970). Genetics may play a critical role in explaining this variability. Indeed, longitudinal and twin studies do demonstrate the role of genetics in the HVR (Collins et al., 1978; Kawakami et al., 1982). Studies utilizing inbred strains of mice have also demonstrated the effect of genetics on the response to hypoxia (Tankersley et al., 1994 & 2000). Two strains of mice in these studies were identified as having extreme responses to hypoxia. The DBA/2J strain demonstrated the highest HVR, whereas the A/J strain demonstrated the lowest HVR (Tankersley et al., 1994). In another study analyzing the role of genetics in a mouse model of sleep-induced hypoxia, the DBA/2J strain demonstrated an increased sensitivity to hypoxia during sleep, compared to that of the A/J strain (Rubin et al., 2004). In order to elucidate a potential explanation that may contribute to this difference in hypoxic sensitivity, we examined CB morphology and volume in adult DBA/2J and A/J strains (Yamaguchi et al., 2003). Results demonstrated a significantly larger volume as well as an increased glomus cell quantity in the CB of DBA/2J strain compared to that of the A/J strain. A question arises whether these differences exist from early neonatal ages. Or, developmental plasticity may contribute to these strain differences. In this study, we analyzed CB volume, glomus cell quantity, and ventilation during development in the DBA/2J and A/J strains of mice. We also introduced a glimpse into the genetic influence on chemoreceptor development with emphasis on the role of glial-cell-line-derived neurotrophic factor (GDNF).

2. METHODS

2.1 Tissue Preparation

Based on methodology previously reported (Yamaguchi et al., 2003), weight-matched, male, DBA/2J and A/J strains ages new-born (1D), one-week (1W), four-weeks (4W) and six-weeks (6W) were anesthetized with ketamine (100-150 mg/kg, i.p.) and pentobarbital (100-150 mg/kg, i.p.). CBs were bilaterally harvested. One CB was used for volumetric analysis and the other used for glomus cell immunohistochemistry for tyrosine hydroxylase (TH).

2.2 Volumetric Analysis

CBs were fixed in 4% paraformaldehyde, embedded in plastic, sectioned at 3 µm on glass slides, and stained with toluidine blue. Slides were subsequently analyzed under a light microscope (Olympus BH-2, Japan) and volumetric measurements were determined as follows: 1) Each section of CB tissue was delineated and area measurements were calculated utilizing Image Pro software. 2) Volume of the whole carotid body (μm^3) was calculated as: $\sum$ {the area of each section (μm^2) x 3 µm (the thickness of each section)}.

2.3 Glomus Cell Immunohistochemistry

For glomus cell quantitification, immunohistostaining for tyrosine hydroxylase (TH) was used (Yamaguchi et al., 2003). CBs were fixed in zinc fixative, embedded in paraffin, sectioned at 4-5 µm, and mounted on a poly-L-lysine coated slide. Sections were deparaffinized with a series of xylene and ethanol washes. Endogenous peroxidase was quenched with 1% H_2O_2 in PBS, and endogenous biotin was blocked with an avidin biotin blocking kit (Vector, CA). Other nonspecific binding was blocked with normal goat serum (1:75) and casein (CAS block; Zymed Laboratories, CA). A polyclonal antibody against TH (Chemicon International, CA; dilution: 1/500-1/1000; made in the rabbit) was applied overnight at 4°C followed by an application of biotinylated anti-rabbit IgG made in the goat (1:2000) for 1 h at room temperature. As negative control, normal rabbit serum was used instead of anti-TH antibody. Subsequently, standard avidin biotin peroxidase techniques were applied using VECTASTAIN Elite ABC kits (Vector). As a chromogen, Vector® SG (Vector) was used. Between each step the slides were washed in 0.1M phosphate buffer saline (PBS). TH immunoreactivity was examined using a light microscope. The largest section of each carotid body was selected for analysis. Carotid body area of the section and TH positive area were calculated utilizing Image Pro Software.

2.4 Ventilatory Function

Ventilatory function was assessed by whole body plethysmography under unanesthetized and unrestrained conditions. Chamber temperature was monitored and was maintained within 30-32°C. Compressed air was humidified and directed through the chamber at a flow rate ~ 500ml/min. Ventilatory functions were measured for 1 minute (for newborn) or 2 minutes for 1-week-old mice. Inspired gases were air for baseline and 15% O_2/air for mild hypoxia. Breathing frequency (f) and tidal volume were recorded on a strip chart recorder (model 7D polygraph, Grass). Due to extremely small tidal volumes, we analyzed only f.

2.5 RT-PCR for GDNF

The CBs were obtained from deeply anesthetized male mice as described above. Total RNA of the CBs from each mouse was separately isolated in TRIzol, and genomic DNA was digested with RQI DNase I (Invitrogen) for 15 min at room temperature. First strand cDNA was obtained using SuperScript III^{TM} (Invitrogen), and multiplex PCR was performed following the manufacture's instruction (Quiagen). The primer pairs used were β-actin and GDNF (final concentrations: 0.5 μM for β-actin and 5 μM for GDNF). The conditions for PCR amplification were as follows. The template was denatured at 95 °C for 15 min followed by 35 cycles of denaturation at 94 °C for 30 sec, annealing of primers at 60 °C for 1.5 min, extension at 72 °C for 1.5 min. Final cycle was 72 °C for 10 min. Positive tissue control was cDNA from the SCG. The negative reaction controls were total RNA and water as templates for PCR.

3. RESULTS

3.1 Volumetric Analysis

CBs and glomus cells at 1D (Figure 1, left) and 1W animals demonstrated very similar morphology in both strains. However, CBs of A/J mice at 4W begin to exhibit their adult phenotype of distorted glomus cell morphology. In regard to volume measurements, the CBs of both strains were similar in volume at birth, but as they age to 4W, the CB of DBA/2J mice grows in volume and the growth of AJ's CB was stunted (Figure 1, right).

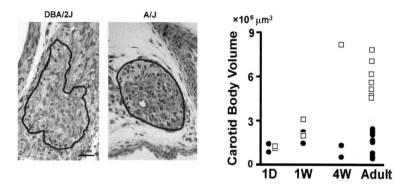

Figure 1. **Left panel:** The largest parts of the CBs in 1-day-old mice. The CB is outlined. **Right panel:** The summary indicates developmental differences in the growth of the CB in these two strains of mice. Open square, DBA/2J mice; closed circle, A/J mice.

3.2 Glomus Cell Immunohistochemistry

At 1D TH positive ratio (TH positive area / CB area) of glomus cells in both strains of mice was significantly greater than their adult counterparts. (Table 1). In the DBA/2J strain, TH positive ratio decreased from 1D to 6W and appears to plateau through adult (8-24wks). However, the A/J strain exhibited a continuous decline in TH positive ratio through adult (Table 1).

Table 1. TH positive ratio (%)

	1-Day	1-Week	4-Weeks	6-Weeks	8-24 Weeks
A/J	24.6 ± 3.1	22.3 ± 2.8	15.5 ± 1.2	13.5 ± 3.5	5.6 ± 0.1
DBA/2J	31.1 ± 2.6	29.4 ± 2.3	23.6 ± 2.1	20.6 ± 1.6	22.2 ± 1.9

Data are presented as mean ± SE.

3.3 Ventilatory Function

At 1D, DBA/2J mice showed a steady f (1.80Hz ±0.06, n=5) with a slight increase in response to mild hypoxia. The A/J mice demonstrated predominantly apneic breathing with a very low f. At 1W, however, A/J mice exhibited a robust, baseline frequency (5.18Hz ±0.20, n=5) with a significant increase in f to mild hypoxia (5.82Hz±0.14, $P<0.05$). In 1W DBA/2J mice baseline f increased compared to newborn animals (3.43Hz±0.18, n=6). In response to hypoxia, f significantly increased (3.99Hz±0.109, $P<0.05$).

3.4 RT-PCR for GDNF

GDNF mRNA is expressed at a higher ratio in the CB of DBA/2J mice compared to that of A/J mice at 1 week and 4-week of age (Fig. 2, left). In order

Genetic Regulation of Chemoreceptor Development

to quantify overall expression, relative ratios of GDNF versus β-actin controls were calculated using Kodak Electrophoresis Documentation and Analysis System 290 (Fig. 2, right).

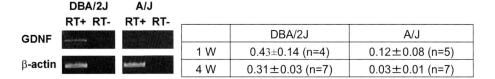

	DBA/2J	A/J
1 W	0.43±0.14 (n=4)	0.12±0.08 (n=5)
4 W	0.31±0.03 (n=7)	0.03±0.01 (n=7)

Figure 2. **Left panel:** Agarose gel electrophoresis of RT-PCR amplified products of mRNA for β-actin and GDNF. The CBs were obtained from 4-week-old mice. **Right panel:** GDNF mRNA levels were expressed relative to β-actin. The values at 4 week of age are significantly different ($p<0.01$), and those at 1 week of age are almost significantly different ($p=0.08$).

4. DISCUSSION

Our results demonstrate that at the neonatal stage (1D) CBs of both strains are similar in volume, TH positive ratio, and overall morphological structure. However, overall CB features of the two strains diverge during postnatal development. The CB volume of DBA/2J mice increased from 1D to adult while that of A/J mice remained constant. In regard to the TH positive ratio, both strains demonstrated a decline in this ratio through 6W. The TH ratio of DBA/2J mice appears to plateau at 6W while that of A/J mice continues declining up to adulthood. TH staining data in conjunction with CB volume data indicate an increase in glomus cell quantity in the DBA/2J strain compared to a decrease in the A/J stain.

In whole-body plethysmography experiments of 1D animals, we discovered that they differ in their basic respiratory patterns. Thus, similar morphology of the CB and of glomus cells in newborn mice does not appear to be reflected in their ventilation. On the other hand, the CBs of DBA/2J and A/J mice respond to hypoxia at 1W. Many studies indicate that the response of the carotid body to hypoxia is minimal at birth, and then gradually reaches the adult level (Carroll, 2003). Our data agree with those obtained from rats and larger animals. It remains to be seen whether developmental differences in the CB after 1W would contribute to ventilatory responses to hypoxia.

GDNF supports the development and survival of dopaminergic neurons, spinal motoneurons, subpopulations of peripheral ganglion neurons (Baloh et al., 2000). The expression of mRNA for GDNF in the rat carotid body has been shown (Nostrat et al., 1996; Toledo-Aral et al., 2003). Because GDNF content in the near-term rat carotid body was reduced by prenatal exposure to cocaine (Lipton et al., 1999), which impairs HVR (Lipton et al., 1996), we have speculated that GDNF plays a role in the strain differences in carotid body development. Our RT-PCR analysis shows that GDNF mRNA is expressed more in the CB of DBA/2J mice than that of A/J mice. Hence, its absence may be a contributing factor in the decline of glomus cells in the developing CB of A/J mice.

These results in two inbred strains of mice demonstrate the fact that genetics play a critical role in the development of the CB as well as the development of the HVR.

The study was supported by HL 72293 and AHA 0255358N.

REFERENCES

Baloh, R.H., Enomoto,H., Johnson,E.M., and Milbrandt,J., The GDNF family ligands and receptors - implications for neural development, Curr Opin Neurobiol, 10 (2000) 103-110.

Carroll,J.L., 2003. Developmental plasticity in respiratory control. *J. Appl. Physiol.* 94: 375-389.

Eisele JH, Wuyam B, Savourey G, Eterradossi J, Bittel JH, Benchetrit G. Individuality of breathing patterns during hypoxia and exercise. *J Appl Physiol* 72: 2446-2453, 1992.

Fitzgerald RS, Shirahata M. Systemic responses elicited by stimulating the carotid body: primary and secondary mechanisms. 1997. In: Carotid Body Chemoreceptors. (Gonzalez C., ed) pp.171-202, Barcelona, Springer-Verlag.

Gonzalez C., Almaraz L., Obeso A., Rigual R., 1994. Carotid body chemoreceptors: from natural stimuli to sensory discharges. *Physiol. Rev.* 74: 829-889.

Collins D.D., Scoggin C.H., Zwillich C.W., Weil J.V., 1978. Hereditary aspects of decreased hypoxic response. *J. Clin. Invest.* 62: 105-110.

Lipton, J.W., Davidson, T.L., Carvey, P.M., Weese-Mayer, D.E., 1996. Prenatal cocaine: Effect on hypoxic ventilatory responsiveness in neonatal rats. *Resp. Physiol.* 106: 161-169.

Lipton, J.W., Ling, Z., Vu, T.Q., Robie, H.C., Mangan, K.P., Weese-Mayer, D.E., Carvey, P.M., 1999. Prenatal cocaine exposure reduces glial cell line-derived neurotrophic factor (GDNF) in the striatum and the carotid body of the rat: implications for DA neurodevelopment. Brain Res. Dev. Brain Res. 118: 231-235.

Kawakami Y., Yoshikawa T., Shida A., Asanuma Y., Murao M., 1982. Control of breathing in young twins. *J. Appl. Physiol.* 52: 537-542.

Nosrat, C.A., Tomac, A., Lindqvist, E., Lindskog, S., Humpel, C., Stromberg, I., Ebendal, T., Hoffer, B.J., and Olson, L., 1996. Cellular expression of GDNF mRNA suggests multiple functions inside and outside the nervous system. *Cell Tissue Res.* 286: 191-207.

Rubin A.E., Polotsky V.Y., Balbir A., Krishnan J.A., Schwartz A.R., Smith P.L., Fitzgerald R.S., Tankersley C.G., Shirahata M., O'Donnell C.P., 2003. Differences in sleep-induced hypoxia between A/J and DBA/2J mouse strains. *Am. J. Resp. Crit. Care Med.* 168: 1520-1527.

Tankersley C.G,. Elston R.C., Schnell A.H., 2000. Genetic determinants of acute hypoxic ventilation: patterns of inheritance in mice. *J. Appl. Physiol.* 88: 2310-2318.

Tankersley C.G., Fitzgerald R.S., Kleeberger S.R., 1994. Differential control of ventilation among inbred strains of mice. Am. *J. Physiol.* 267: R1371-R1377.

Toledo-Aral, J.J., Mendez-Ferrer, S., Pardal, R., Echevarria, M., and Lopez-Barneo,J., 2003. Trophic restoration of the nigrostriatal dopaminergic pathway in long-term carotid body-grafted parkinsonian rats. *J. Neurosci.* 23: 141-148.

Vizek M., Pickett C.K., Weil J.V., 1987. Interindividual variation in hypoxic ventiatory response: potential role of carotid body. *J. Appl. Physiol.* 63: 1884-1889.

Weil J.V., Bryne-Quinn E., Sodal I.E., Friesen W.O., Underhill B., Filley G.F., Grover R.F., 1970. Hypoxic ventilatory drive in normal man. *J. Clin. Invest.* 49: 1061-1072.

Yamaguchi S., Balbir A., Schfield B., Coram J., Tankersley C.G., Fitzgerald R.S., O'Donnell C.P., Shirahata M., 2003, Structural and functional differences of the carotid body between DBA/2J and A/J strains of mice. *J. Appl. Physiol.* 94:1536-1542.

Genetic Influence on Carotid Body Structure in DBA/2J and A/J Strains of Mice

SHIGEKI YAMAGUCHI[1,2], ALEXANDER BALBIR[1], MARIKO OKUMURA[1], BRIAN SCHOFIELD[1], JUDITH CORAM[1], CLARKE G. TANKERSLEY[1], ROBERT S. FITZGERALD[1], CHRISTOPHER P. O'DONNELL[3], MACHIKO SHIRAHATA

Department of Environmental Health Sciences, The Johns Hopkins Bloomberg School of Public Health, Baltimore, USA[1], Department of Anesthesiology, Dokkyo University School of Medicine, Tochigi, Japan[2], Department of Medicine, University of Pittsburgh Medical Center, Pittsburgh, USA[3]

1. INTRODUCTION

The carotid body is a major chemosensory organ for hypoxia, hypercapnia and acidosis in the arterial blood (Fitzgerald and Shirahata 1997; Gonzalez et al., 1994). During hypoxia, the neural output from the carotid body increases and reflexly modifies several variables in the respiratory system. A prominent response is an increase in ventilation, but the hypoxic ventilatory response (HVR) among individuals varies widely (Eisele et al., 1992; Vizek et al., 1987). Studies in humans (Collins et al., 1978; Kawakami et al., 1982; Nishimura et al., 1991; Thomas et al., 1993) suggest that genetic factors significantly contribute to those differences. Similarly, studies in mice (Tankersley et al., 1994) and rats (Weil et al., 1998) clearly indicate that genetic determinants robustly influenced HVR. Among several inbred strains of mice the DBA/2J mice demonstrated the highest HVR and the A/J mice the lowest HVR (Tankersley et al., 1994). The differences in HVR between these two strains of mice may be closely related to the structural differences of the carotid body (Yamaguchi et al., 2003). The size of the carotid body and the quantity of glomus cells in the DBA/2J mice are significantly larger than those in the A/J mice. Those differences were clearly segregated between the strains, suggesting that genetic factors strongly influence the observed phenotypic differences between the DBA/2J and A/J mice. The purpose of the current study was to confirm that the morphological characteristic differences in the carotid body between the DBA/2J and A/J mice are controlled by genetic factors. Thus, we generated the first-filial progeny (F1) by a crossing the DBA/2J (female) and A/J (male) strains of mice, and examined the morphological differences of the carotid body in the DBA/2J, A/J and their F1 mice.

2. METHODS

2.1 Experimental Animals

We used the DBA/2J and A/J strains of mice and their first-filial progeny (8 to 26-week old). DBA/2J and A/J strains of mice were initially purchased from Jackson Laboratories (Bar Harbor, ME), then bred in the Johns Hopkins Bloomberg School of Public Health animal center together with F1. All mice were housed in a facility where the temperature and light cycles (12 hour cycle) were controlled. All animal protocols were approved by the Animal Care and Use Committee of the Johns Hopkins University.

2.2 Morphology and Immunocytochemistry

Mice were deeply anesthetized with ketamine (100 –150 mg/kg, i.p.) and pentobarbital (100 –150 mg/kg, i.p.). The heart was removed to avoid bleeding in the neck, and both carotid bifurcations with carotid bodies were harvested immediately. One carotid body was used for morphological assessment, and the other was used for immunostaining for tyrosine hydroxylase (TH). Procedures for morphological studies were similar to those described before (Yamaguchi et al., 2003). In short, the carotid bifurcation was fixed in buffered 4% formalin, embedded in plastic, sectioned at 3 µm thickness, mounted on glass slides, and stained with toluidine blue. Structural differences among each mouse were examined with a standered bright field microscope. The size of the carotid bodies was measured by using a computer imaging program (Sonic Image Beta 4.02, Scion, Frederick, MD). The carotid body in each section was delineated, and the area of the carotid body was calculated. Subsequently, the volume of the whole carotid body (μm^3) was estimated as: $\Sigma\{$the area of each section (μm^2) × 3 µm (the thickness of each section)$\}$.

The quantity of glomus cells in the carotid body from each mouse was estimated by the immunocytochemical signal for TH. The carotid bifurcation was immersed in the zinc fixative, embedded in paraffin, sectioned at 4–5 µm, and mounted on a poly-L-lysine-coated slide. After deparaffinization, the sections were treated in boiling 0.01 M citric acid buffer (pH 8.0) for 5 min to retrieve antigens (Shi et al., 1997), then, endogeneous peroxidase was quenched with 1% H_2O_2 in PBS. Endogenous biotin was blocked (avidin biotin-blocking kit: Vector Lab.), and other nonspecific binding was blocked with normal goat serum (1:75) and casein (CAS block, Zymed Laboratories). The sections were incubated with a polyclonal antibody against TH (Chemicon International; dilution: 1:500 to 1:1,000; made in the rabbit) overnight at 4 °C followed by an application of biotinylated anti-rabbit IgG made in the goat (1:2,000) for 1 h at room temperature. As negative control, normal rabbit serum was used instead of anti-TH antibody. Subsequently, standard avidin-biotin-peroxidase techniques were applied. As a chromogen, Vector SG (Vector) was used. Between each step, the slides were washed in 0.1 M PBS. The largest section of each carotid body was selected, when possible. The area of the carotid body and the stained area for TH were measured by the imaging software (Scion Image Beta 4.02). Then, TH-positive ratios were determined: TH-positive ratio (%) = (the stained area for TH / the area of the carotid body) × 100.

All data are presented as means±SE. The Kruskal-Wallis test was used to evaluate the significant differences among strains. Differences were considered to be statistically significant when $P < 0.05$.

3. RESULTS

3.1 Morphology

We detected four major characteristics in morphology of the carotid body in F1 mice. First, the boundary of the carotid body of F1 mice was not smooth (Fig. 1B). The carotid body and other tissues, including connective tissues, bundles of nerves, and large vessels, were entangled. These characteristics are different from DBA/2J mice, but similar to A/J mice (Fig. 1A). Second, many glomeruli, which contained several glomus cells surrounded by sheath cells, were observed in F1 mice, similar to the DBA/2J mice. Third, many glomus cells in the carotid body of the F1 mice showed typical large and round nuclei and abundant cytoplasm, similar to the DBA/2J mice (Fig. 1B). Fourth, the carotid body volume of F1 mice was intermediate that of the DBA/2J and A/J mice (Table 1).

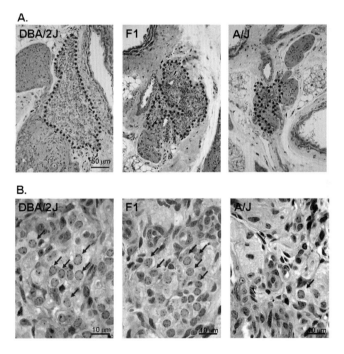

Figure 1. **A:** Histology of the carotid bodies in the DBA/2J, A/J and F1 mice. Each section shows the largest part of the carotid body in each mouse. Dotted borders outline the entire carotid body of each strain. **B:** With higher magnification, many typical glomus cells (abundant cytoplasm with a large, round nucleus) are observed in the DBA/2J and F1, but only a few were evident in the A/J mice. Some glomus cells are indicated by arrows.

Table 1. The volume of the carotid body.

Mice	DBA/2J	F1	A/J
Volume (10^6 μm^3)	5.9 ± 1.0	3.5 ± 0.8	1.6 ± 0.6

The values were significantly different from each other.

3.2 Immunohistochemistry

In agreement with the light microscopic observation (Fig. 1B), TH immunoreactivity is easily observed in the cytoplasm of glomus cells in the DBA/2J and F1 mice. Fewer TH-positive cells were seen in the A/J mice. Glomus cell quantity was expressed in two ways. TH positive ratio presents the density of glomus cells within the carotid body. The ratios for DBA/2J, F1 and A/J mice were 23.3±3.7, 20.9±2.3, and 5.6±1.7%, respectively. The ratio of F1 mice is significantly different from that of A/J mice, but not from that of DBA/2J mice. We also calculated glomus cell number index (= carotid body volume × TH ratio) which is roughly proportional to the total number of glomus cells. The index for the F1 mice (74±12, n=8; arbitrary unit) was significantly lower than that for DBA/2J mice (139±7, n=8) and higher than that for A/J mice (9±2, n=8).

4. DISCUSSION

The structural characteristics and the size of the carotid body can be influenced by environmental and genetic factors. For example, as environmental factors, hypoxia (Barer et al., 1972), hypertension (Habeck, 1991), and high altitude (Lahiri, 2000) have been known to increase the size of the carotid body and change its cellular components. In this study, we demonstrated distinctive differences in morphological characteristics between DBA/2J, A/J, and F1 mice. F1 mice inherited the morphological characteristics of the carotid body from the DBA/2J and A/J mice. Inheritance patterns suggest that several genetic factors may separately regulate the morphology of the carotid body and glomus cells. Together with our previous study (Yamaguchi et al., 2003), we recognized at least four prominent phenotypes in the morphology of the carotid body as mentioned above. F1 mice resemble to DBA/2J in the "typical" morphology of glomus cells and clearly delineated glomeruli. These may be reflected by similar TH ratio of F1 and DBA/2J mice. The demarcation of the carotid body of F1 mice is similar to A/J mice. On the other hand, the volume of the carotid body is intermediate between the two parental strains. Thus, F1 mice inherited several genes for determining characteristics of the carotid body, glomeruli and glomus cells from their parental strains. These results agree with our hypothesis which states the morphological characteristics are controlled by genetic factors and several determinants separately regulate the carotid body and glomus cells.

ACKNOWLEDGEMENT

The study was supported by HL 72293, HL 61596 and AHA0255358N.

REFERENCES

Barer G.R., Edwards C., Jolly A.I., 1972. Changes in the ventilatory response to hypoxia and in carotid body size in chronically hypoxic rats. *J. Physiol.* 221: P27-P28.
Eisele J.H., Wuyam B., Savourey G., Eterradossi J., Bittel J.H., Benchetrit G., 1992. Individuality of breathing patterns during hypoxia and exercise. *J. Appl. Physiol.* 72: 2446-2453.
Fitzgerald R.S., Shirahata M., 1997. Systemic responses elicited by stimulating the carotid body: primary and secondary mechanisms. In: *Carotid Body Chemoreceptors,* (Gonzalez C., ed.) pp.171-202, Barcelona, Springer-Verlag.
Gonzalez C., Almaraz L., Obeso A., Rigual R., 1994. Carotid body chemoreceptors: from natural stimuli to sensory discharges. *Physiol. Rev.* 74: 829-889.
Habeck, J.O., 1991. Peripheral arterial chemoreceptors and hypertension. *J. Auton. Nerv. Syst.* 34:1-7.
Kawakami Y., Yoshikawa T., Shida A., Asanuma Y., Murao M., 1982. Control of breathing in young twins. *J. Appl. Physiol.* 52: 537-542.
Lahiri S., Rozanov C., Cherniack N.S., 2000. Altered structure and function of the carotid body at high altitude and associated chemoreflexes. *High Alt. Med. Biol.* 1: 63-74.
Nishimura M., Yamamoto M., Yoshioka A., Akiyama Y., Kishi F., Kawakami Y., 1991. Longitudinal analyses of respiratory chemosensitivity in normal subjects. *Am. Rev. Respir. Dis.* 143:1278-1281.
Shi S.R., Cote R.J., Taylor C.R., 1997. Antigen retrieval immunohistochemistry: past, present, and future. *J. Histochem. Cytochem.* 45:327-343.
Tankersley C.G., Fitzgerald R.S., Kleeberger S.R., 1994. Differential control of ventilation among inbred strains of mice. *Am. J. Physiol.* 267: R1371-R1377.
Thomas D.A., Swaminathan S., Beardsmore C.S., McArdle E.K., MacFayden U.M., Goodenough P.C., Carpenter R., Simpson H., 1993. Comparison of peripheral chemoreceptor responses in monozygotic and dizygotic twin infants. *Am. Rev. Res. Dis.* 148: 1605-1609.
Vizek M., Pickett C.K., Weil J.V., 1987. Interindividual variation in hypoxic ventiatory response: potential role of carotid body. *J. Appl. Physiol.* 63: 1884-1889.
Weil J.V., Stevens T., Pickett C.K., Tatsumi K., Dickinson M.G., Jacoby C.R., Rodman D.M. 1998. Strain-associated differences in hypoxic chemosensitivity of the carotid body in rats. *Am. J. Physiol.* 274: L767-L774.
Yamaguchi S., Balbir A., Schfield B., Coram J., Tankersley C.G., Fitzgerald R.S., O'Donnell C.P., Shirahata M., 2003. Structural and functional differences of the carotid body between DBA/2J and A/J strains of mice. *J. Appl. Physiol.* 94:1536-1542.

The Effect of Hyperoxia on Reactive Oxygen Species (ROS) in Petrosal and Nodose Ganglion Neurons during Development (Using Organotypic Slices)

D. J. KWAK, S. D. KWAK AND E. B. GAUDA

Department of Pediatrics, Johns Hopkins University School of Medicine, Baltimore, MD, 21287-3200, USA

1. INTRODUCTION

Peripheral arterial chemoreceptors, within the carotid body (CB), are critical in maintaining respiratory and cardiac hemostasis by uniquely sensing changes in O_2 tension. The components of the peripheral arterial chemoreceptors are found within the CB, but the physiologic effects of activation of these chemoreceptors are widespread and significant (Gonzalez et al., 1994). The CB is located in the bifurcation of the carotid artery and consists of three major neuronal components that include: 1) type I chemosensory cells, also known as glomus cells, which contain neurotransmitters and autoreceptors; 2) type II cells, which are similar to supportive glial cells; and 3) chemoafferent nerve fibers from the carotid sinus nerve, a branch of the IX cranial nerve, with cell bodies in the petrosal ganglion (PG) (Verna 1997).

Numerous studies, in several mammalian models, support a role for peripheral arterial chemoreceptors in contributing to stable ventilation at a critical period of development during early postnatal life, which establishes rhythmogenesis that is sustained throughout life (for review, Gauda et al. 2004; Carroll 2003). Chronic hyperoxia, during early postnatal development (PND), depresses ventilatory responses to subsequent acute hypoxia in newborn animals and in premature infants (Ling et al., 1997; Ling et al., 1997). Furthermore, histological evidence suggests that hyperoxia is cytotoxic to the CB and its associated chemoafferent neurons (Erickson et al. 1998). In other model systems, hyperoxia is associated with increased production of reactive oxygen species (ROS), including superoxide (O_2-), hydroxyl radical (OH•), and hydrogen peroxide (H_2O_2) (for review, Jamieson et al., 1986). If the balance between intracellular oxidants and antioxidants is disturbed, ROS contribute to cellular damage via lipid peroxidation, enzyme inactivation, and protein and nucleic acid oxidation, resulting in apoptosis or necrosis. We hypothesize that hyperoxia leads to an increased production of ROS and subsequent cytotoxic response in PG neurons.

2. METHODS

CBs and petrosal-nodose ganglia (PG/NG) from Sprague-Dawley rat pups (n=10 and 9, respectively, for each age group) at day of life (DOL) 5-6 and 17-18 were rapidly removed, embedded in 3% Agar or 3% low melting point agarose, and sectioned at 45 microns.

2.1 Organotypic Slice Culture

Tissue slices were incubated at 37° C (21% O_2/5% CO_2) overnight. Tissues were exposed to hypoxia (8% O_2), normoxia (21% O_2) or hyperoxia (95% O_2) for 4 hours in the presence or absence of the ROS-sensitive fluorescent indicator, 5-(and 6)-chloromethyl-2', 7'-dichlorodihydrofluorescein diacetate, acetyl ester (CM-H_2DCFDA, 2µM, Molecular Probes, Eugene, OR). Tissue slices were also exposed to hyperoxia in the presence or absence of the antioxidants: N-Acetyl-L-cysteine (NAC, 20mM, Sigma, St. Louis, MO) and 4-((9-acridinecarbonyl) amino)-2,2,6,6-tetramethylpiperidin-1-oxyl, free radical (TEMPO-9-AC, 10mM, Molecular Probes, Eugene, OR). Propidium iodide (PI, 500nM, Molecular Probes, Eugene, OR) was used to determine the effect of hyperoxia on cell death in PG/NG neurons in the presence or absence of antioxidants. Sections were fixed in 2% paraformaldehyde and then mounted onto slides using Anti-Fade to apply cover slips.

2.2 Data Analysis

CM-H_2DCFDA fluorescent signals were examined with fluorescent microscopy and analyzed using IPLAB image program. After subtracting for background fluorescence and normalization, gray level intensities for 200 neurons of the PG/NG complex were measured. At each developmental age and O_2 tension exposure, a group mean was determined and differences were analyzed by one-way ANOVA and posthoc analysis. Significance was set at $P < 0.05$.

3. RESULTS

In tissues removed from rats at DOL 5-6 and DOL 17-18, increasing O_2 tension augmented the level of ROS production in PG/NG neurons. In rats at DOL 5-6, normoxic exposure increased ROS production in PG/NG neurons, by 19.2% in comparison to hypoxic exposure ($P<0.01$) with a further increase of 54.8% from normoxia to hyperoxia ($P<0.001$), as shown in the figure. The greatest increase in fluorescence intensity was observed from hypoxia to hyperoxia, which was 84.7% ($P<0.001$). However, in rats at DOL 17-18 PG/NG neurons had a decreased response to changes in O_2 tension, as compared to the younger animals. In rats at DOL 17-18, normoxic exposure increased ROS production in PG/NG neurons by 8.2% in comparison to hypoxic exposure ($P<0.01$) with a further increase of 31.2% from normoxia to hyperoxia ($P<0.001$). In the older animals the greatest increase in fluorescence intensity, was observed from hypoxia to hyperoxia, which was 42.2% ($P<0.001$), in contrast to the 84.7% observed in the younger animals, see figure. In a subset of rat pups (n=3 each age group), tissues were exposed to hyperoxia in the presence and absence of the antioxidants, NAC and TEMPO-9-AC. There was a decrease of 18% ± 3.3 and 10.9% ± 2.2 in hyperoxia-induced ROS production in NG/PG neurons in tissues from rats at DOL 5-6 and DOL 17-18 respectively. PI was used to indicate the level of cell death of PG/NG neurons as a result of oxygen exposure in both age groups. There was a greater increase in cell death, evidenced by a greater level of fluorescent intensity in tissues from the younger animals, in each oxygen tension in compared to tissues from the older animals, $P<0.05$, ANOVA.

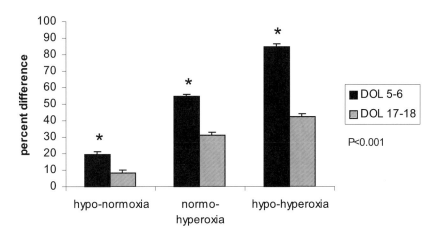

Bar graph depicting developmental increases in relative percent difference in grey-level intensity from one oxygen tension level to another. The percent differences were greater for the younger animals than the older animals, with the greatest relative difference demonstrated in the hypo-hyperoxia comparison. $p<0.001$, ANOVA

4. DISCUSSION

Perinatal hyperoxic exposure severely depress responses to subsequent hypoxia, as recorded from carotid sinus nerves of kittens (Hanson et al. 1989) and adult rats (Vidruk et al., 1996), suggesting impairment of CB chemoreflex function. Chronic hyperoxic exposure in newborn rats decreases the number of unmyelinated axons in the carotid sinus nerve, and causes degeneration of chemoafferent neurons and marked hypoplasia of the CB (Erickson et al. 1998). However, less is known about the effects of ROS-induced hyperoxia on the CB, and the PG/NG in newborn animal models.

Using the novel technique of organotypic slices of the CB and PG/NG complex, our data show a direct correlation between the level of O_2 tension exposure and ROS production in NG/PG neurons. Younger rat pups have a more augmented production of ROS in PG/NG neurons in response to hyperoxia than do older rat pups. Cell death, as measured by PI fluorescence, increased with increasing O_2 tension in the younger animals. Increased ROS production within chemoafferent cell bodies may account for histological evidence of cytotoxicity and ablation of hypoxic chemosensitivity in these cells, in newborns exposed to hyperoxia (Calder et al., 1994). We speculate that blunted chemoreceptor responses in newborn infants with chronic lung disease (Katz-Salamon et al., 1995) may be, in part, related to hyperoxia-induced cytotoxicity of key cells and neurons in the peripheral arterial chemoreceptors.

ACKNOWLEDGEMENTS

This investigation was supported by grants from NIDA DA 013940-03 and NIH HL 072748. I would like to thank Gabrielle L. McLemore, PhD, for her assistance in editing the manuscript. Also, I would like to thank Reed Z. Cooper, MS, for his help in dealing with technical issues, and Ariel V. Mason for her help in conducting the actual study.

REFERENCES

Calder NA, Williams BA, Smyth J, Boon AW, Kumar P, Hanson MA. Absence of ventilatory responses to alternating breaths of mild hypoxia and air in infants who have had bronchopulmonary dysplasia: implications for the risk of sudden infant death. Pediatr Res. 1994 Jun; 35(6):677-81.
Carroll JL, Developmental plasticity in respiratory control. J Appl Physiol. 2003. 94(1):375-89.
Erickson JT, Mayer C, Jawa A, Ling L, Olson EB Jr, Vidruk EH, Mitchell GS, Katz DM. Chemoafferent degeneration and carotid body hypoplasia following chronic hyperoxia in newborn rats. J Physiol. 1998. 509 (Pt 2):519-26.
Gauda EB, McLemore GL, Tolosa J, Marston-Nelson J, Kwak D. Maturation of peripheral arterial chemoreceptors in relation to neonatal apnoea. Semin Neonatol. 2004. 9(3):181-94.
Gonzalez C, Almaraz L, Obeso A, Rigual R. Carotid body chemoreceptors: from natural stimuli to sensory discharges. Physiol Rev. 1994. 74(4):829-98.
Hanson, MA, Eden, GJ, Nijhuis, JG, Moore, PJ. 1989. Peripheral chemoreceptors and other oxygen sensors in the fetus and newborn. In *Chemoreceptors and Reflexes in Breathing: Cellular and Molecular Aspects*, ed. Lahiri S, Forster RE, Davies RO, Pack AI, pp. 113-120. Oxford University Press, New York.
Jamieson D, Chance B, Cadenas E, Boveris A. The relation of free radical production to hyperoxia.
 Annu Rev Physiol. 1986. 48:703–719.
Katz-Salamon M, Jonsson B, Lagercrantz H. Blunted peripheral chemoreceptor response to hyperoxia in a group of infants with bronchopulmonary dysplasia. Pediatr Pulmonol. 1995 20(2):101-6.
Ling L, Olson EB Jr, Vidruk EH, Mitchell GS. Integrated phrenic responses to carotid afferent stimulation in adult rats following perinatal hyperoxia. J Physiol. 1997 1;500 (Pt 3):787-96.
Ling L, Olson EB Jr, Vidruk EH, Mitchell GS. Phrenic responses to isocapnic hypoxia in adult rats following perinatal hyperoxia. Respir Physiol. 1997 109(2):107-16.
Pagano A and Barazzone-Argiroffo C. Alveolar cell death in hyperoxia-induced lung injury. Ann N Y Acad Sci. 2003. 1010:405-16.
Verna A. The mammilian carotid body: morphological data. In *The Carotid Body Chemorecptors*, ed. Gonzales, C 1997, pp. 1-29. Springer-Verlag, Heidelberg, Germany.
Vidruk EH, Olson EB Jr., Ling L, Mitchell GS. Carotid sinus nerve responses to hypoxia and cyanide are attenuated in adult rats following perinatal hyperoxia. Physiologist 1996. 39:190.
Wenninger JM, Olson EB, Wang Z, Keith IM, Mitchell GS, Bisgard GE. Carotid sinus nerve responses and ventilatory acclimatization to hypoxia in adult rats following 2 weeks of postnatal hyperoxia. Respir Physiol Neurobiol. 2005. Jun 21; [Epub ahead of print]

Carotid Body Volume in Three-Weeks-Old Rats Having an Episode of Neonatal Anoxia

CHIKAKO SAIKI[1], MASAYA MAKINO[2], AND SHIGEJI MATSUMOTO[1]

[1]*Department of Physiology,* [2] *Department of Pediatric Dentistry, Nippon Dental University, School of Dentistry at Tokyo, Tokyo, Japan*

1. INTRODUCTION

The development of oxygen chemosensitivity in carotid chemoreceptor cells, i.e. type I cell (glomus cell), is reported to continue postnatally (Wasicko et al., 1999), and it has been suggested that environmental experiences such as episode of hypoxia and chronic hypoxia during critical period of maturation may result in long-term alterations in the structure or function of the respiratory control neural network (Carroll, 2003). In the previous studies, we have observed no apparent effect on the hypoxic ventilatory response (HVR) in the day 7 newborn rats, which had daily episode of anoxia from day 1 to day 6 (day 0 = day of birth) (Saiki and Mortola, 1994), but significantly higher HVR in the 3-weeks-old rats, which had an episode of anoxia on day 3-4 after birth (Saiki and Matsumoto, 1999). These results suggest that the severity of anoxia and the timing of the anoxic episode as well as the assessment of HVR may be important factors, and that an episode of anoxia during the neonatal period has long-lasting effects on the control of ventilation in rats. Because no further information is available on the effects, including carotid body chemoreceptors, we examined whether or not an episode of anoxia in neonatal period induces changes in the carotid body and glomus cell structures in the three-weeks-old rats.

2. MATERIALS AND METHODS

On day 3 (day 0 = date of birth), rats were exposed to anoxia (100%N_2) for 20 min (EXP rats). By switching 100%N_2 to air, all EXP rats were autoresuscitated. After the recovery, they were reared by their dam for 3 weeks with the litter mates, which exposed to air, instead of 100%N_2, on day 3 (CONT rats). The procedure was similar to the HVR measurements described in a previous study (Saiki and Matsumoto, 1999).

On day 25, the CONT and EXP rats were anesthetized with sodium pentobarbital (45 mg/kg, i.p.), and perfused with 0.01M phosphate buffer saline (PBS), followed by 4% paraformaldehyde solution. The carotid bodies were excised, infiltrated with 30% sucrose in PBS, and embedded and frozen in O.C.T.

compound, then stored at -80°C until use. All experiments with animals were performed in accord with "Guiding Principles for the Care and Use of Animals in the Fields of Physiological Sciences" published by the Physiological Society of Japan.

The frozen sections (10μm) were cut through the entire bifurcation and thaw-mounted onto silan-coated microscope slides, and were processed for immunohistochemistry according to the immunofluorescence method. The sections were air dried, rinsed in PBS for 30 min (10 min x 3 times), and soaked in dilution buffer (2% bovine serum albumin in PBS with 10% normal goat serum) for 1-2 hours at room temperature, and incubated at 4°C overnight with the primary antisera, i.e. polyclonal rabbit anti-PGP9.5 (1:750-1000, Neuromics) or rabbit anti-tyrosine hydroxylase (TH; 1:750-1000, Chemicon International). After rinsed in PBS for 30min (10min x 3 times), the sections were transferred for 1-2 hours to secondary antibodies (Alexa Fluor® 568 goat anti-rabbit IgG, 1:1000, Molecular Probes) at room temperature, rinsed in PBS, and coverslipped with mounting medium containing DAPI (4·6 diamidino-2-phenylindole, a counterstain for DNA) (Vector laboratories). Serial sections of the carotid body immunostained for PGP9.5 were used for carotid body volume measurements. The volume of the carotid body and the area occupied by glomus cells were determined using image analysis software, and values were calculated based on the cross-sectional areas of the carotid body compartment, section thickness and total number of sections containing the carotid body. In addition, in the sections through around the center of the carotid body (in total, 24 sections from 12 carotid bodies) were used to evaluate the volume proportion of glomus cells and vessels, and this examination was performed in the sections immunostained not only for PGP9.5, but also for TH to see whether or not different results between CONT and EXP rats could be obtained by PGP9.5 and TH. The immunoreactivity for both PGP9.5 and TH is known to detect glomus cells in the carotid body (Kent and Rowe, 1992; Erickson et al., 1998; Kusakabe et al., 1998). Values were expressed as mean±SEM. Statistical comparisons between the control and experimental values were determined using the two-tailed t test.

3. RESULTS

3.1 Somatic Growth

Figure 1 shows the individual data points for body weight of CONT and EXP rats between days 3 and 25 after birth. At any age, body weight was similar between CONT and EXP rats. At times of the study (i.e. day 25), body weights of CONT and EXP rats were 78±2g and 76±2g, respectively (n=8 in each rat group).

3.2 Immunohistochemical Results

The immunoreactivity for PGP9.5 was similar between CONT and EXP rats (Figure 2). The calculated total carotid body volume ($10^6 \mu m^3$) of 4 CBs was not significantly different between CONT and EXP rats [28.5 ± 1.3 (n=4) and 25.9 ± 2.1 (n=4), respectively], and the volume proportions (%) of glomus cells and vessels, which were examined in 12 sections from 6 CBs, between CONT and

EXP rats [37.5 ± 2.0 (n=6) and 41.9 ± 2.1 (n=6) for glomus cells, and 11.1 ± 0.6 (n=6) and 9.6 ± 1.4 (n=6) for vessels] were not significantly different.

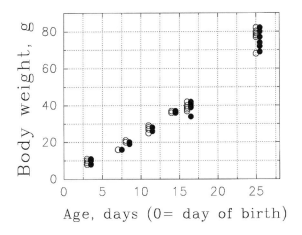

Figure 1. Body weight vs. postnatal age in CONT rats (open symbols, 32 points from 8 rats of 3 litters) and EXP rats (solid symbols, 32 points from 8 rats of 3 litters).

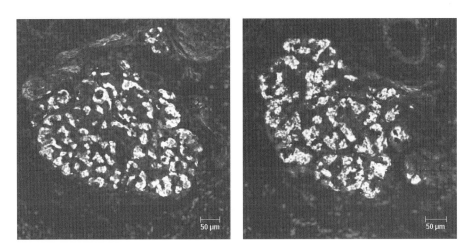

Figure 2. PGP9.5 immunoreactivities in CONT (A) and EXP (B) rats carotid bodies.

The immunoreactivity for PGP9.5 was also used to estimate the diameter of glomus cell nuclei (long axis and short axis) and the number of glomus cells per $10^6 \mu m^3$ glomus tissue (Figure 3). We obtained 16-18 cells from 4 CBs in each rat group. The calculated diameter of glomus cell nuclei (μm) was not significantly different between CONT and EXP rats [7.7 ± 0.3 (n=18) and 7.9 ± 0.2 (n=16) for long axis, and 5.4 ± 0.3 (n=18) and 5.8 ± 0.2 (n=16) for short axis]. The number of glomus cells in $10^6 \mu m^3$ glomic tissue was calculated from 23 PGP9.5 positive clusters from 4 CBs in each rat group and the values between CONT and EXP rats [1653 ± 71 (n=23) and 1613 ± 69 (n=23)] was not significantly different.

The immunoreactivity for TH was similar to that for PGP9.5, and was similar between CONT and EXP rats. Although TH was not used for the calculated total carotid body volume, the volume proportions (%) of glomus cells between CONT and EXP rats [31.3 ± 3.1 (n=6) and 31.6 ± 3.7 (n=6)] and those of vessels between CONT and EXP rats [9.1 ± 1.2 (n=6) and 9.4 ± 0.8 (n=6)] were not significantly different.

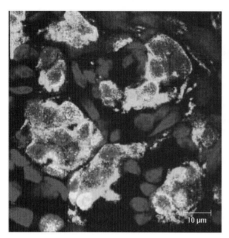
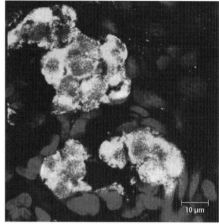

Figure 3. PGP9.5 immunoreactivities (indicated by arrows) in CONT (A) and EXP (B) rat carotid bodies. Nuclei are stained with DAPI.

4. DISCUSSION

At three weeks after the episode, we found that the volume of carotid body and the area occupied with glomus cells and/or vessels were not significantly influenced by an episode of neonatal anoxia. The results are consistent with a previous observation showing that a significant increase in ventilation occurred only at acute hypoxia but not in normoxia in the developing rats with an anoxic episode in neonatal period (Saiki and Matsumoto, 1999). These results suggest that the long-lasting effects of one episode of neonatal anoxia may become apparent exclusively at hypoxic exposure at their weaning age, i.e. about 25 days after birth.

Long-lasting effects of neonatal hypoxia in the first week after birth are well known, and in the cases, the rats at 7 weeks of age increase their resting ventilation but reduce the HVR (Okubo and Mortola, 1988; 1990). Although our previous findings are opposite to the results obtained in neonatal hypoxia, a recent study has reported that intermittent hypoxia augments carotid body and ventilatory response in neonatal rat pups (Peng et al., 2004) and the results suggest that episodes of hypoxia during neonatal period may also act to facilitate carotid body sensory response to hypoxia. In another study, chronic intermittent hypoxia was reported to evoke long-lasting facilitation of carotid body sensory activity with no changes in carotid body morphology in adult rats (Peng et al., 2003). These observations suggest that the influence of neonatal hypoxia on the HVR may be not uniform and that the difference in the severity of hypoxia and the timings of the hypoxic episode as well as the assessment of HVR may bring different results on the HVR in developing animals. In addition, it should be considered that the influence of hypoxia and anoxia (i.e. 0% O_2) on the respiratory control system may be different.

In conclusion, although further examinations are needed to clarify how hypoxic episodes change the ventilatory control mechanisms, the earlier studies and our present results suggest the possibility that hypoxic ventilatory responses of the rat having an episode of neonatal anoxia may be facilitated three weeks after the episode, but those ventilatory alterations could occur exclusively at hypoxia without changing their carotid body structures.

REFERENCES

Carroll J.L. Plasticity in respiratory motor control. Invited Review: Developmental plasticity in respiratory control. J Appl Physiol 2003; 94:375-389.

Erickson J.T., Mayer C., Jawa A., Ling L., Olson E.B., Vidruk E.H., Mitchell G.S., Katz D.M. Chemoafferent degeneration and carotid body hypoplasia following chronic hyperoxia in newborn rats. J Physiol 1998; 509: 519-526.

Kent C., Rowe H.L. The immunolocalisation of ubiquitin carboxyl-terminal hydrolase (PGP9.5) in developing paraneurons in the rat. Dev Brain Res 1992; 68:241-246.

Kusakabe T., Hayashida Y., Matsuda H., Gono Y., Powell F.L., Ellisman M.H., Kawakami T., Takenaka T. Hypoxic adaptation of the peptidergic innervation in the rat carotid body. Brain Res 1998; 806:165-174.

Okubo S, Mortola, J.P. long-term respiratory effects of neonatal hypoxia in the rat. J Appl Physiol 1988; 64:952-958.

Okubo S, Mortola, J.P. Control of ventilation in adult rats hypoxic in the neonatal period. Am J Physiol 1990; 259:R836-R841.

Peng Y-J., Overholt J.L., Kline D., Kumar G.K., Prabhakar N.R. Induction of sensory long-term facilitation in the carotid body by intermittent hypoxia: implications for recurrent apneas. PNAS 2003; 100: 10073-10078.

Peng Y-J., Rennison J., Prabhakar N.R. Intermittent hypoxia augments carotid body and ventilatory response to hypoxia in neonatal pups. J Appl Physiol 2004; 97: 2020-2025.

Saiki C., Matsumoto S. Effect of neonatal anoxia on the ventilatory response to hypoxia in developing rats. Pediatr Pulmonol 1999; 28:313-320.

Saiki C., Mortola J.P. Ventilatory control in infant rats after daily episodes of anoxia. Pediatr Res 1994; 35:490-493.

Wasicko M.J., Sterni L.M., Bamford O.S., Montrose M.H., Carroll, J.L. Resetting and postnatal maturation of oxygen chemosensitivity in rat carotid chemoreceptor cells. J Physiol 1999; 514:493-503.

The Effect of Development on the Pattern of A1 and A2a-Adenosine Receptor Gene and Protein Expression in Rat Peripheral Arterial Chemoreceptors

ESTELLE B. GAUDA, REED Z. COOPER, [2]DAVID F. DONNELLY, ARIEL MASON AND GABRIELLE L. McLEMORE

Department of Pediatrics, Division of Neonatology Johns Hopkins Medical Institutions. Baltimore, Maryland 21287. [2]Department of Pediatric Pulmonary Medicine, Yale University School of Medicine, New Haven, CT, USA

1. INTRODUCTION

The peripheral arterial chemoreceptors in the carotid body (CB) are the first step in a closed–loop feedback control system that acts to normalize arterial oxygen and carbon dioxide levels by rapidly modulating ventilation. Type I cells in the CB are excitable and contain O_2- sensitive K^+ channels (Gonzalez et al., 1995; Montoro et al., 1996; Wyatt et al., 1995). Reduction of K+ conductance in response to hypoxia is the signal that triggers Type I cell depolarization, Ca^{2+} entry, and secretion of neurotransmitters that bind to receptors on the first order sensory nerve endings of the carotid sinus nerve with cell bodies in the petrosal ganglion {(Gonzalez et al., 1994;Gonzalez et al., 1992). These first order sensory neurons (chemoafferents) project to second order neurons within the nucleus tractus solitarii (nTS), which send projections to the muscles of respiration. While the cascade of molecular and cellular events occurs in multiple CB preparations from multiple mammalian species, key aspects of the cascade are still unknown, particularly identification of the specific oxygen sensor within the Type I cell that initiates the cascade and the specific excitatory neurotransmitter systems that are involved in chemoexcitation. Furthermore, in multiple immature mammalian species, including human infants, hypoxic chemosensitivity matures during the first several weeks of postnatal life. Specific mechanisms mediating that maturation are unknown.

Adenosine (ADO) is an ubiquitous molecule that is released from metabolically active cells by facilitated diffusion or is generated extracellularly by degradation of released ATP (Zimmermann & Braun, 1996). ADO levels increase in response to hypoxia. ADO modifies cellular function by binding to specific cell-surface receptors. All four identified ADO-Rs (A1-R, A2a-R, A2b-R, and A3-R) are members of the superfamily of G protein-coupled receptors (GPCRs). A1- and A3-Rs interact with pertussis toxin-sensitive G proteins (Gi and Go), inhibit adenylyl cyclase (AC), and hyperpolarize cells by G protein-coupled K^+ channels, whereas A2a- and A2b-Rs interact with G proteins and activate AC (Fredholm et al., 2000). In response to hypoxia there is an initial increase in ventilation followed by gradual decline. While the role of ADO in the gradual decline, known as the hypoxic " roll-off," is through ADO inhibiting central networks modulating ventilation in both adult and immature models, the role of ADO on the initial increase in ventilation, is through ADO exciting influence on peripheral arterial chemoreceptor in adult models. Studies in newborn rabbits suggest that ADO may also have an inhibitory role at the level of the peripheral arterial chemoreceptors. Using semi-quanitative in situ hybridization (ISHH) immunohistochemistry, and real time quantitative polymerase chain reaction, we tested the hypothesis that the ADO-R profile in the peripheral arterial chemoreceptors shifts from an inhibitory profile to an excitatory profile with postnatal maturation.

2. METHODS

Tissues were removed from Sprague-Dawley rats within the first 3 weeks of postnatal development. All animals were briefly anesthetized with halothane and decapitated. The bifurcation of the carotid artery with the carotid body and the nodose/petrosal ganglion (NPG) complex were quickly removed "en bloc", placed in embedding media and quick frozen on dry ice. The tissue was later sectioned and processed for either ISHH or immunohistochemistry. Tissues samples used for real-time PCR were homogenates of the NPG from 6-8 animals at each of the ages examined.

2.1 ISHH

Radioactive ISHH was performed as previously described Gauda et al. (Gauda *et al*, 2000) and briefly summarized. Tissue blocks were cut in to 12 µM sections on a cryostat. Serial sections were thaw-mounted onto gelatin-chrome, alum-subbed slides. Slide-mounted sections were fixed in 4% paraformaldehyde, acetylated in fresh 0.25% acetic anhydride in 0.1M triethanolamine, dehydrated in ascending series of alcohols, delipidated in chloroform, and then rehydrated in a descending series of alcohols. ^{35}S-UTP-antisense ribonucleotide probes were used for the detection of mRNAs for A2a Rs and A1- Rs. The antisense nucleotide probes were constructed from complementary DNAs (cDNA) 100-426 base pairs of the rat A2a-R gene , (Fink et al., 1992) and 396-842 base pairs of the rat A1- R gene (Weaver, 1996). Probes were labeled with $1.2\text{-}1.5 \times 10^6$ dpm of labeled probes were added to 100 µl of hybridization buffer (Gauda et al., 2000). Hybridization was performed at 55^0 C overnight. The slides were then washed, treated with RNAase A (20mg/ml), washed in 60^0C in 0.2 X SSC, for 10 mins x 4, then rinsed in deionized water and air dried, dipped in Kodak photographic emulsion, dried and exposed in the dark at -20^0C for 8-12 weeks. After exposure, the slides were thawed at room temperature and developed with Dektol and counterstained with thionin and coverslips applied with Permount.

Slides were qualitatively and semi-quantitatively analyzed for the pattern of expression of the silver grains in the carotid body and PNG complex and for the change in the mean number of cluster silver grains over ganglion cells for the entire NPG complex for animals at postnatal day 5, 15 and 22. Silver grains generated by ^{35}S in the emulsion were analyzed by counting algorithm of the

NIH image analysis program. The darkfield image was captured and digitized at 400 X.

2.2 Single and Double-labeled Immunofluorescence

The pattern of A1-R and A2A-R immunoreactivity (-ir) in the CB and NPG was determined with immunofluorescence. Frozen, slide-mounted, 12μm sections of the CB and NPG complex were thawed at room temperature then fixed in 4% paraformaldehyde in 0.9% normal saline for 10min, washed three times for 5mins with 1X TBS, pH 7, permeabilized in 100% ice cold acetone for 10mins, and subsequently washed three times with TBS. Tissue endoperoxidases were quenched in 2% hydrogen peroxide for 5mins. The slides were then washed with TBS and non-specific binding was blocked by incubating the slides in 3% BSA containing 0.3% Triton X-100 (Sigma-Aldrich, Inc., St. Louis, MO) for 60mins at room temperature. Afterwards slides were incubated in primary antibodies followed by secondary antibodies as outlined. All primary and secondary antibodies were diluted in 1 X Tris-buffered saline (TBS) containing 0.3% Triton X-100 and 3% BSA.

Slides were incubated in primary anti-rat A2a-R monoclonal antibody (1.5 μg/ml; Upstate, Lake Placid, NY), overnight at room temperature, washed three times for 5mins with 1X TBS, then incubated with rhodamine-conjugated anti-mouse secondary antibody (1:400; Santa Cruz Biotechnology, Inc., Santa Cruz, CA). Next, the slides were washed with 1X TBS and incubated with anti- rat A1-R- polyclonal antibody raised in rabbit (1:400 Sigma Aldrich, Inc. St. Louis, MO) in 1X TBS for 2 hrs at room temperature, washed 3X and then incubated with Alexa Flour® 488 goat anti-rabbit secondary antibody (1ug/ml; Molecular Probes, Inc., Eugene, OR) for 2hrs at room temperature. Lastly, slides were washed with 1X TBS and coverslips applied with ProLong® antifade reagent (Molecular Probes, Inc., Eugene, OR) and examined with fluorescent microscopy.

Fluorescent images were examined with a fluorescent microscope (Nikon Eclipse E-400) captured at 40 X with an attached CCD camera (Photometrics CoolSnap FX). The images where then stored in an image analysis program (IPLab version 3.5). The fluorescent cells were detected using filters for rhodamine (abs/em 555/580) and for FITC (abs/em 494/519). Sequential capturing of both fluorescent images with each filter set was done for each tissue section to determine co-expression of A1-R and A2a-R-ir. Background fluorescence was subtracted by comparing images with primary antibody with sections that were processed without primary antibody. Slides were qualitatively analyzed for the pattern of A1 and A2a-R –ir in the CB and NPG complex in animals at postnatal day 3 and 14.

3. RESULTS

During postnatal development, A2a-R mRNA was abundantly expressed in the CB of all the animals at each postnatal age while A1-R mRNA was not detected in the CB of any of the animals examined (data not shown). In contrast, A1-R mRNA was highly expressed in essentially all the ganglion cells in the NPG complex for animals at each postnatal age as seen qualitatively in

Figure 1 for animals at 0, 3 and 14 PNDs. As shown in Figure 2, A1-R mRNA (A ,C) and A2a-R mRNA (B,D) expression in the NPG complex for animals at 6 (A,B) and 16 (C,D) PNDs shows more extensive expression pattern for the A1-R mRNA than that of the A2a-R at both postnatal ages. Similar findings were seen for animals at PND 0, 14 and 22.. With developmnt A2a-mRNA expression was characterized by greater number of cluster of silver grains in the ganglion cells of the NPG complex while the number of clusters for A1 mRNA remained constant with development (Figures 3 and 4). The increase in expression of A2a-mRNA was confirmed by real- time PCR which showed a two fold increase in A2a-mRNA expression in the pooled homogenates of the NPG from 6 animals at PND 5 to PND 22 each. These data suggested to us that the ratio of A1-R to A2a-R mRNA expression decreased with postnatal development, shifting the ratio to a more excitatory receptor profile for the binding of ADO in response to hypoxia.

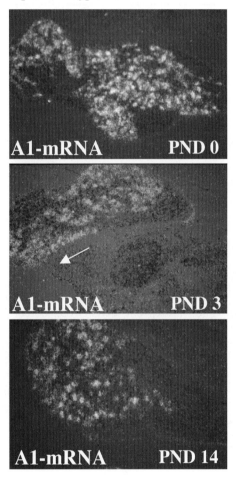

Figure 1. Representative low power photomicrograph depicting A1-mRNA expression as multiple white dots (silver grains) throughout the NPG complex for three animals at PNDs 0, 3 and 14. Also note the absence of A1-mRNA expression in the CB (arrow) shown in the photomicrograph of a representative animal at PND 3. PND is postnatal day and CB is carotid body.

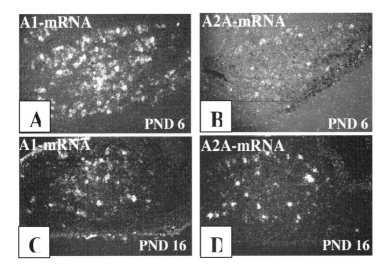

Figure 2. Low power photomicrograph depicting A1-mRNA and A2a-mRNA expression (multiple white dots) on serial sections of the NPG complex from 2 animals at PNDs 6 and 16.

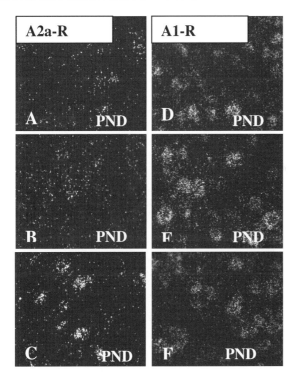

Figure 3. High power darkfield photomicrograph depicting the effect of postnatal development on the expression pattern of A2a-R (A-C) and A1-R mRNA (D-F) expression in ganglion cells in the petrosal ganglion in 3 animals at PNDs 5, 15 and 22. Note the increased clusters of silver grains for A2a-R mRNA while the expression pattern for A1-R mRNA remains consistent with development.

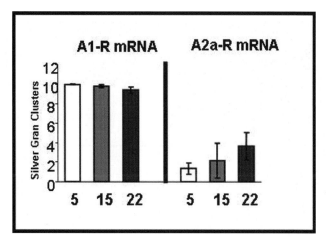

Figure 4. Bar graph showing the effect of development on the number of clusters of silver grains over ganglion cells per NPG complex for A1-R and A2a-R mRNAs. N=5 for each age group.

The pattern of A1-R and A2a-R protein expression was also determined in tissue sections of the CB and NPG complex during postnatal development. A2a-R –ir was abundantly expressed throughout the CB at all gestational ages as shown for one representative animal at DOL 3. The intensity of the A2aR-ir in the CB was high and appeared consistent during development. The pattern of A1-R- ir appeared lacey throughout the CB suggesting to us that the localization of A1-R-ir may be in nerve fibers innervating the CB (arrows, Figure 5). The intensity A1-R -ir in the carotid body was greater in animals at PND 3 than at PND 14 as shown in Figure 5. A2a-R-ir and A1-R-ir was also abundantly expressed in the NPG complex at both ages. Of interest, the pattern of A2a-ir localized to the nucleus of the ganglion cells while the A1-R –ir localized to the cytoplasm of the same cells (Figure 6). Confocal fluorescent microscopy confirmed the nuclear localization of the A2a-R- ir to in ganglion cells while nuclear localization of A2a-R- ir in type I cells in the CB was demonstrated with immunoelectromicroscopy (data not shown).

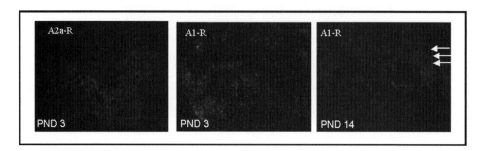

Figure 5. High power photomicrographs of immunoflourescence microscopy showing the expression pattern for A2a-R immunoreacivity in the CB from one animal at PND 3 and A1-R immunoreactivity for two animals at PND 3 and PND 14. Note the lacey and linear (arrows) pattern of expression for A1-R immunoreactivity.

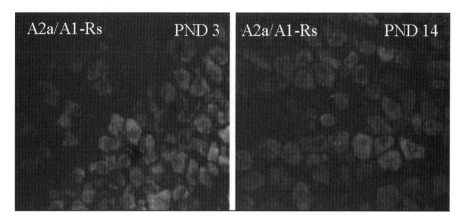

Figure 6. High power photomicrographs of double-labeled immunoflourescence microscopy of the petrosal ganglion cell bodies showing the expression pattern in for A2a-R (red) and A1-R immunoreactivity (green) for two animals at PND 3 and PND 14. Note the cytoplasmic expression of the A1-R-ir and the nuclear pattern of expression for A2a-R-ir. Also note that the intensity of the A2a-R –ir is greater in the animal at PND 14.

4. DISCUSSION

Using immunohistochemical techniques we describe the pattern of gene and protein expression for A1 and A2a-R in peripheral arterial chemoreceptors during postnatal development. We confirm and extend our previous findings (Gauda et al., 2000) by demonstrating that 1) gene expression for inhibitory A1-Rs is not present in type cells in the CB but is abundantly expressed in essentially all cells in the NPG which contains the cell bodies for chemoafferents, 2) A1-R mRNA expression in the NPG complex remains constant throughout development while A2a-R mRNA expression increases during development, 3) immunoreactivity for inhibitory A1-Rs is present in the CB and the pattern of expression suggest that the A1-R –ir is within the nerve fibers innervating the CB, and lastly 4) A2a-R-ir is present in cell bodies of chemoafferents but localization within in the cells is nuclear which differs from the cytoplasmic localization of A1-R -ir.

Several lines of evidence suggest that ADO plays an excitatory role in hypoxic chemosensitivity via binding to postsynaptic A2a-Rs. Hypoxic induced ventilation is increased following intracarotid injections of either exogenous ADO or enzymatic blockers that inhibit the degradation of endogenous ADO (Monteiro & Ribeiro, 1987; Monteiro & Ribeiro, 1989). Carotid sinus nerve denervation or exposure to exogenous A2a-R antagonists abolished ADO-mediated hypoxic ventilatory responses (Monteiro & Ribeiro, 1987; Monteiro & Ribeiro, 1989) in adult rats. In adult cats, the dose-dependent increase in chemoexcitation was blocked by theophylline, a nonspecific adenosine receptor antagonist (McQueen & Ribeiro, 1981). Experiments in a superfused adult CB preparation have also shown that exogenous ADO induces chemoexcitation in response to hypoxia (Runold et al., 1990). We and others have shown that -ir and mRNA for excitatory A2a-Rs have been identified in the CB, whereas mRNA for the inhibitory A1- and A3-Rs have not been detected by ISHH (Gauda et al., 2000, Kobayashi et al., 2000) or by PCR (Kobayashi et al., 2000).

While the central inhibitory effects of ADO are well described for newborns, the role of ADO in modulating hypoxic chemosensitivity in peripheral arterial chemoreceptors is less well described even though non-specific ADO receptor blockers are one of the most common classes of drugs used for the treatment of apnea of prematurity in human infants. Studies in newborn rabbits suggest that ADO at the level of peripheral arterial chemoreceptors is also inhibitory via an A1-R mechanism (Runold M et al., 1989). By demonstrating the presence of –ir and the absence of mRNA for A1-Rs in the CB and the presence of both mRNA and -ir for A1-Rs in the petrosal ganglion, our data corroborate the findings of (Runold M et al., 1989). We speculate that ADO functions as a major inhibitory neuromodulator that acts through postsynaptic A1-Rs in immature animals. The decrease in the ratio of postsynaptic A1-Rs to A2a-Rs on chemoafferents with postnatal development may in part contribute to the maturation of hypoxic chemosensitivity that occurs with postnatal development in newborn animals and human infants.

ACKNOWLEDGMENTS

This work was supported by National Institutes of Drug Abuse RO1 DA-13940. We would also like to thank Carol Cook for her technical expertise with the immunoelectronmicroscopy and Debbie Flock for her technical expertise in immunoblotting.

REFERENCES

Fink, J.S., Weaver, D.R., Rivkees, S.A., Peterfreund, R.A., Pollack, A.E., Adler, E.M., & Reppert, S.M. 1992, Molecular cloning of the rat A2 adenosine receptor: selective co- expression with D2 dopamine receptors in rat striatum. *Brain Res.Mol.Brain Res.*, 14, 186-195.

Fredholm, B.B., Arslan, G., Halldner, L., Kull, B., Schulte, G., & Wasserman, W. 2000, Structure and function of adenosine receptors and their genes. *Naunyn Schmiedebergs Arch.Pharmacol.*, 362, 364-374.

Gauda, E.B., Northington, F.J., Linden, J., & Rosin, D.L. 2000, Differential expression of A_{2A}, A_1- adenosine and D_2-dopamine receptor genes in rat peripheral arterial chemoreceptors during postnatal development. *Brain Res.*, 872, 1-10.

Gonzalez, C., Almaraz, L., Obeso, A., & Rigual, R. 1992, Oxygen and acid chemoreception in the carotid body chemoreceptors. *Trends Neurosci.*, 15, 146-153.

Gonzalez, C., Dinger, B.G., & Fidone, S.J. (1994) Mechanisms of carotid body chemoreception. Regulation of Breathing (ed. by J. A. Dempsey & A. I. Pack), pp. 391-470. Marcel Dekker, Inc.

Gonzalez, C., Lopez-Lopez, J.R., Obeso, A., Perez-Garcia, M.T., & Rocher, A. 1995, Cellular mechanisms of oxygen chemoreception in the carotid body. *Respir.Physiol*, 102, 137-147.

Kobayashi, S., Conforti, L., & Millhorn, D.E. 2000, Gene expression and function of adenosine A_{2A} receptor in the rat carotid body. *Am.J.Physiol Lung Cell Mol.Physiol*, 279, L273-L282.

McQueen, D.S. & Ribeiro, J.A. 1981, Effect of adenosine on carotid chemoreceptor activity in the cat. *British Journal Of Pharmacology*, 74, 129-136.

Monteiro, E.C. & Ribeiro, J.A. 1987, Ventilatory effects of adenosine mediated by carotid body chemoreceptors in the rat. *Naunyn Schmiedebergs Arch.Pharmacol*, 335, 143-148.

Monteiro, E.C. & Ribeiro, J.A. 1989, Adenosine deaminase and adenosine uptake inhibitions facilitate ventilation in rats. *Naunyn Schmiedebergs Arch.Pharmacol.*, 340, 230-238.

Montoro, R.J., Urena, J., Fernandez-Chacon, R., Alvarez, T., & Lopez-Barneo, J. 1996, Oxygen sensing by ion channels and chemotransduction in single glomus cells. *J Gen.Physiol*, 107, 133-143.

Runold, M., Lagercrantz, H, Prabhakar N.R., & Fredholm B.B. Role of adenosine in hypoxic ventilatory depression. J Appl Physiol [2], 541-546. 1989. Ref Type: Journal (Full)

Runold, M., Cherniack, N.S., & Prabhakar, N.R. 1990, Effect of adenosine on isolated and superfused cat carotid body activity. *Neuroscience Letters*, 113, 111-114.

Weaver, D.R. 1996, A_1-adenosine receptor gene expression in fetal rat brain. *Developmental Brain Research*, 94, 205-223.

Wyatt, C.N., Wright, C., Bee, D., & Peers, C. 1995, O_2-sensitive K^+ currents in carotid body chemoreceptor cells from normoxic and chronically hypoxic rats and their roles in hypoxic chemotransduction. *Proc.Natl.Acad.Sci.U.S.A*, 92, 295-299.

Zimmermann, H. & Braun, N. 1996, Extracellular metabolism of nucleotides in the nervous system. *J.Auton.Pharmacol.*, 16, 397-400.

A Comparative Study of the Hypoxic Secretory Response between Neonatal Adrenal Medulla and Adult Carotid Body from the Rat

A.J. RICO, S. P. FERNANDEZ, J. PRIETO-LLORET, A. GOMEZ-NIÑO, C. GONZALEZ AND R. RIGUAL

Departamento de Bioquímica y Biología Molecular y Fisiología /(IBGM). Universidad de Valladolid /(CSIC). Facultad de Medicina. Universidad de Valladolid

1. INTRODUCTION

Perinatal adrenal medulla responds to hypoxia by increasing the release of catecholamine (CA) that are responsible for metabolic, cardio-circulatory and respiratory mechanisms crucial to the adaptation to the extrauterine life (3, 11). In neonatal rat, hypoxia elicits directly this secretory response since splachnic innervation of AM is not mature until the second week of postnatal life (10).

Despite of the similarities in the proposed hypoxic transduction cascades in CB chemoreceptor cells (1) and in perinatal adrenomedullary chromaffin cells (5, 12) it has never been compared the final product of both transduction cascades, i.e., the release of CA, which in the case of the CB would act as neurotransmitters, and in the case of the AM they would act as hormones. This comparison would allow making straightforward conclusions on the thresholds and on the overall gain of both transduction cascades in both organs.

2. MATERIAL AND METHODS

Rats of 1-2 days and 3 months old were anaesthetized with sodium pentobarbitone (60 mg/kg i.p.). After a longitudinal incision in the abdomen, adrenal glands were removed and placed in a lucite chamber filled with ice cold Tyrode-bicarbonate solution (in mM: NaCl, 116; KCl, 5; $CaCl_2$, 2; $MgCl_2$, 1.1; $NaHCO_3$, 23; glucose, 5; pH = 7.4) equilibrated with 95%O_2-5%CO_2. Adrenal glands were gently cleaned of adrenal cortex with fine forceps under a dissecting microscope (see Rigual et al., 2002). Carotid artery bifurcations were removed and CB cleaned of surrounding connective tissue in the same conditions (see ref 13). After dissection, in order to facilitate the penetration of the carbon fiber electrode inside the tissue, CB from 3 months old rats were incubated in a dilute enzymatic solution (Tyrode-bicarbonate equilibrated with 95%O_2-5%CO_2 containing 0.1% collagenase (Worthington type I) for 5 min at 37°C. Tissues were transferred to a thermostatic lucite recording chamber and superfused by gravity with Tyrode-bicarbonate control solution equilibrated with 20%O_2, 5%CO_2 and 75%N_2 (3-4ml/min, 37° C).

Both tissues, neonatal AM and adult CB were placed side to side in the recording chamber and were impaled with paired carbon electrodes (single 5-μm; Dagan Corporation) under microscope. Free CA concentration inside tissue was used as an index of the secretory response. This experimental setting allows the recording for both preparations simultaneously to assure identical recording conditions The electrodes were attached to an EI-400 potentiostat (Ensman Instrumentation). Recordings were undertaken with a fixed voltage (0.5 V amperometric mode). Currents, proportional to free tissue CA concentrations, were sampled at 5 Hz, digitalized and recorded by computer (Digidata 1322, Axoscope 9; Axon Instruments) for offline analysis (Microcal Origin). After a superfusion period of 30-40 min the amperometric recording was stable and preparations were subjected to different experimental protocols detailed in the Results section. Prior to and after the recording from the tissues the carbon fiber electrode was advanced into the bath chamber and calibrated by switching between normal Tyrode and Tyrode containing 10 μM epinephrine.

3. RESULTS

With one isolated adult CB and one isolated neonatal AM (1-2days) placed in the recording chamber side by side, we stimulated the preparations with severe (solution equilibrated with 0 %O2; PO2≈15-20mm Hg in the chamber; 3min) and mild hypoxia (solution equilibrated with 5% O2; PO2≈ 48mmHg; 3min) and to depolarizing solution (30mM KCl; 1min). Figure 1A shows a record from a sample experiment.

The intensity of the secretory responses evoked by hypoxia in adult CB and in neonatal AM was similar, but the response to 30mM KCl was much higher in neonatal AM than in the CB. Figure 1B shows mean results from 4 experiments. The responses evoked by hypoxia are quite similar in the CB and AM: mild hypoxia produced a release response in CB that yielded a free CA tissue concentration of 2.8 ± 1.0 μM, and an AM tissue concentration of 3.9 ± 1.0 μM; severe hypoxia also rose free CA tissue concentrations by comparable magnitudes: 11.5 ± 4.4 μM in the CB and 12.5 ± 3.3 μM in the AM (mean ± SEM; n=4). However the depolarizing solution (30mM KCl) evoked responses were markedly different. In the CB concentration of free tissue CAs rose to 6,6 ± 5,1 μM and in AM they rose to 399,9 ± 97,7 μM. When the CA released by hypoxia are calculated as percentage of that the evoked by high K+, it was found that for severe hypoxia (0%O_2) the percentage were 372.0 ±128.0% and 3,6 ± 0.9% for adult CB and neonatal AM, respectively. Since the measurement of the release is made as free tissue CA concentration, the absolute values of free CA are greatly determined by the stored concentrations of CA in each tissue (CA content/mg wet tissue). Therefore in Figure 1C we have represented the ratios of CA release/stored CA tissue concentration in both organs. For high K+ as stimulus the ratio is comparable in both tissues (20.6 ± 16.0 for adult CB and 24.9 ± 6.1 for neonatal AM), while for mild and severe hypoxia the ratios are completely different: 8.9 ± 3.1 in the CB and 0.2 ± 0.1 in the AM, for 5% O2,

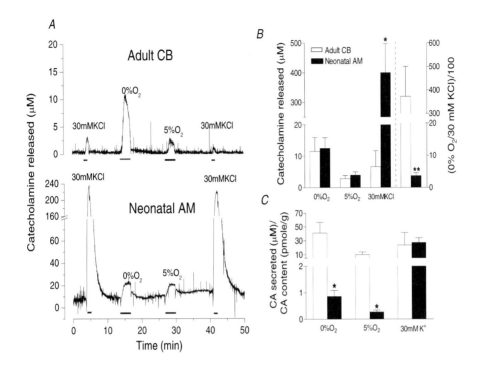

Figure 1. Release of CA in response to hypoxia from *in vitro* superfused adult CB and neonatal (1-2 days) AM. A, representative example of the effects of 3 min mild (5%O_2), severe (0%O_2) hypoxic and depolarizing stimuli (30mM KCl, 1min) on free-CA tissue concentration from adult CB and neonatal AM (1-2days old). B, upper part on the left represents averaged results of the free-CA tissue concentration peaks evoked by hypoxic and depolarizing stimuli applied on adult CB and neonatal AM as in part A; upper part on the right represents the quotient between the free-tissue CA peaks evoked by hypoxic stimuli and that evoked by depolarizing stimulus of [(CA secreted by hypoxia / CA secreted by K^+) x 100]. C, shows ratios between free- tissue CA peaks evoked by stimuli and CA content in the organs. Values represent means± SEM of 4-7 data from 4 experiments as in A. *$P < 0.05$, **$P < 0.01$.(Student's unpaired *t* test).

and 36.3 ± 13.8 for the CB and 0.8 ± 0.2 for the AM, for 0% O2. Both comparisons indicate that the transduction cascade coupling the hypoxic stimulus to exocytosis is ≈30-100times more efficient in the adult CB than in the neonatal AM, while the efficiency of the stimulus-secretion coupling for high external K+ stimuli is comparable in the adult CB and in the neonatal AM.

4. CONCLUSION

AM from 1-2 days old AM secretes CA in response to hypoxia, in increasing amounts as the intensity of hypoxia increases. The hypoxic threshold for the neonatal AM response is a PO_2 of 50 mmHg or higher and in the same range to that described earlier for *in vitro* preparations of the adult rabbit CB (6, 9) and very similar to that found for CB in vivo (PO_2=70-75 mmHg) (2). However, as we have already mentioned in the Results, the gain of the hypoxic transduction cascade to activate the exocytotic machinery is much lower in the neonatal AM, while the efficiency of the stimulus-secretion coupling for high external K^+ depolarization is comparable in the adult CB and in the neonatal AM. This comparison is extremely interesting when it is considered that the magnitude of Ca^{2+} currents is over ten times larger in the rat adrenomedullary cells than in the CB chemoreceptor cells (4 *vs* 7), even though the size of both cells is nearly identical. These facts would imply that chemoreceptor cells should express some mechanism, presently unknown, that facilitates the coupling of Ca^{2+} to exocytotic machinery during high K^+ depolarization and much more strongly during hypoxic stimulation.

ACKNOWLEDGEMENTS

WE wish to thank María de los Llanos Bravo for technical assistance. Supported by Grants from the DGICYT BFI2003/16271 and BFU2004-06394 and by grants of ICiii-FISS (Red Respira and Grant PI042462).

REFERENCES

1. Gonzalez, C., Almaraz, L., Obeso, A., and Rigual, R. Oxygen and acid chemoreception in the carotid body chemoreceptors. *Trends Neurosci.* 15:146-53, 1992.
2. Gonzalez, C., Almaraz, L., Obeso, A. and Rigual, R. Carotid body chemoreceptors: from natural stimuli to sensory discharges. *Phsiol. Rev.* 74:829-898, 1994.
3. Lagercrantz, H., and Slotkin, T.A. The "stress" of being born. *Sci. Am.* 254:100-107, 1986.
4. Lopez-Lopez, J.R., and Peers, C. Electrical Properties of chemoreceptor cells. In: The carotid body chemoreceptors, (C. Gonzalez eds) Springer-Verlag , NY, 1997, pp. 41-78.
5. Mochizuki-Oda, N., Takeuchi, Y., Matsumura, K., Oosawa, Y., and Watanabe, Y. Hypoxia-Induced catecholamine release and intracellular Ca^{2+} increase via supression of K^+ channels in cultured rat adrenal chromaffin cells. *J. Neurochem.* 69:377-387, 1997.
6. Perez-Garcia, M.T., Almaraz, L., and Gonzalez, C. Cyclic AMP modulates differentially the release of dopamine induced by hypoxia and other stimuli and increases dopamine synthesis in the rabbit carotid body. *J. Neurochem.* 57: 1992-2000, , 1991.
7. Prakriya, M., and Lingle, C.J. BK channel activation by brief depolarizations requires Ca2+ channels in rat chromaffin cells. *J. Neurophysiol.* 81: 2267-2278, , 1999.
8. Rigual. R., Almaraz L., Gonzalez C., and Donnelly, D.F. Developmental changes in chemoreceptor nerve activity and catecholamine secretion in rabbit cartodid body: posible role of Na^+ and Ca^{2+} currents. *Plugers Arch.* 439:463-470, 2000.
9. Rigual, R., Montero, M., Rico, A.J., Prieto-Lloret, J., Alonso, M.T., and Alvarez, J. Modulation of secretion by the endoplasmic reticulum in mouse chromaffin cells. *Eur. J. Neurosci.* 167:1690-1696, 2002.

10. Seidler, F.J., and Slotkin, T.A. Adrenomedullary function in the neonatal rat: response to acute hypoxia. *J. Physiol.* 358:1-16, 1985.
11. Slotkin, T.A., and Seidler, F.G. Adrenomedullary catecholamine release in the fetus and newborn: secretory mechanisms and their role in stress and survival. *J. Dev. Physiol.* 10:1-16, 1988.
12. Thompson, R.J., Jackson, A., and Nurse, C.A. Developmental loss of hypoxic chemosensitivity in rat adrenomedullary chromaffin cells. *J. Physiol.* 498: 503-510, 1997.
13. Vicario, I., Rigual, R., Obeso, A., and Gonzalez, C. Characterization of the synthesis and release of catecholamine in the rat carotid body in vitro. *Am. J. Physiol. Cell Physiol.* 278:490-499, 2000.

In Search of the Acute Oxygen Sensor
Functional proteomics and acute regulation of large-conductance, calcium-activated potassium channels by hemeoxygenase-2

[1]PAUL J KEMP, [2]CHRIS PEERS, [1]DANIELA RICCARDI, [3]DAVID E. ILES, [1]HELEN S. MASON, [2]PHILLIPPA WOOTTON AND [1]SANDILE E. WILLIAMS

[1]*School of Biosciences, Cardiff University, Cardiff, CF10 3US, UK;* [2]*Institute for Cardiovascular Research, Worsley Building, University of Leeds, Leeds LS2 9JT UK;* [3]*School of Biology, Manton Building, University of Leeds, Leeds LS2 9JT UK.*

1. INTRODUCTION

Detecting and reacting to acute perturbation in the partial pressure of atmospheric oxygen (pO_2), particularly hypoxia, is a fundamental adaptive mechanism which is conserved throughout the animal kingdom. In mammals, a number of cellular systems respond, often co-operatively as oxygen availability becomes compromised, with the express aim of maximising oxygen uptake by the lungs and of optimising its delivery to the metabolically most active tissues. Thus, during hypoxia, ventilation rate and depth are increased to maximize air flow across the gaseous exchange surface, local lung perfusion rates become rapidly matched to local alveolar ventilation and systemic arteriolar dilatation ensures that tissue and cerebral blood flow become swiftly optimized. Central to many of the oxygen-sensitive responses is hypoxic inhibition of large conductance, Ca^{2+}-activated potassium (BK, maxiK or *slo*) channels. Thus, BK channels are strongly implicated as critical components of the acute O_2 signalling cascade in; a) carotid body chemoreceptors (Peers, 1990; Riesco-Fagundo et al., 2001), where low arterial pO_2 is detected by BK channels and the resulting depolarizing signal is ultimately transduced into increased ventilation; b) fetal and postnatal pulmonary arteriolar myocytes, where BK channels may contribute to both persistent prenatal (Cornfield et al., 1996) and acute postnatal hypoxic pulmonary vasoconstriction (Peng et al., 1999; Cornfield et al., 1996) in order to match ventilation to perfusion; c) neonatal adrenomedullary chromaffin cells (Thompson & Nurse, 1998), where hypoxic inhibition of BK channels induces the huge surge in catecholamine secretion crucial for preparing the newborn's lung for air-breathing by activating alveolar fluid reabsorption and surfactant secretion and; d) central neurones (Liu et al., 1999; Jiang & Haddad, 1994b; Jiang & Haddad, 1994a), where hypoxic depression of BK channel activity may contribute to the excitotoxicity which results from increased neuronal excitability.

Although a number of mechanisms of hypoxic regulation of K^+ channels in general, and BK channels in particular, have been proposed see (Lopez-Barneo et al., 2001) for recent review, the nature of the sensor is still unclear. Indeed, there are varying reports which have suggested involvement of either cytosolic factors e.g. (Wyatt & Peers, 1995), direct channel modulation e.g. (Liu et al., 1999; Riesco-Fagundo et al., 2001; Jiang & Haddad, 1994a) and membrane-delimited regulation via activation/inhibition of associated BK channel protein partners (Lewis et al., 2002). To investigate the later, we have employed HEK293 cells stably co-expressing identified α- and β-subunits of a human BK channel in order to examine the O_2-sensitivity of these recombinant K^+ channels at the single channel level. Furthermore, we have employed a sequential strategy (termed functional proteomics) to determine; 1) potential protein partners of BK α-subunit; 2) the role of particular protein partners in both BK regulation and hypoxic inhibition of BK activity and; 3) the effect of specific protein knockdown using post transcriptional gene suppression on BK regulation by hypoxia. Finally, we have used these data to inform experiments aimed at determining the role of hemeoxygenase-2 in hypoxic inhibition of BK channels natively expressed in rat carotid body glomus cells (Williams et al., 2004a).

Funded by The Wellcome Trust and The British Heart Foundation

2. METHODS

2.1 Membrane Preparation and Immunoprecipitation

Cell pellets were homogenised in 1 ml ice-cold homogenization buffer and centrifuged at 1,000g for 5 minutes. The supernatants were transferred to fresh tubes and further spun at 16,000g for 30 minutes at 4°C. Pellets were solubilised by addition of Triton X-100. Suspensions were incubated with antibodies at 4°C for 1 hour before further incubation with Protein G beads for 4 hours. The beads were pelleted at 7,000g.

2.2 Electrophoresis and Protein Visualisation

For single dimension SDS-PAGE, pelleted immunoprecipitates were taken up in sample buffer, heated to 100°C for 3 minutes and loaded onto SDS-polyacrylamide gels comprising 4% stacking and 10% resolving gels For 2-D electrophoresis, the pelleted immunoprecipitates were resuspended in 250 µl rehydration buffer, sonicated for 20 minutes and loaded onto 11 cm, pI 4 - 7, IPG strips. After focussing, the IPG gel strips were equilibrated for 15 minutes in 15 ml reducing solution and 15 minutes in alkylating solution and then loaded onto 8 – 18% precast gels.

2.3 In-gel Trypsin Digestion and Peptide Mass Mapping

Bands of interest were excised from the gels, dehydrated in 0.5 ml acetonitrile for 10 minutes and dried in air. Proteins were reduced by incubation in 50 µl 10 mM DTT, 50 mM NH_4HCO_3 for 1 h at 56°C. Upon cooling, the DTT solution was replaced by 55 mM iodoacetamide, 50 mM NH_4HCO_3. The gel was then hydrated at 4°C for 45 minutes in trypsinisation buffer (20 ng/µl trypsin, 50 mM NH$_4$HCO$_3$). Digestion was carried out overnight at 37°C. 1.5 μl of the digest was mixed with 1.5 μl MALDI matrix solution (10 mg/ml α-cyano-4-hydroxycinnamic acid in 50% v/v ethanol/50% v/v acetonitrile) and dried onto a MALDI target. The sample was analysed on a Micromass TofSpec MALDI/TOF-mass spectrometer. Monoisotopic peptide fingerprints were used to search databases.

2.4 siRNA Design, and Transfection into HEK293 Cells

Two short interfering (si) RNAs were designed to target nucleotides 212 - 232 (HO-2 siRNA1) and 481 - 501 (HO-2 siRNA2) of the HO-2 coding sequence and the following oligonucleotide templates were designed (underlined bases are complimentary to the HO-2 coding sequence):

Antisense$_1$; 5'- AGCACACGACCGGGCAGAAACCTGTCTC -3';
Sense$_1$; 5'- AATTTCTGCCCGGTCGTGTGCCCTGTCTC -3'.
Antisense$_2$; 5'- AGTACGTGGAGCGGATCCACCCTGTCTC -3';
Sense$_2$; 5'- AAGTGGATCCGCTCCACGTACCCTGTCTC -3'.

For transfections, 125 pmol duplex, Cy3-labelled siRNA was diluted into 490 μl Optimem and 10 μl Lipofectamine. The mixture was incubated at room temperature for 20 minutes before being added to the cells. Cells were cultured or a further 48 hours before electrophysiological analyses

2.5 Electrophysiology – Inside-out Patches

BK channels were recorded from inside-out patched of wild type HEK293, BKαβ HEK 293 and rat carotid body glomus cells (prepared as described previously (Hatton et al., 1997)). Pipette solution contained: 135 mM NaCl, 5 mM KCl, 1.2 mM MgCl$_2$, 2.5 mM CaCl$_2$, 5 mM HEPES, 10 mM glucose. Osmolarity of the solution was adjusted to 300 mOsm.L^{-1}with sucrose. pH was adjusted to 7.4 with 1M NaOH. Bath solution contained 10 mM NaCl, 117 mM KCl, 2 mM MgCl$_2$, 11 mM HEPES, 0.15 mM EGTA, 0.1 mM CaCl$_2$. The pH was adjusted to pH 7.2 using KOH. Free ionised Ca^{2+} [Ca^{2+}]$_i$ was 336 nM. Normoxic solutions (pO$_2$ ~ 150 mmHg) were bubbled with medical air and hypoxic (15-25 mmHg) solutions were bubbled with N$_2$. Patches were held at +20 mV (-Vp).

3. RESULTS AND DISCUSSION

Proteins which potentially associate with the α-subunit of recombinant human (BKα) were immunoprecipitated with a specific BKα antibody from lysates of HEK293 cells stably co-expressing both human BKα$_1$ (KCNMA1) and BKβ$_1$ (KCNMB1) and separated by 2-D (Figure 1A, right panel) and 1-D (Figure 1B, right lane) gel electrophoresis. Parallel immunoprecipitation experiments were performed on untransfected, wild type HEK293 cells for comparison purposes (Figure 1A, left panel; Figure 1B, left lane). Of the unique proteins that immunoprecipitated from the stable BKαβ cell line, peptide mass mapping using mass spectroscopy of trypsin digests consistently identified gamma glutamyl transpeptidase (GGT) and hemeoxygenase-2 (HO-2) as potential protein partners. Although GGT associates directly with BKα, we have recently demonstrated that it is not involved in hypoxic inhibition of BK channels (Williams et al., 2004b).

Western blot of BKαβ cell lysates demonstrated the expected BKα subunit immunoreactivity only in transfected cells, whereas there was constitutive expression of HO-2 in both cell lines (Figure 1C). Biochemical interaction between BKα subunit and HO-2 was confirmed by co-immunoprecipitation of BKα with an HO-2 antibody, and *vice versa*, only in cells stably co-expressing either BKαβ (Figure 1D) or BKα alone (Figure 1E). Specificity of this interaction was demonstrated by the inability of BKα to immunoprecipitate either endothelial nitric oxide synthase (eNOS) or the α-subunit of the Na^+/K^+-ATPase (Figure 1D) despite both proteins being abundantly expressed (Figure 1C).

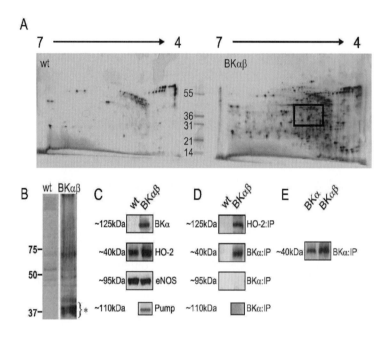

Figure 1. Hemeoxygenase-2 as a BKα protein partner. (A) 2-D gel electrophoresis of proteins immunoprecipitated with a BKα antibody from wild type (wt) and BKαβ HEK293 cells. Boxed area indicates location of protein spots selected for MALDI/TOF analysis. (B) SDS-PAGE of immunoprecipitates from wt and BKαβ cells. Bands removed for MALDI/TOF analysis are indicated by the asterisk. Linear pH gradients and/or molecular weight markers (in KDa) are shown. (C) Western blot analyses from lysates of wt and BKαβ cells show HO-2, endothelial nitric oxide (eNOS) and α-subunit of Na^+/K^+-ATPase (pump) are constitutively expressed. (D) Western blot identification of BKα and HO-2 following immunoprecipitation (IP) with the antibodies shown to the right (top two blots). Neither eNOS nor the pump immunoprecipitated with BKα (lower two blots). (E) Western blot identification of HO-2 following IP with the BKα antibody using lysates from BKαβ cells and BKα cells (with no β subunit). Adapted from Williams et al., 2004a.

In Search of the Acute Oxygen Sensor

141

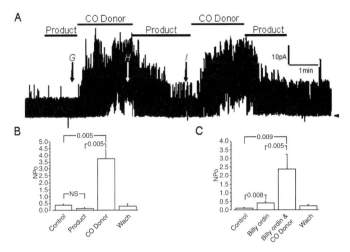

Figure 2. Hemeoxygenase metabolites activate BK channels. (A) Exemplar current recording from an inside-out patch excised from a BKαβ cell. Periods of application of CO-donor and its control (Product) are shown above the trace. (B) mean NPo plot showing effect of CO-donor. (C) Mean NPo plot showing additive effects of biliverdin and CO-donor. Patch potential (-Vp) = +20 mV, $[Ca^{2+}]_i$ = 335 nM, in this and all subsequent figures. P values are shown above bars and are from ANOVA/Bonferroni post hoc test. Adapted from Williams et al., 2004a.

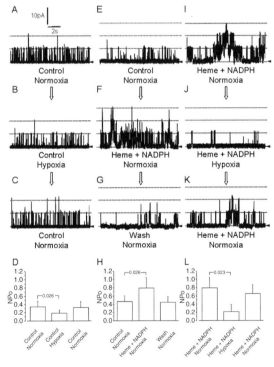

Figure 3. Hemeoxygenase substrates augment BKαβ channel activity and hypoxic inhibition. Exemplar traces and mean NPo plots indicating modest hypoxic channel inhibition in untreated patches (A-D), increased baseline channel activity by 1nM heme/1 μM NADPH (E-H) and augmentation of the hypoxic inhibition in the continued presence of heme/NADPH (I-L). All traces from inside-out patches excised from BKαβ cells. Student's t-test. Adapted from Williams et al., 2004a.

In the presence of O_2 and NADPH, hemeoxygenases catalyse the breakdown of heme to biliverdin, iron and CO (Prabhakar, 1999). BKαβ channel activity was robustly and reversibly activated by 30 μM of the chemical CO-donor, $[Ru(CO)_3Cl_2]_2$; 30 μM of the breakdown product of this compound, $RuCl_2(DMSO)_4$, which does not release CO and, therefore, acts as control, did not affect channel activity indicating that CO strongly activates BK channels in inside-out patches (Figure 2A). Normalised NPo was increased 15-fold by the CO-donor (Figure 2B; n = 13) 10 μM biliverdin evoked a more modest, but significant 4-fold activation of BKαβ channel activity (Figure 2C; n = 12). In patches treated sequentially with biliverdin and the CO-donor, the activation was additive with the CO-donor causing a further increase to 28-fold above control (Figure 2C; n=5). Wild type HEK293 cells did not display BK currents (data not shown but see (Lewis et al., 2002)) and no activation was observed upon addition of the CO-donor (data not shown).

Consistent with earlier reports (Lewis et al., 2002; Williams et al., 2004b), acute hypoxia resulted in a modest depression in NPo of inside-out patches excised from BKαβ cells (Figure 3A-2D; n = 14). In the presence of O_2, addition of the HO-2 co-substrates, heme (1nM) and NADPH (1μM), evoked a large increase in patch NPo (Figure 3E-2H; n = 15). 1nM heme alone has been previously shown not to modulate recombinant BKα channel activity (Tang et al., 2003), an observation which we have extended to BKαβ since NPo in the absence 0.252 ± 0.236 or presence 0.312 ± 0.233 of 1nM heme were not significantly different from each other (P > 0.25, n = 8, data not shown). Importantly, in the continued presence of the HO-2 co-substrates, hypoxia evoked a dramatic decrease in channel activity of over 70% suggesting that the enzymatic activity of HO-2 confers a significant enhancement to the O_2 sensing ability of the HO-2/BK protein complex (Figure 3I-L; n = 10). Thus, O_2 sensing by human BKαβ channels consists of two components of which the HO-2-dependent part is quantitatively more important.

Selective knock-down of HO-2 protein at 48 h was achieved by transfecting cells with siRNA species. Successful transfection of Cy3-labelled siRNA, designed against either a scrambled human GAPDH coding sequence or the human HO-2 coding sequence was followed using fluorescence microscopy. No knock-down of HO-2 immunoreactivity was observed using the scrambled siRNA. In complete contrast, almost total loss of HO-2 immunoreactivity was achieved with the specific HO-2 siRNA (data not shown). Identification of successfully transfected cells was achieved by observing Cy3 fluorescence prior to seal formation. The NADPH/heme-dependent hypoxic suppression seen in untreated cells was maintained following 48h incubation with the scrambled, control siRNA (Figure 4A-D; n = 10). Following post-transcriptional gene suppression of HO-2 for 48 h with HO-2 siRNA, mean patch NPo was dramatically depressed and NADPH/heme-dependent hypoxic suppression was completely absent (Fig 4E-H; n = 7). However, the CO-donor was able to rescue this loss-of-function in all patches tested (Fig. 4I-L; n = 3).

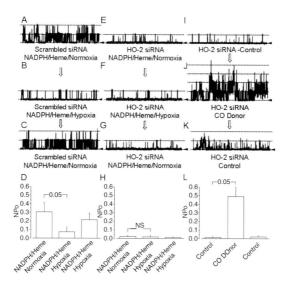

Figure 4. Modulation of heme/NADPH-dependent hypoxic inhibition in BKαβ cells following protein knock-down of HO-2 by siRNA. Exemplar traces and mean NPo plots NADPH/heme-dependent hypoxic channel inhibition in scrambled siRNA treated patches (A-D), almost complete loss of channel activity by in HO-2 treated patches (E-H) and rescue of channel activity by the CO-donor in HO-2 treated patches (I- L). All traces are from inside-out patches excised from BKαβ cells identified as siRNA-positive by Cy3 fluorescence prior to patch clamp. Statistical comparisons made by paired Student's t-test. Adapted from Williams et al., 2004a.

The physiological relevance of this novel enzyme-linked O_2 sensing by large conductance Ca^{2+}-dependent K^+ channels is illustrated in Figure 5 which shows the result of activating hemeoxygenase in inside-out patches excised from the membrane of rat carotid body glomus cells. Consistent with previous data obtained in native carotid body (Riesco-Fagundo et al., 2001), the large conductance K^+ channel was only modestly inhibited by hypoxia (Fig. 5A, B, C and G; n = 7). Similar to the recombinant system, supplying the channel complex with hemeoxygenase substrates (heme and NADPH - Fig 5G) or addition of the CO-donor increased patch NPo (35-fold, n = 7; data not shown). More importantly, NADPH/heme-dependent hypoxic inhibition was greatly augmented suggesting that the HO-2-dependent O_2 system is fully operable in native carotid body glomus cells (Figure 5 D - G; n = 7).

Amongst the numerous proteins which directly associate with the α-subunit of BK, HO-2 is notable in that it is concentrated in neuronal and chemosensing tissues, including carotid body glomus cells, where it is constitutively expressed (Prabhakar, 1999; Maines, 1997; Verma et al., 1993; Prabhakar et al., 1995). Such constitutive expression also holds true for the recombinant system in which we have chosen to study human BK channels, HEK 293 cells (Ahring et al., 1997). HO-1 immunoreactivity was not detected in HEK293 cells and has previously been discounted in rat carotid glomus cells (Prabhakar et al., 1995). Importantly, immunoprecipitation of proteins from BKαβ cells provides direct evidence that HO-2 is associated with the BK α-subunit. That the system is still functionally intact in excised patches suggests strongly that the protein-protein interaction is membrane-delimited. Whether this interaction is direct or whether it occurs via intermediate proteins is uncertain; either way, it is clear that such a co-localization of BKα with HO-2 is necessary for both

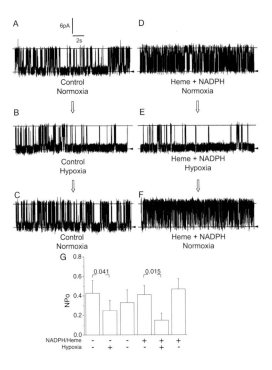

Figure 5. Augmentation of carotid body glomus cell BK channel activity by hemeoxygenase substrates. Exemplar traces indicating the modest hypoxic channel inhibition observed in untreated patches (A-C), increased baseline channel activity by 1nM heme/1 μM NADPH (C-D) and augmentation of the hypoxic inhibition in the continued presence of heme/NADPH (D-F). Corresponding mean NPo values are shown in (G). All traces are from inside-out patches excised from carotid body glomus cells. Patch potential (-Vp) = +20 mV, $[Ca^{2+}]_i$ = 335 nM. Statistical comparisons made by paired Student's t-test. Adapted from Williams et al., 2004a.

basal and O_2-dependent activity. That it is necessary for basal BK activity is demonstrated by the observation that HO-2 knock-out results in a dramatic loss of channel activity which is fully rescued by the HO-2 product, CO. Activation by CO gas has been reported in glomus cells, supporting our suggestion that HO-2 activity is crucial to native BK channel regulation (Riesco-Fagundo et al., 2001). Together with the data presented herein, the presence of HO-2 in the BK channel complex provides a molecular explanation for the observation that HO inhibition results in carotid body excitation (Prabhakar et al., 1995) and that HO-2 knock-out promotes blunted hypoxic ventilatory responses (Adachi et al., 2004). In the proposed model (Williams et al., 2004a), O_2-sensing is conferred upon the BK channel by co-localization with HO-2. In normoxia, tonic HO-2 activity generates CO and biliverdin, both of which maintain the open state probability of the channel at a relatively high level. Our data show that the presence of CO and biliverdin together evoke BK channel activation which is more than additive, this may represent a means by which the normoxic signal is amplified. However, since biliverdin is rapidly broken down to bilirubin by the actions of biliverdin reductase, it seems likely that the physiological messenger is CO. At this juncture, one can only speculate on the mechanism of CO action. In the absence of other second messenger systems (such as gas-activation of guanylate cyclase) an appealing candidate, based on earlier data in native vascular tissue (Wang & Wu, 1997), is conformational regulation through direct interaction of CO with a histidine residue, potentially in the heme-binding domain of BKα (Wood & Vogeli, 1997; Tang et al., 2003). Whatever the molecular nature of the CO effect, cellular CO levels are reduced during hypoxic challenge as HO-2 substrate (O_2) becomes scarce, and rapidly fall below the critical threshold for the maintenance of BK channel activity at the tonically high level. In other words, HO-2 functions as a sensor of O_2 by regulating BK channel activity primarily through the production of CO.

REFERENCES

Adachi T, Ishikawa K, Hida W, Matsumoto H, Masuda T, Date F, Ogawa K, Takeda K, Furuyama K, Zhang Y, Kitamuro T, Ogawa H, Maruyama Y, & Shibahara S (2004). Hypoxemia and blunted hypoxic ventilatory responses in mice lacking heme oxygenase-2. *Biochem Biophys Res Commun* **320**, 514-522.

Ahring PK, Strobaek D, Christophersen P, Olesen S-P, & Johansen TE (1997). Stable expression of the human large-conductance Ca^{2+}-activated K^+ channel α- and β-subunits in HEK 293 cells. *FEBS Lett* **415**, 67-70.

Cornfield DN, Reeve HL, Tolarova S, Weir EK, & Archer S (1996). Oxygen causes fetal pulmonary vasodilation through activation of a calcium-dependent potassium channel. *Proc Natl Acad Sci U S A* **93**, 8089-8094.

Hatton CJ, Carpenter E, Pepper DR, Kumar P, & Peers C (1997). Developmental changes in isolated rat type I carotid body cell K+ currents and their modulation by hypoxia. *J Physiol* **501**, 49-58.

Jiang C & Haddad GG (1994a). A direct mechanism for sensing low oxygen levels by central neurons. *Proc Natl Acad Sci U S A* **91**, 7198-7201.

Jiang C & Haddad GG (1994b). Oxygen deprivation inhibits a K^+ channel independently of cytosolic factors in rat central neurons. *J Physiol* **481**, 15-26.

Lewis A, Peers C, Ashford MLJ, & Kemp PJ (2002). Hypoxia inhibits human recombinant maxi K^+ channels by a mechanism which is membrane delimited and Ca^{2+}-sensitive. *J Physiol* **540**, 771-780.

Liu H, Moczydlowski E, & Haddad GG (1999). O_2 deprivation inhibits Ca^{2+}-activated K^+ channels via cytosolic factors in mice neocortical neurons. *J Clin Invest* **104**, 577-588.

Lopez-Barneo J, Pardal R, & Ortega-Saenz P (2001). Cellular mechanism of oxygen sensing. *Ann Rev Physiol* **63**, 259-287.

Maines MD (1997). The heme oxygenase system: a regulator of second messenger gases. *Annu Rev Pharmacol Toxicol* **37**, 517-554.

Peers C (1990). Hypoxic suppression of K^+ currents in type-I carotid-body cells - selective effect on the Ca^{2+}-activated K^+ current. *Neurosci Lett* **119**, 253-256.

Peng W, Hoidal JR, & Farrukh IS (1999). Role of a novel KCa opener in regulating K^+ channels of hypoxic human pulmonary vascular cells. *Am J Respir Cell Mol Biol* **20**, 737-745.

Prabhakar NR (1999). NO and CO as second messengers in oxygen sensing in the carotid body. *Respir Physiol* **115**, 161-168.

Prabhakar NR, Dinerman JL, Agani FH, & Snyder SH (1995). Carbon monoxide: a role in carotid body chemoreception. *Proc Natl Acac Sci USA* **92**, 1994-1997.

Riesco-Fagundo AM, Perez-Garcia MT, Gonzalez C, & Lopez-Lopez JR (2001). O_2 modulates large-conductance Ca^{2+}-dependent K^+ channels of rat chemoreceptor cells by a membrane-restricted and CO-sensitive mechanism. *Circ Res* **89**, 430-436.

Tang XD, Xu R, Reynolds MF, Garcia ML, Heinemann SH, & Hoshi T (2003). Haem can bind to and inhibit mammalian calcium-dependent Slo1 BK channels. *Nature* **425**, 531-535.

Thompson RJ & Nurse CA (1998). Anoxia differentially modulates multiple K^+ currents and depolarizes neonatal rat adrenal chromaffin cells. *J Physiol* **512**, 421-434.

Verma A, Hirsch DJ, Glatt CE, Ronnett GV, & Snyder SH (1993). Carbon monoxide: a putative neural messenger. *Science* **259**, 381-384.

Wang R & Wu L (1997). The chemical modification of KCa channels by carbon monoxide in vascular smooth muscle cells. *J Biol Chem* **272**, 8222-8226.

Williams SE, Wootton P, Mason HS, Bould J, Iles DE, Riccardi D, Peers C, & Kemp PJ (2004a). Hemoxygenase-2 is an oxygen sensor for a calcium-sensitive potassium channel. *Science* **306**, 2093-2097.

Williams SE, Wootton P, Mason HS, Iles DE, Peers C, & Kemp PJ (2004b). siRNA knock-down of gamma-glutamyl transpeptidase does not affect hypoxic K^+ channel inhibition. *Biochem Biophys Res Commun* **314**, 63-68.

Wood LS & Vogeli G (1997). Mutations and deletions within S8-S9 interdomain region abolish complementation of N- and C-terminal domains of Ca^{2+}-activated K^+ (BK) channels. *Biochem Biophys Res Comm* **240**, 623-628.

Wyatt CN & Peers C (1995). Ca^{2+}-activated K^+ channels in isolated type-I cells of the neonatal rat carotid-body. *J Physiol* **483**, 559-565.

Does AMP-activated Protein Kinase Couple Inhibition of Mitochondrial Oxidative Phosphorylation by Hypoxia to Pulmonary Artery Constriction?

A. MARK EVANS[1], KIRSTEEN J.W. MUSTARD[2], CHRISTOPHER N. WYATT[1], MICHELLE DIPP[1], NICHOLAS P. KINNEAR[1], D. GRAHAME HARDIE[2]

[1]Department of Biomedical Sciences, School of Biology, Bute Building, University of St Andrews, St. Andrews, Fife. KY16 9TS, UK. [2] Division of Molecular Physiology, School of Life Sciences, Wellcome Trust Biocentre, University of Dundee, Dow Street, DD1 5EH, UK.

1. INTRODUCTION

Pulmonary arteries constricts in response to hypoxia and thereby aid ventilation-perfusion matching in the lung[1]. Although O_2-sensitive mechanisms independent of mitochondria may also play a role, it is generally accepted that relatively mild hypoxia inhibits mitochondrial oxidative phosphorylation and that this underpins, at least in part, cell activation[2-6]. Despite this consensus, the mechanism by which inhibition of mitochondrial oxidative phosphorylation couples to Ca^{2+}-dependent vasoconstriction has remained elusive. To date, the field has focussed on the role of the cellular energy status (ATP)[7], reduced redox couples[3] and reactive oxygen species[4,5], respectively, but investigation of these hypotheses has delivered conflicting data and failed to unite the field[8].

Recently, the AMPK cascade has come to prominence as a sensor of metabolic stress that appears to be ubiquitous throughout eukaryotes[9,10]. AMPK complexes are heterotrimers comprising a catalytic α subunit and regulatory β and γ subunits[9], which monitor the cellular AMP/ATP ratio as an index of metabolic stress[9]. Binding of AMP to two sites in the γ subunits triggers activation of the kinase via phosphorylation of the α subunit at Thr-172, an effect antagonized by high concentrations of ATP[9,10,12]. This phosphorylation is catalyzed by upstream kinases (AMPK kinases) the major form of which is a complex between the tumour suppressor kinase, LKB1, and two accessory subunits, STRAD and MO25[14,15]. Given that inhibition of mitochondrial oxidative phosphorylation by hypoxia would be expected to promote a rise in the AMP/ATP ratio[9] we considered the proposal[16] that AMPK activation may mediate, in part, pulmonary artery constriction by hypoxia.

2. MATERIALS AND METHODS

2.1 Tissue and Cell Isolation

All experiments were performed under the UK Animals (Scientific Procedures) Act 1986. Pulmonary arteries were excised from male Wistar rats (150-300g) after cervical dislocation and smooth muscle cells were obtained as described previously[17].

2.2 AMPK Isoforms and Activities

AMPK subunit protein expression was analysed using pre-cast 4-12% Bis-Tris gels in MOPS buffer. Proteins were transferred to nitrocellulose membranes using an Xcell II Blot Module and probed with antibodies against AMPK subunits[18]. Isoform-specific AMPK activities were determined by immunoprecipitating 100 mg of tissue lysate with antibodies raised against $\alpha 1$, $\alpha 2$, $\gamma 1$, $\gamma 2$ or $\gamma 3$ subunits bound to protein-G Sepharose beads, and quantified using the *AMARA* peptide and [γ-^{32}P] ATP substrates[19]. Adenine nucleotide content of arterial smooth muscle lysates was determined by capillary electrophoresis[20].

2.3 Immunocytochemistry

Cells were fixed using ice cold methanol (15 min), permeabilised with 0.3% Triton X-100 and incubated overnight at 4°C with antibodies against the AMPK $\alpha 1$ subunit (1:500). Coverslips were washed with blocking solution and incubated (1hr, 22°C, dark) with FITC-conjugated secondary antibodies (1:200; excitation 490 nm, emission 518 nm). Images were acquired using a Deltavision microscope system (Applied Precision), on an Olympus IX70 microscope using a 60x, 1.40 n.a., oil immersion objective and photometric CH300 CCD camera. Single or multiple Z sections (0.2 µm) were taken through a cell. Images were deconvolved and analysed off-line via Softworx (Applied Precision).

2.4 Ca^{2+} Imaging

Intracellular Ca^{2+} concentration was reported by Fura-2 fluorescence ratio (F340 / F380 excitation; 510 nm emission). Emitted fluorescence was recorded at 22°C with a sampling frequency of 0.02 Hz using a Hamamatsu 4880 CCD camera via a Zeiss Fluar 40x, 1.3 n.a. oil immersion lens and Leica DMIRBE microscope. Background subtraction was performed on-line. Analysis was via Openlab (Improvision, UK)[17]. Pharmacological agents were applied extracellularly by a micro-superfusion system via a flow pipe positioned close to the cell under investigation, as described previously[17].

2.5 Tension Recording

Records were from pulmonary artery branches (i.d. 300 – 400 µm; 2-3 mm in length) via small vessel myographs (AM10, Cambustion Biological, Cambridge) as described previously[17]. Experimental chambers were filled with PSS-B: 118 $NaCl$, 4 KCl, 1 $MgSO_4$, 1.2 NaH_2PO_4, 24 $NaHCO_3$, 2 $CaCl_2$, 2

MgCl$_2$, 5.6 glucose, pH 7.4, 37°C, and bubbled with 75% N$_2$, 20% O$_2$, 5% CO$_2$ (normoxia:150-160 Torr;) or 93% N$_2$, 2% O$_2$, 5% CO$_2$ (hypoxia:16-21 Torr) via a gas-mixer (Columbus, USA).

3. RESULTS

3.1 AMPK Subunit Isoforms in Pulmonary Artetrial Smooth Muscle and AMPK Activation by Hypoxia

Western blot analysis in combination with co-immunoprecipitation identified the presence of the α1, α2, β2, γ1 and γ3 subunits of AMPK in smooth muscle lysates from 2nd and 3rd order branches of the pulmonary arterial tree (Figure 1A-B). Anti-γ2 antibodies were not sufficiently specific in Western blotting to confirm the presence of the γ2 subunit. However, they do not cross-react with anti-α1 or -α2 antibodies and immunoprecipitate kinase assays revealed that γ2 accounted for 40% and γ1 for 60% of the total AMPK activity, with γ3 accounting for an insignificant fraction (Figure 1B (i); n = 3, 32 arteries, 8 animals). Furthermore, anti-α1 and -α2 antibodies showed that α1 accounted for 80-90%, and α2 only 10-20%, of the total catalytic activity (Figure 1B (ii)). Thus, AMPK α1β2γ1 predominates over other combinations (i.e. α1β2γ2, α2β2γ1 and α2β2γ2).

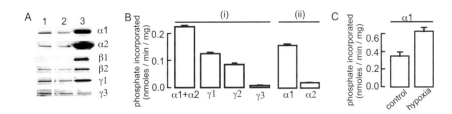

Figure 1. AMPK subunit isoforms in pulmonary versus systemic arterial smooth muscle and AMPK activation in pulmonary arterial smooth muscle by hypoxia. **A,** Western blot of AMPK-α1, -α2, -β1, -β2, -γ1, and –γ3 expression: 1 and 2, pulmonary arterial smooth muscle; 3, rat liver. **B(i)** AMPK activity immunoprecipitated from pulmonary arterial smooth muscle with anti-α1+α2 (total), -γ1, -γ2 and –γ3 antibodies and (ii) AMPK activity immunoprecipitated from pulmonary and systemic arterial smooth muscle with anti-α1 and –α2 antibodies. **C,** activation of AMPK-α1 in pulmonary arterial smooth muscle by switching from normoxia (154-160 Torr; 1hr) to hypoxia (16-21 Torr, 1hr).

Capillary electrophoresis analysis on pulmonary arterial smooth muscle lysates (32 arteries, 8 animals) showed that the AMP/ATP ratio rose from 0.040 under normoxia (155-160 Torr, 2 hr) to 0.083 under hypoxic conditions (16-21 Torr, 1 hr; following 1 hr normoxia). Most significantly, immunoprecipitate kinase assays demonstrated that the increase in the AMP/ATP ratio was associated with a concomitant, 2-fold increase in AMPK-α1 activity (Figure 1D; n = 3, 32 arteries, 8 animals).

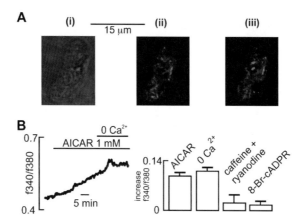

Figure 2. AMPK activation elicits cADPR-dependent SR Ca^{2+} release in pulmonary arterial smooth muscle. **Ai**, brightfield image of a pulmonary arterial smooth muscle cell; **(ii)** z-section showing staining by antibodies to AMPK-α1; **(iii)** 3D reconstruction. B, effect on Fura-2 fluorescence ratio (F340/F380) in an isolated pulmonary arterial smooth muscle cell of AICAR (1 mM) with and without extracellular Ca^{2+} (+ 1mM EGTA), and the mean increase in the Fura-2 fluorescence ratio ± s.e.m. (n ≥ 7) in the presence and absence of Ca^{2+}, ryanodine (10 µM) + caffeine (10 mM) and 8-bromo-cADPR (100 µM).

3.2 AMPK Activation Initiates cADPR-dependent Ca^{2+} Release from the Sarcoplasmic Reticulum

Our previous studies have established that cADPR-dependent Ca^{2+} release from smooth muscle SR stores is required for the full expression of HPV[21,22]. Consistent with a role in this process, immunocytochemistry showed that AMPK-α1 was distributed throughout the cytoplasm in pulmonary arterial smooth muscle (Figure 2A). To determine whether or not AMPK activation mimicked the effects of hypoxia, we employed AICAR (1 mM). AICAR is taken up into cells, metabolised to yield the AMP mimetic ZMP (AICAR monophosphate) and thereby selectively activates AMPK without affecting the cellular AMP/ATP ratio[23]. AICAR (1 mM) induced an increase in intracellular Ca^{2+} concentration in isolated pulmonary arterial smooth muscle cells (20-22 °C), as reported by an increase in the Fura-2 fluorescence ratio (F340/F380) by 0.10 ± 0.01 (n = 22; Figure 2B). This increase remained unaffected upon removal of extracellular Ca^{2+}, but was virtually abolished upon: (1) block of SR stores by pre-incubation of cells with ryanodine (10 µM) and caffeine (10 mM), (2) block of the Ca^{2+} mobilising messenger cADPR[24,25] using a selective cADPR antagonist, 8-bromo-cADPR (100 µM)[22].

3.3 AMPK Activation Elicits Pulmonary Artery Constriction Via the Mechanisms that Underpin HPV

To ascertain whether or not AMPK plays a functional role in HPV, we compared the effects of hypoxia and AMPK activation by AICAR on isolated pulmonary arteries at 37 °C. Consistent with the maintained phase of constriction by hypoxia[21,22], AICAR (1 mM) induced a slow, sustained constriction of pulmonary artery rings from 1.3 ± 0.4 to 3.2 ± 0.9 mN mm^{-1} (n = 4; Figure 3A, left panel). This was reversed rapidly on washing, consistent with ZMP being metabolised rapidly at 37°C[23]. Removal of the pulmonary artery endothelium reduced the constriction in response to AICAR (1 mM; n = 4; right panel) and hypoxia (16-21 Torr) by approximately 29% and 28%, respectively (Figure 3B). Furthermore, the endothelium-dependent component of constriction by AICAR (1 mM; n = 4) and hypoxia (16-21 Torr; n = 3), respectively, was abolished upon removal of extracellular Ca^{2+}. In contrast, constriction mediated by mechanisms intrinsic to the smooth muscle was not, although it is notable that this component of constriction was attenuated in the absence of extracellular Ca^{2+}. Significantly, therefore, block of SR Ca^{2+} stores with caffeine (10 mM) and ryanodine (10 μM) or blockade of cADPR with 8-bromo-cADPR (100 μM) completely abolished the constriction of pulmonary arteries, with or without endothelium, by both AICAR (1 mM) and hypoxia (16-21 Torr). Thus, the partial dependence of smooth muscle constriction on extracellular Ca^{2+} must be due to SR store depletion-activated Ca^{2+} influx / store-refilling, which is not mediated by AMPK activation by AICAR nor hypoxia *per se* (Figure 3B).

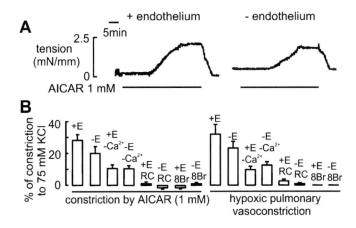

Figure 3. AMPK activation by AICAR replicates hypoxic pulmonary vasoconstriction. **A**, records show pulmonary artery constriction in response to AMPK activation by AICAR (1 mM) with (left panel) and without (right panel) the endothelium. **F**, mean constriction ± s.e.m. (n ≥ 3) of isolated pulmonary artery rings by AICAR (1 mM; left panel) and hypoxia (right panel) under the following conditions: E = endothelium, RC= ryanodine and caffeine, 8Br = 8-bromo-cADPR.

4. DISCUSSION

Our findings suggest that hypoxia increases the AMP/ATP ratio in pulmonary arterial smooth muscle and consequently activates AMPK. We identified a number of possible AMPK subunit isoform combinations in pulmonary arterial smooth muscle, although it seems likely that the α1β2γ1 combination predominates. Most significantly, however, we find that activation of AMPK by AICAR mimicked precisely the mechanism of pulmonary artery constriction by hypoxia: (1) cADPR-dependent SR Ca^{2+} release in the smooth muscle (2) a secondary component of constriction driven by calcium influx into the pulmonary artery endothelium[17,21,22].

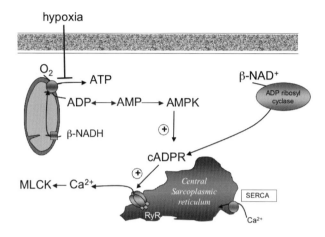

Figure 4. Proposed mechanism of chemotransduction by AMPK in pulmonary arterial smooth muscle cells.

We propose, therefore, that AMPK may act as a primary metabolic sensor in O_2-sensing cells, and the primary effector of HPV. This novel role for AMPK may therefore unite for the first time the mitochondrial and Ca^{2+} signalling hypotheses for chemotransduction by hypoxia[8]. This process would be exquisitely sensitive to metabolic stress by hypoxia because of the triple mechanism by which AMPK is activated: adenylate kinase converts any rise in the cellular ADP/ATP ratio into a rise in the AMP/ATP ratio[11], whilst AMP binding to the γ subunits of AMPK not only promotes phosphorylation by LKB1 but also inhibits dephosphorylation by protein phosphatases and causes allosteric activation of the phosphorylated enzyme[9,10]. The acute sensitivity of O_2-sensing cells to relatively small changes in pO_2 may, therefore, be conferred by: (1) The expression of specific AMPK isoforms in a given cell type, (2) The subcellular distribution of specific subunit combinations and (3) The reliance on mitochondrial oxidative phosphorylation for ATP production.

ACKNOWLEDGEMENTS

Studies funded by The Wellcome Trust (Grant Ref. 070772), the European Commission (LSHM-CT-2004-005272).

REFERENCES

1. von Euler US, Liljestrand G. Observations on the pulmonary arterial blood pressure in the cat. *Acta Physiol. Scand.* 1946;12, 301-320.

2. Rounds S, McMurtry IF. Inhibitors of oxidative ATP production cause transient vasoconstriction and block subsequent pressor responses in rat lungs. *Circ. Res.* 1981;48:393-400.
3. Archer SL, Will JA, Weir EK. Redox status in the control of pulmonary vascular tone. *Herz* 1986;11:127-141.
4. Killilea DW, Hester R, Balczon R, Babal P, Gillespie MN. Free radical production in hypoxic pulmonary artery smooth muscle cells. *Am. J. Physiol.* 2000;279:L408-L412.
5. Waypa GB, Chandel NS, Schumacker PT. Model for hypoxic pulmonary vasoconstriction involving mitochondrial oxygen sensing. *Circ. Res.* 2001; 88:1259-1266.
6. Leach RM, Hill HS, Snetkov VA, Robertson TP, Ward JPT. Divergent roles of glycolysis and the mitochondrial electron transport chain in hypoxic pulmonary vasoconstriction of the rat. Identity of the hypoxic sensor. *J. Physiol.* 2001;536:211-224.
7. Buescher PC, Pearse DB, Pillai RP, Litt MC, Mitchell MC, Sylvester JT. Energy state and vasomotor tone in hypoxic pig lungs. *J. Appl. Physiol.* 1991;70: 1874-1881.
8. Gonzalez C, Sanz-Alfayate G, Agapito T, Gomez-Nino A, Rocher A, Obeso A. Significance of ROS in oxygen sensing in cell systems with sensitivity to physiological hypoxia. *Respiratory Physiol. Neurobiol.* 2002;132:17-41.
9. Hardie DG, Scott JW, Pan DA, Hudson ER. Management of cellular energy by the AMP-activated protein kinase system. *FEBS Lett.* 2003;546:113-120.
10. Hardie DG. The AMP-activated protein kinase pathway – new players upstream and downstream. *J. Cell Sci.* 2004;117:5479-5487.
11. Hardie DG, Hawley SA. AMP-activated protein kinase: the energy charge hypothesis revisited. *BioEssays* 2001;23:1112-1119.
12. Hawley SA, Selbert MA, Goldstein EG. 5'-AMP activates the AMP-activated protein kinase cascade, and Ca2+/calmodulin activates the calmodulin-dependent protein kinase I cascade, via three independent mechanisms. *J. Biol. Chem.* 1995;270:27186-27191.
13. Hawley SA, Boudeau J, Reid JL, Mustard KJ, Udd L, Makela TP, Alessi DR, Hardie DG. Complexes between the LKB1 tumor suppressor, STRAD□/□ and MO25□/□ are upstream kinases in the AMP-activated protein kinase cascade. *J. Biol.* 2003;2: 28.
14. Shaw RJ, Bardeesy N, Manning BD, Lopez L, Kosmatka M, DePinho RA, Cantley LC. The LKB1 tumor suppressor negatively regulates mTOR signaling. *Cancer Cell* 2004;6:91-99.
15. Woods A, Johnstone SR, Dickerson K, Leiper FC, Fryer LG, Neumann D, Schlattner U, Wallimann T, Carlson M, Carling D. LKB1 is the upstream kinase in the AMP-activated protein kinase cascade. *Curr. Biol.* 2003;*13*:2004-2008.
16. Evans AM 2004 Hypoxia cell metabolism and cADPR accumulation. In: Yuan X-J (ed) Hypoxic pulmonary vasoconstriction: cellular and molecular mechanisms. Kluwer Academic Publications. p 313-338.
17. Dipp M, Nye PCG, Evans AM. Hypoxic release of calcium from the sarcoplasmic reticulum of pulmonary artery smooth muscle. *Am. J. Physiol.* 2001;281:L318-L325.
18. Durante PE, Mustard KJ, Park SH, Winder WW, Hardie DG. Effects of endurance training on activity and expression of AMP-activated protein kinase isoforms in rat muscles. *Am. J. Physiol.* 2002;283:E178-86.
19. Cheung PC, Salt IP, Davies SP, Hardie DG, Carling D. Characterization of AMP-activated protein kinase gamma-subunit isoforms and their role in AMP binding. *Biochem. J.* 2000;346, 659-669.
20. Schrauwen P, Hardie DG, Roorda B, Clapham JC, Abuin A, Thomason-Hughes M, Green K, Frederik PM, Hesselink MK. Improved glucose homeostasis in mice overexpressing human UCP3: a role for AMP-kinase? *Int J Obes Relat Metab Disord.* 2004;28:824-828.
21. Wilson HL, Dipp M, Thomas JM, Lad C, Galione A, Evans AM. ADP-ribosyl cyclase and cyclic ADP-ribose hydrolase act as a redox sensor: A primary role for cyclic ADP-ribose in hypoxic pulmonary vasoconstriction. *J. Biol. Chem.* 2001;276:11180-11188.
22. Dipp M, Evans AM. cADPR is the primary trigger for hypoxic pulmonary vasoconstriction in the rat lung *in-situ*. *Circ. Res.* 2001;89:77-83.

23. Corton JM, Gillespie JG, Hawley SA, Hardie DG. 5-Aminoimidazole-4-carboxamide ribonucleoside: a specific method for activating AMP-activated protein kinase in intact cells? *Eur. J. Biochem.* 1995;229:558-565.
24. Lee HC, Walseth TF, Bratt GT, Hayes RN Clapper DL. Structural determination of a cyclic metabolite of NAD with intracellular calcium-mobilizing activity. *J. Biol. Chem.* 1989;264:1608-1615.
25. Galione A, Lee HC, Busa WB. Ca^{2+}-induced Ca^{2+} release in sea urchin egg homogenates: modulation by cyclic ADP-ribose. *Science* 1991;253:1143-1146.

Function of NADPH Oxidase and Signaling by Reactive Oxygen Species in Rat Carotid Body Type I Cells

L. HE[1], B. DINGER[1], C. GONZALEZ[2], A. OBESO[2] AND S. FIDONE[1]

[1]Department of Physiology, University of Utah School of Medicine, Salt Lake City, Utah 84108 USA, [2]Departamento de Bioquimica y Biologia Molecular y Fisiologia/IBGM Facultad de Medicina. Universidad de Valladolid/CSIC. 47005Valladolid. Spain

1. INTRODUCTION

O_2-sensing in the carotid body occurs in neuroectoderm-derived type I glomus cells, where hypoxia elicits a complex chemotransduction cascade involving membrane depolarization, Ca^{2+} entry and the release of excitatory neurotransmitters. Efforts to understand the exquisite O_2-sensitivity of these cells have focused primarily on the relationship between PO_2 and the activity of K^+-channels. An important hypothesis developed by Acker and his colleagues suggests that coupling between local PO_2 and the open-closed state of K^+-channels is mediated by reactive oxygen species (ROS) generated by a phagocytic-like multisubunit enzyme, NADPH oxidase (Nox)(1). According to this scheme, ROS production will occur in proportion to the prevailing PO_2, and a subset of K^+-channels which control the E_M, should close as ROS levels decrease. In O_2-sensitive cells contained in lung neuroepithelial bodies (NEB), experiments have confirmed that ROS levels decrease in hypoxia, and that E_M and K^+-channel activity are indeed controlled by ROS produced by an Nox isoform similar, if not identical to the enzyme expressed in phagocytic cells that use ROS as part of an extracellular killing mechanism activated in response to invading micro-organisms(8; 15).

ROS-generating phagocytic Nox is a complex enzyme comprised of two trans-membrane and four cytosolic subunits. The large 91 kD glycoprotein (gp91^{phox} ; phox: phagocytic oxidase) and a 22 kD protein (p22^{phox}) form a membrane bound cytochrome b558(2). Immunologic stimulation initiates a protein kinase C (PKC)-dependent process in which cytosolic subunits, including p67^{phox}, p40^{phox}, p47^{phox} and a small GTPase (Rac-1 or Rac-2), unite at the membrane to form the active enzyme. An electron is then transferred from NADPH (produced in the pentose phosphate shunt [PPS] pathway of glucose metabolism) to O_2, thus forming O_2 X plus $NADP^+$. In non-phagocytic cells, recent studies have demonstrated Nox activity consistent with ROS involvement in intracellular signaling(3; 21). For example, vascular endothelial cells contain Nox in which cytochrome b558 and the cytosolic subunits are preassembled on cytoskeletal elements in the perinuclear region, where they are engaged in constitutive production of intracellular ROS(13). Moreover, homologs of gp91^{phox} have been cloned (designated Nox1-Nox5, plus Duox 1 and Duox2) and sequenced from a variety of cell types, suggesting that multiple forms of the enzyme are adapted to function in specific signaling pathways in diverse tissues(12). These alternative forms of Nox have likewise been proposed as O_2-sensors which produce ROS in proportion to local levels of PO_2 (19).

Although ROS have been proposed as intracellular mediators of chemotransduction events in O_2-sensitive type I cells in the carotid body, efforts to determine the precise mode(s) of involvement of Nox in low-O_2 sensing have resulted in conflicting views(16). A fundamental tenet of this hypothesis suggests that H2O2 derived from O2 X by the action of SOD would facilitate K+-channel activity. Indeed, application of low concentrations of H_2O_2 to the *in vitro* rat carotid body/CSN preparation has been shown to depress chemoreceptor nerve activity(1), and an inhibitor of the oxidase, diphenyleneiodonium (DPI), alters nerve activity evoked by hypoxia(6). Furthermore, certain subunits common to the phagocytic and non-phagocytic forms of the enzyme, including p22^{phox}, gp91^{phox}, p47^{phox}, and p67^{phox} (phox: phagocytic oxidase), have been localized to type I cells by immunocytochemical staining techniques(11).

Opposing these positive findings are multiple observations which question the validity of the Nox hypothesis of chemotransduction. First, recent studies in the rat have shown that the heme-binding molecule, carbon monoxide (CO), activates K$^+$-channels, and blocks the inhibitory effects of hypoxia, suggesting that a heme-protein mediates the effects of hypoxia on channel activity(14). Second, although early studies demonstrated that maxiK and TASK-1 channels fail to respond to hypoxia following excision of membrane patches from type I cells(4; 23), more recent studies have obtained opposite results indicating that cytosolic factors (i.e., subunits of NADPH oxidase) are not required for low-O_2 inhibition of the voltage-dependent K$^+$-current(20). Third, pharmacological inhibitors of Nox fail to evoke catecholamine release from carotid bodies, suggesting that hypoxia excites type I cells via a mechanism independent of Nox(16). Finally, our studies using genetically modified mice lacking the gp91^{phox} subunit (Nox2) demonstrated normal inhibition of I_K by hypoxia in type I cells, as well as normal hypoxia-evoked Ca^{2+}-responses and CSN activity(10).

None of the above cited studies of Nox in type I cells have addressed the relationship between PO_2 and ROS production, nor have they investigated possible target molecules in the chemotransduction machinery which could be affected by ROS. We have embarked on a series of experiments to examine fundamental issues of ROS biology in the carotid body. In a recent study of mouse type I cells, we established that hypoxia activates Nox, thus increasing ROS production, a finding which is inconsistent with the fundamental notion of the Nox hypothesis(9). Moreover, this response was absent in Nox deficient cells in which the p47^{phox} subunit had been gene-deleted. The role of ROS in chemotransduction was indicated in experiments which demonstrated increased hypoxia-induced depression of K$^+$-current, and elevated hypoxia-evoked chemoreceptor activity in p47^{phox} gene-deleted preparations. These findings indicate that ROS facilitate K$^+$-channel activity following cell depolarization, thereby modulating cell excitability. In the present study, we have applied similar techniques to rat type I cells to examine ROS production and the effect of Nox inhibition on K^+-current.

2. METHODS

Dissociation and culture of carotid body type I cells. Carotid bodies were removed from anesthetized young adult rats (~100g), cleaned of connective tissue and cut into pieces, which were placed on a coverslip and incubated in 100 µl of Ham's F-12 medium (Ca^{2+}-and Mg^{2+}-free) containing collagenase and trypsin (0.2%; 30 min) in a CO_2 incubator. Tissue fragments were rinsed in F-12 medium (Ca^{2+}- and Mg^{2+}-free) and transferred to poly-L-lysine coated glass coverslips, where they were triturated with a polished Pasteur pipette in 100 µl of F-12 medium containing 10% fetal calf serum and 5 µg/ml insulin. Coverslips containing isolated cells were maintained in a CO_2 incubator for later recording (2-6 hours).

Perforated Whole-Cell Patch-Clamp Recordings. Coverslips containing carotid body type I cells were positioned in a 0.3 ml flow-chamber mounted on the stage of an inverted microscope and perfused (0.5 ml/min) at 35-36.5 °C. Bath solution contained in mM: NaCl, 141; KCl, 4.7; $MgCl_2$, 1.2, $CaCl_2$, 1.8; glucose, 10; HEPES, 10, (pH 7.40) and was routinely air-equilibrated (PO_2=120 mmHg). Pipette solution (K-glutamate, 145; KCl, 15; $MgCl_2$, 2; HEPES, 20; pH 7.2 and 37 °C) also contained nystatin (150-200 µg/ml); pipette resistances varied from 2 to 10 MΩ. Hypoxic solutions have a PO_2 of 30-32 mmHg, similar to that in CB tissue during moderate hypoxia (5). Whole cell voltage-dependent K^+ currents (I_K) were evoked by step voltage changes from a holding potential of -70 mV, recorded (Axopatch 200A patch-clamp amplifier and a CV 201A headstage; Axon Instruments, Inc.), displayed on an oscilloscope and digitized with a DigiData 1200 computer interface for analysis (PCLAMP, 5.0 software; Axon Instruments, Inc.). The series resistance (typically 40 mΩ) was not compensated. Junction potentials (2-4 mV) were canceled at the onset of the experiment. Capacitance and leakage currents were subtracted.

Measurement of ROS production. Chemoreceptor type I cells attached to coverslips were loaded with 5µM dihydroethidium (DHE) in the CO_2 incubator (30 min). Coverslips were superfused in a recording chamber (Tyrode solution; perfusion rate, 1.0-1.2 ml/min; 32-34°C) mounted into a Zeiss/Attofluor workstation equipped with an excitation wavelength selector and an intensified charge-coupled device camera system. DHE fluorescence (excitation at 535 nm) was recorded at 645 nm (sampling rate 100 msec/4 sec). Data were collected and analyzed using Attofluor Ratiovision software. Much of the fluorescent oxidized form of DHE is cell trapped; fluorescence gradually increases due to basal ROS production, and any stimulus which elevates ROS production increases the signal slope (22).

3. RESULTS

Assessment of ROS production in a rat type I cells is shown in figure 1. Basal fluorescence established at $PO_2 \sim 120$ Torr was flat or increased only slighty. The introduction of hypoxic media equilibrated at $PO_2 \sim 24\text{-}26$ Torr elicited slow increase in fluorescence, consistent with elevated ROS production.

The slope of the fluorescence signal increased substantially as hypoxia continued for several minutes. Moreover, superfusion with hypoxic media containing the specific NADPH oxidase inhibitor, 4-(2-aminoethyl)-benzenesulfonyl fluoride (AEBSF, 3 mM (7)), elicited a marked decrease in the fluorescence emission, and following washout of the drug, the signal again indicated increased ROS production which subsided upon re-introduction of normoxic media. Subsequently, exposure to the cytochrome oxidase inhibitor, azide (5 μM) elicited a large increase in fluorescence, demonstrating ROS production from a viable cell containing metabolically active mitochondria.

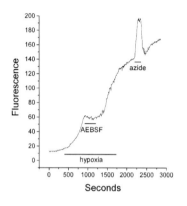

Figure 1. Effect of hypoxia on fluorescence in a type I cell loaded with dihydroethidum (DHE). The NADPH oxdiase inhibitor, AEBSF (3 mM), attenuates hypoxia-evoked fluorescence. Azide (5 μM), a cytochrome oxidase inhibitor, elicits a large increase in fluorescence, consistent with reactive oxygen species production in mitochondria.

The records in figure 2A demonstrate the effect of 100 μM H_2O_2 on hypoxia-induced depression of K^+-current. In accord with the notion that the open-state of K^+-channels in type I cells is facilitated by an oxidative shift in the redox environment, H_2O_2 hampered current depression in hypoxia. Figure 2B shows that the Nox inhibitor, AEBSF enhances the hypoxia-evoked current depression, further implicating endogenous ROS in regulation of channel function.

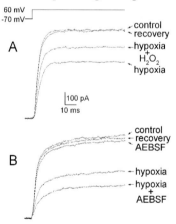

Figure 2. (A) H_2O_2 (100 μM) hampers hypoxia-evoked depression of voltage-dependent K^+-currents in rat carotid body type I cell. (B) Current depression in hypoxia is enhanced in the presence of the NADPH oxidase inhibitor, AEBSF (3 mM).

4. DISCUSSION

The current observations from rat carotid body type I cells confirm our recent studies in mouse, which demonstrated increased AEBSF-sensitive ROS production induced by hypoxia, as well as enhanced hypoxia induced depression of K^+-current after p47^{phox} gene-deletion or Nox inhibition(9). These findings support the notion that ROS produced by Nox modulate type I cell activity by facilitating the open-state of voltage-dependent K^+-channels. In earlier studies, we showed that hypoxia triggers ouabain-sensitive glucose uptake in type I cells, consistent with increased Na/K ATPase activity following cell depolarization(17). Because the primary source of the Nox cofactor, NADPH, is via the PPS pathway, the enzyme may be activated in hypoxia due to increased metabolic flux, resulting in enhanced ROS production and modulation of cell excitability. Indeed, preliminary experiments in our laboratory suggest that hypoxia-evoked ROS production in type I cells is attenuated in the presence of PPS inhibitors. In conclusion, our findings indicate that Nox homologs are adapted to diverse roles in the O_2-sensing machinery contained in arterial versus airway chemoreceptors(18).

ACKNOWLEDGEMENTS

Supported by USPHS Grants NS 12636 and NS 07938; and Spanish DGICYT grant (BFI2001-1713) and Red Respira-Separ (Fiss).

REFERENCES

1. Acker H, Bolling B, Delpiano MA, Dafau E, Gorlach A and Holtermann G. The meaning of H_2O_2 generation in carotid body cells for PO_2 chemoreception. *J Auton Nerv Syst* 41(1-2): 41-51, 1992.
2. Babior BM. NADPH oxidase: An update. *Blood* 93(5): 1464-1476, 1999.
3. Brar SS, Kennedy TP, Sturrock AB, Huecksteadt TP, Quinn MT, Murphy TM, Chitano P and Hoidal JR. NADPH oxidase promotes NF-κB activation and proliferation in human airway smooth muscle. *Am J Physiol Lung Cell Molec Physiol* 282: L782-L795, 2002.
4. Buckler KJ. Background leak K^+-currents and oxygen sensing in carotid body type 1 cells. *Resp Physiol* 115: 179-187, 1999.
5. Buerk DG, Nair PK and Whalen WJ. Evidence for second metabolic pathway for O_2 from $PtiO_2$ measurements in denervated cat carotid body. *J Appl Physiol* 67: 1578-1584, 1989.
6. Cross AR, Henderson L, Jones OTG, Delpiano MA, Hentschel J and Acker H. Involvement of an NAD(P)H oxidase as a pO_2 sensor protein in the rat carotid body. *Biochem J* 272: 743-747, 1990.
7. Diatchuk V, Lotan O, Koshkin V, Wikstroem P and Pick E. Inhibition of NADPH oxidase activation by 4-(2-aminoethyl)-benzenesulfonyl fluoride and related compounds. *J Biol Chem* 272(2): 13292-13301, 1997.
8. Fu XW, Wang D, Nurse CA, Dinauer MC and Cutz E. NADPH oxidase is an O_2 sensor in airway chemoreceptors: Evidence from K^+ current modulation in wild-type and oxidase-deficient mice. *PNAS* 97(8): 4374-4379, 2000.
9. He L, Dinger B, Sanders K, Hoidal J, Obeso A, Stensaas L, Fidone S and Gonzalez C. Effect of p47phox gene-deletion on reactive oxygen species (ROS) production and oxygen sensing in mouse carotid body chemoreceptor cells. *Am J Phsiol* in press: 2005.
10. He L, Chen J, Dinger B, Sanders K, Sundar K, Hoidal J and Fidone S. Characteristics of carotid body chemosensitivity in NADPH oxidase-deficient mice. *Am J Physiol Cell Physiol* 282: C27-C33, 2002.

11. Kummer W and Acker H. Immunohistochemical demonstration of four subunits of neutrophil NAD(P)H oxidase in type I cells of carotid body. *J Appl Physiol* 78(5): 1904-1909, 1995.
12. Lambeth JD. Nox/Duox family of nicotinamide adenine dinucleotide (phosphate) oxidases. *Current Opinion in Hematology* 9: 11-17, 2002.
13. Li J-M and Shah AJ. Intracellular localization and preassembly of the NADPH oxidase complex in cultured endothelial cells. *J Biol Chem* 277(22): 19952-19960, 2002.
14. Lopez-Lopez JR and Gonzalez C. Time course of K^+ current inhibition by low oxygen in chemoreceptor cells of adult rabbit carotid body: effects of carbon monoxide. *FEBS Lett* 299: 251-254, 1992.
15. O'Kelly I, Lewis A, Peers C and Kemp PJ. O_2 sensing by airway chemoreceptor-derived cells: Protein kinase C activation reveals functional evidence for involvement of NADPH oxidase. *J Biol Chem* 275(11): 7684-7692, 2000.
16. Obeso A, Gomez-Nino A and Gonzalez C. NADPH oxidase inhibition does not interfere with low PO_2 transduction in rat and rabbit CB chemoreceptor cells. *Am J Physiol* 276(Cell Physiol.45): C593-C601, 1999.
17. Obeso A, Gonzalez C, Rigual R, Dinger B and Fidone S. Effect of low O_2 on glucose uptake in rabbit carotid body. *J Appl Physiol* 74(5): 2387-2393, 1993.
18. Peers C and Kemp PJ. Acute oxygen sensing: Diverse but convergent mechanisms in airway and arterial chemoreceptors. *Respir Res* 2: 145-149, 2001.
19. Porwol T, Ehleben W, Brand V and Acker H. Tissue oxygen sensor function of NADPH oxidase isoforms, and an unusual cytochrome aa3; producing reactive oxygen species. *Resp Physiol* 128: 331-348, 2001.
20. Riesco-Fagundo AM, Pérez-García MT, Gonzalez C and López-López JR. O_2 modulates large-conductance Ca^{2+}-dependent K^+ channels of rat chemoreceptor cells by a membrane-restricted and CO-sensitive mechanism. *Circ Res* 89: 430-436, 2001.
21. Szöcs K, Lassè B, Sorescu D, Hilenski LL, Valppu L, Couse TL, Wilcox JN, Quinn MT, Lambeth JD and Griendling KK. Upregulation of Nox-Based NAD(P)H oxidases in restenosis after carotid injury. *Arterioscler Thromb Vasc Biol* 22: 21-27, 2002.
22. Tarpey MM and Fridovich I. Methods of detection of vascular reactive species: Nitroc oxide, superoxide, hydrogen peroxide and peroxynitrite. *Circ Res* 89: 224-236, 2001.
23. Wyatt CN and Peers C. Ca^{2+}-activated K^+ channels in isolated type I cells of the neonatal rat carotid body. *J Physiol* 483.3: 559-565, 1995.

Hypoxemia and Attenuated Hypoxic Ventilatory Responses in Mice Lacking Heme Oxygenase-2
Evidence for a Novel Role of Heme Oxygenase-2 as an Oxygen Sensor

YONGZHAO ZHANG[1], KAZUMICHI FURUYAMA[1], TETSUYA ADACHI[1], KAZUNOBU ISHIKAWA[2], HAYATO MATSUMOTO[2], TAKAYUKI MASUDA[3], KAZUHIRO OGAWA[4], KAZUHISA TAKEDA[1], MIKI YOSHIZAWA[1], HIROMASA OGAWA[5], YUKIO MARUYAMA[2], WATARU HIDA[6], AND SHIGEKI SHIBAHARA[1]

[1]*Department of Molecular Biology and Applied Physiology, Tohoku University School of Medicine, Sendai, Japan,* [2]*First Department of Internal Medicine, Fukushima Medical University, Fukushima, Japan,* [3]*Division of Pathology, School of Health Sciences, Tohoku University, Sendai, Japan,* [4]*Laboratory of Molecular Pharmacology, Tohoku University School of Medicine, Sendai, Japan,* [5]*Department of Cardiovascular Medicine, Tohoku University School of Medicine, Sendai, Japan, and* [6]*Health Administration Center, Tohoku University, Sendai, Japan.*

1. INTRODUCTION

All nucleated cells depend on heme for their survival, as heme senses or uses oxygen. In fact, heme is a prosthetic moiety of various hemoproteins such as hemoglobin, myoglobin, and cytochromes. Accordingly, heme must be synthesized and degraded within an individual cell, because heme cannot be recycled among different cells, except for senescent erythrocytes, which are phagocytosed by macrophages in the reticuloendothelial system (for review, Shibahara, 2003). Heme, derived from hemoproteins, is broken down by heme oxygenase, which catalyzes the oxidative breakdown of heme, generating biliverdin, carbon monoxide (CO), and iron (Fig. 1). These heme degradation products are important bioactive molecules (For review, Shibahara, 2003 and references therein). Bilirubin functions as a chain-breaking antioxidant. CO represents a direct marker for heme catabolism, and binds to hemoglobin to form carboxyhemoglobin, which is transported to the lungs and is excreted in exhaled air. CO has received much attention because of its physiological functions similar to those of NO. Iron is transported to the entire tissues, especially bone marrow, and is reutilized for erythropoiesis and heme biosynthesis.

2. TWO ISOZYMES ARE BETTER THAN ONE

Heme oxygenase consists of two isozymes: heme oxygenase-1 (HO-1) and heme oxygenase-2 (HO-2) (Shibahara et al., 1985; Maines et al., 1986). Human HO-1 and HO-2 share 43% amino acid sequence identity (Yoshida et al., 1988;

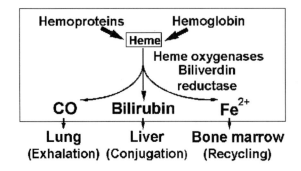

Figure 1. Heme catabolism in mammalian cells. In a typical nucleated cell, heme is derived from turnover of hemoproteins and is cleaved by heme oxygenase (either HO-1 or HO-2) to generate CO, biliverdin and ferrous iron. Reduction of biliverdin to bilirubin represents the final step of the heme breakdown reaction. Iron is used for heme synthesis in the cell or transported via transferrin to other tissues, mainly the bone marrow. Bilirubin is transported to the liver, where bilirubin is conjugated and excreted into bile. CO is transported to the lung and exhaled. Note that hemoglobin derived from senescent erythrocytes is a major source of heme in macrophages.

McCoubrey et al., 1992; Ishikawa et al., 1995). Human HO-1 consists of 288 amino acids and contains no cysteine residues (Yoshida et al., 1988) (Figure 2). In contrast, human HO-2 of 316 amino acids contains two copies of cysteine and proline (CP motif), a potential heme-binding site, which are not involved in heme breakdown reaction (McCoubrey et al., 1997). HO-2 may therefore possess a hitherto unknown function besides heme catabolism. In addition, there is a remarkable difference in the regulation of expression of the two HO isozymes. Expression of HO-1 is inducible (Shibahara et al., 1987) or repressible (Takahashi et al., 1999; Nakayama et al., 2000; Kitamuro et al., 2003) depending on cellular microenvironments, whereas expression levels of HO-2 are largely unchanged (Ewing and Maines, 1991; Shibahara et al., 1993; Takahashi et al., 1996). Such a difference suggests separate roles of HO-1 and HO-2. These topics are discussed in a comprehensive review (Shibahara, 2003).

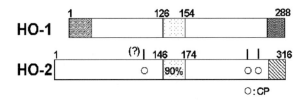

Figure 2. Structures of human HO-1 and HO-2. The conserved catalytic domain is shown as stippled, and the number indicates the amino acid sequence identity to human HO-1. Human HO-2 contains three copies of CP motif, and one of them, marked with (?), is not conserved in mouse and rat HO-2. In contrast, human HO-1 contains no Cys residues.

HO-2 deficient mice are fertile and survive normally for at least one year (Poss et al., 1995), and thus show milder phenotypes than those of HO-1 deficient mice, which are characterized by infertility and prenatal lethality (Poss and Tonegawa 1997). Subsequent studies revealed that ejaculatory abnormalities and reduced mating behavior in male HO-2 (-/-) mice (Burnett et al., 1998) and increased susceptibility to hyperoxic lung damage (Dennery et al., 1998). These results indicate that the function of HO-2 is not completely compensated by HO-1. In fact, recent studies have identified a novel role of HO-2 for oxygen sensing (Adachi et al., 2004; Williams et al., 2004).

3. UNEXPECTED ABNORMALITIES IN HO-2$^{-/-}$ MICE

3.1 Hypoxemia

Hypoxemia is a common manifestation of various respiratory diseases and is caused by one of four mechanisms: a decrease in O_2 pressure in inspired gas, hypoventilation, shunting (vascular or intra-pulmonary), and ventilation-perfusion mismatching. HO-2$^{-/-}$ mice exhibited hypoxemia, as judged by lower arterial oxygen tension (PaO_2) and lower oxygen contents compared to those of the wild-type mice, but normal $PaCO_2$ (Adachi et al., 2004). Carboxyhemoglobin level, which represents the amount of *in vivo* heme degradation, is also decreased in HO-2$^{-/-}$ mice. There are no significant differences in hemoglobin levels and pH between HO-2$^{+/+}$ and HO-2$^{-/-}$ mice. In addition, there are no noticeable morphological changes in bronchioles, alveoli, and pulmonary arterioles of the mutant lung. These findings suggest that hypoxemia may be due to ventilation-perfusion mismatch. There are at least two types of the chemoreceptors in the lung, which are responsible for ventilation-perfusion matching: airway neuroepithelial bodies that are clusters of amine- and peptide-producing cells distributed throughout the airway mucosa (Youngson et al., 1993; Kemp et al., 2002) and pulmonary artery smooth muscle cells (Prabhakar, 1998). These chemoreceptors initiate vasoconstriction of the pulmonary arterioles by sensing hypoxia, thereby maintaining ventilation-perfusion matching. We therefore propose that the function of neuroepithelial bodies and/or pulmonary artery smooth muscle cells may be partially impaired in HO-2$^{-/-}$ mice, which leads to ventilation-perfusion mismatch and eventually causes hypoxemia.

3.2 Attenuated Hypoxic Ventilatory Responses

We analyzed the lung function of unanesthetized HO-2$^{-/-}$ mice by whole-body plethysmography (Mizusawa et al., 1994), which allowed us to record the basal breathing patterns of freely moving HO-2$^{-/-}$ mice. There are no noticeable differences under basal conditions between HO-2$^{+/+}$ and HO-2$^{-/-}$ mice in the following parameters: respiratory frequency (f), tidal volume (TV/g), and minute ventilation (VE/g) (Adachi et al., 2004). Thus, HO-2$^{-/-}$ mice maintain normal ventilation, thereby excluding hypoventilation as a cause of hypoxemia.

To evaluate the function of neuroepithelial bodies and/or pulmonary arteries in unanesthetized mice, we assessed the function of the carotid body that senses reduced arterial oxygen tension to enhance ventilation. In this context, a common mechanism involving K^+ channels has been implicated in oxygen sensing at glomus cells, neuroepithelial bodies, and pulmonary artery smooth muscle cells (Prabhakar, 1998; Kemp, et al., 2002). Accordingly, we analyzed the immediate ventilatory responses of freely moving HO-2$^{-/-}$ mice by whole-body plethysmography; we measured the initial phase of ventilatory responses to acute hypoxia (10% O_2) or acute hypercapnia (10% CO_2), which were detected within 20 sec after challenge (Adachi et al., 2004). Notably, the tidal volume (TV/g) and minute ventilation (VE/g) to acute hypoxia were significantly lower in HO-$2^{-/-}$ mice, indicating that acute hypoxic responses are attenuated in HO-$2^{-/-}$ mice. In contrast, the hypercapnic ventilatory responses are maintained in HO-$2^{-/-}$ mice from those in HO-$2^{+/+}$ mice. These results suggest the normal function of the central chemoreceptor and respiratory controller for hypercapnic chemoreception in the HO-$2^{-/-}$ brain, but indicate the impaired function of the oxygen-chemosensors in the carotid body and/or in the lung.

3.3 Hypertrophied Pulmonary Venous Myocardium

There are two unexpected changes in the pulmonary veins of HO-$2^{-/-}$ mice: thickening of the vascular walls of the pulmonary veins and over-expression of HO-1 protein in the thickened venous walls (Adachi et al., 2004). The entire vascular walls of the major pulmonary veins are thicker in HO-$2^{-/-}$ mice than those in the HO-$2^{+/+}$ mice. The thickening of the venous walls is due to the hypertrophy of the pulmonary venous myocardium, which represents the extension of atrial myocardium into the vascular walls of the pulmonary veins (Klavins, 1963). In addition, HO-1 protein is detected in the pulmonary venous myocardium in HO-$2^{+/+}$ mice, and is over-expressed in the hypertrophied venous myocardium of HO-$2^{-/-}$ mice. The thickening of the venous myocardium may represent the adaptation to hypoxemia (Wagenvoort and Wagenvoort, 1976). Importantly, there are no detectable changes in pulmonary arteries and arterioles in HO-$2^{-/-}$ mice, despite the hypoxemia, which could induce remodeling of the pulmonary arteries and cause pulmonary hypertension. In fact, we did not notice any changes in the size and weight of their hearts, suggesting that HO-$2^{-/-}$ mice may maintain normal blood pressure of the pulmonary artery.

4. EXPRESSION OF HO-2 IN HUMAN LUNG CELLS

In contrast to the systemic arteries that exhibit hypoxic vasodilatation, the pulmonary arteries constrict in response to hypoxia, which is an essential mechanism that optimizes the oxygenation of pulmonary blood at alveoli. An important question remains to be answered is a cellular basis for the ventilation-perfusion mismatch in HO-$2^{-/-}$ mice, as we were unable to detect the expression of HO-2 protein in the wild-type lung by immunohistochemical analysis. To explore the expression of HO-2 in the lung cells, we analyzed its expression in primary cultures of smooth muscle cells derived from human pulmonary artery, A549 human lung cancer cells, and H146 human small cell lung cancer cells. HO-2 mRNA and protein are expressed in these cell types (data not shown), supporting the notion that HO-2 is expressed in pulmonary artery smooth muscle cells and neuroepithelial bodies.

5. IMPLICATIONS

Hypoxic vasoconstriction of the pulmonary artery is essential for ventilation-perfusion matching, which ensures efficient oxygenation of pulmonary blood and maintains normal PaO_2. Conversely, ventilation–perfusion mismatch is the most common cause of hypoxemia, and is seen in a variety of disease processes that affect the airways or the pulmonary parenchyma, such as chronic obstructive pulmonary disease, asthma, and pneumonia, which account for common causes of disability and death in the developed world. Recently, it has been shown that human HO-2 represents a part of a large conductance, calcium-sensitive potassium channel (the BK channel); namely, HO-2 interacts with BK α-subunit and functions as an oxygen sensor for a BK channel (Williams et al., 2004). Hypoxia inhibits the BK channel through HO-2, thereby enhancing ventilation. However, this hypoxic inhibition of the BK channel might be impaired in $HO-2^{-/-}$ mice.

In summary, $HO-2^{-/-}$ mice exhibit hypoxemia with normal $PaCO_2$ and intact alveolar architecture, hypertrophy of the myocardium localized within the pulmonary venous walls, and attenuated hypoxic ventilatory responses (Adachi et al., 2004). The presence of hypoxemia with normal $PaCO_2$ and intact alveolar architecture suggests that loss of HO-2 may impair oxygen sensing at the lung chemoreceptors, which leads to ventilation-perfusion mismatching. Taken together with the fact that HO-2 functions as a component of the BK channel (Williams et al., 2004), we propose that HO-2 is involved in oxygen sensing and may be responsible for the hypoxic pulmonary vasoconstriction.

ACKNOWLEDGEMENTS

This work was supported in part by Grants-in-Aid for Scientific Research (B) (to S.S.) and for Exploratory Research (to S.S.) from the Ministry of Education, Science, Sports, and Culture of Japan, and by the 21st Century COE Program Special Research Grant "the Center for Innovative Therapeutic Development for Common Diseases" from the Ministry of Education, Science, Sports and Culture of Japan.

REFERENCES

Adachi T., Ishikawa K., Hida W., Matsumoto H., Masuda T., Date F., Ogawa K., Takeda K., Furuyama K., Zhang Y., Kitamuro T., Ogawa H., Maruyama Y., Shibahara S. Hypoxemia and blunted hypoxic ventilatory responses in mice lacking heme oxygenase-2. Biochem. Biophys. Res. Commun. 320: 514-522, 2004.

Burnett A.L., Johns D.G., Kriegsfeld L.J., Klein S.L., Calvin D.C., Demas G.E., Schramm L.P., Tonegawa S., Nelson R.J., Snyder S.H., Poss K.D. Ejaculatory abnormalities in mice with targeted disruption of the gene for heme oxygenase-2. Nat. Med. 4: 84-87, 1998.

Dennery P.A., Spitz D.R., Yang G., Tatarov A., Lee C.S., Shegog M.L., Poss K.D. Oxygen toxicity and iron accumulation in the lungs of mice lacking heme oxygenase-2, J. Clin. Invest. 101: 1001-1011, 1998.

Ishikawa K., Takeuchi N., Takahashi S., Matera K.M., Sato M., Shibahara S., Rousseau D.L., Ikeda-Saito M., Yoshida T. Heme oxygenase-2. Properties of the heme complex and purified tryptic peptide of human heme oxygenase-2 expressed in Escherichia coli. J. Biol. Chem., 270: 6345-6350, 1995.

Kemp P.J., Lewis A., Hartness M.E., Searle G.J., Miller P, O'Kelly I., Peers C. Airway chemotransduction: from oxygen sensor to cellular effector, Am. J. Respir. Crit. Care Med. 166: S17-S24, 2002.

Kitamuro T., Takahashi K., Ogawa K., Udono-Fujimori R., Takeda K., Furuyama K., Nakayama M., Sun J., Fujita H., Hida W., Hattori T., Shirato K., Igarashi K., Shibahara S. Bach1 functions as a hypoxia-inducible repressor for the heme oxygenase-1 gene in human cells. J. Biol. Chem. 278: 9125-9133, 2003.

Klavins J.V. Demonstration of striated muscle in the pulmonary veins of the rat, J. Anat. 97: 239-341, 1963.

Maines M.D., Trakshel G.M., Kutty R.K. Characterization of two constitutive forms of rat liver microsomal heme oxygenase. Only one molecular species of the enzyme is inducible, J. Biol. Chem. 261: 411-419, 1986.

McCoubrey W.K.Jr., Ewing J.F., Maines M.D. Human heme oxygenase-2: characterization and expression of a full-length cDNA and evidence suggesting that the two HO-2 transcripts may differ by choice of polyadenylation signal. Arch. Biochem. Biophys. 295: 13-20, 1992.

McCoubrey W.K.Jr., Huang T.J., Maines M.D. Heme oxygenase-2 is a hemoprotein and binds heme through heme regulatory motifs that are not involved in heme catalysis. J. Biol. Chem. 272: 12568-12574, 1997.

Mizusawa A., Ogawa H., Kikuchi Y., Hida W., Kurosawa H., Okabe S., Takishima T., Shirato K. In vivo release of glutamate in the nucleus tractus solitarius of rat during hypoxia, J. Physiol. (London) 478: 55-65, 1994.

Nakayama M., Takahashi K., Kitamuro T., Yasumoto K., Katayose D., Shirato K., Fujii-Kuriyama Y., Shibahara S. Repression of heme oxygenase-1 by hypoxia in vascular endothelial cells. Biochem. Biophys. Res. Commun. 271: 665-671, 2000.

Poss K.D., Thomas M.J., Ebralidze A.K., O'Dell T.J., Tonegawa S. Hippocampal long-term potentiation is normal in heme oxygenase-2 mutant mice. Neuron 15: 867-873, 1995.

Poss K.D., Tonegawa S. Heme oxygenase 1 is required for mammalian iron reutilization. Proc. Natl. Acad. Sci. USA 94: 10919-10924, 1997.

Prabhakar N.R. Endogenous carbon monoxide in control of respiration, Respir. Physiol. 114: 57-64, 1998.

Shibahara S. The heme oxygenase dilemma in cellular homeostasis: New insights for the feedback regulation of heme catabolism. (Invited Review) Tohoku J. Exp. Med. 200: 167-186, 2003.

Shibahara S., Muller R.M., Taguchi H. Transcriptional control of rat heme oxygenase by heat shock. J. Biol. Chem. 262: 12889-12892, 1987.

Shibahara S., Yoshizawa M., Suzuki H., Takeda K., Meguro K., Endo K. Functional analysis of cDNAs for two types of human heme oxygenase and evidence for their separate regulation. J. Biochem. 113: 214-218, 1993.

Takahashi K., Hara E., Suzuki H., Shibahara S. Expression of heme oxygenase isozyme mRNAs in the human brain and induction of heme oxgenase-1 by nitric oxide donors. J. Neurochem. 67: 482-489, 1996.

Takahashi K., Nakayama M., Takeda K., Fujita H., Shibahara, S. Suppression of heme oxygenase-1 mRNA expression by interferon-γ in human glioblastoma cells. J. Neurochem. 72: 2356-2361, 1999.

Takeda K, Ishizawa S, Sato M, Yoshida T, Shibahara S. Identification of a cis-acting element that is responsible for cadmium-mediated induction of the human heme oxygenase. J. Biol. Chem. 269: 22858-22867, 1994.

Wagenvoort C.A., Wagenvoort N. Pulmonary venous changes in chronic hypoxia, Virchows Arch. A Pathol. Anat. Histol. 372 : 51-56, 1976.

Williams S.E., Wootton P., Mason H.S., Bould J., Iles D.E., Riccardi D., Peers C., Kemp P.J. Hemoxygenase-2 is an oxygen sensor for a calcium-sensitive potassium channel. Science 306: 2093-2097, 2004.

Yoshida T., Biro P., Cohen T., Muller R.M., Shibahara S. Human heme oxygenase cDNA and induction of its mRNA by hemin. Eur. J. Biochem. 171: 457-461, 1988.

Youngson C., Nurse C., Yeger H., Cutz E. Oxygen sensing in airway chemoreceptors, Nature 365: 153-155, 1993.

Regulation of a TASK-like Potassium Channel in Rat Carotid Body Type I Cells by ATP

RODRIGO VARAS AND KEITH J. BUCKLER

University Laboratory of Physiology, Parks Road, Oxford, OX1 3PT, U.K.

1. INTRODUCTION

The carotid body plays a central role in initiating cardiovascular and respiratory responses to hypoxia. Previous work from this laboratory has demonstrated that hypoxia inhibits TASK-like K^+-channels in isolated neonatal rat carotid body type I cells (Buckler et al., 2000). The consequent reduction in background K^+-current leads to type-1 cell membrane depolarisation (Buckler, 1997), voltage-gated calcium entry and thus excitation of the carotid body. The mechanisms by which hypoxia modulates these K^+ channels is still unknown, however the effects of hypoxia are mimicked by a wide range of inhibitors of oxidative phosphorylation. Uncouplers, electron transport inhibitors (e.g. cyanide, rotenone & myxothiazol) and inhibitors of ATP synthase (oligomycin) are all potent stimulants of the carotid body (Anichkov & Belen`kii, 1963; Gonzalez et al., 1994) and potent inhibitors of background K^+-current (Wyatt & Buckler, 2004). Moreover background K^+-current sensitivity to hypoxia is lost in the presence of metabolic inhibitors (Wyatt & Buckler, 2004) suggesting that ATP synthesis may be a prerequisite for the expression of oxygen sensitivity. Further support for the idea that background K^+-channel activity may be dependent upon cellular ATP levels comes from the observation that following patch excision (from the cell attached to the inside out configuration) there is an abrupt rundown of background K^+-channel activity which can be partially reversed by the addition of mM levels of ATP to the intracellular solution (Williams & Buckler, 2004). In this study we have further investigated the action of ATP on background K^+-channels in order to evaluate the potential role for ATP in modulating channel activity and to identify general mechanisms by which ATP might act.

2. METHODS

Carotid bodies were excised from anaesthetized (4% halothane) Sprague-Dawley neonatal rats (11- 15 days old) and put in ice-cold saline solution. The rats were then killed by decapitation whilst still anaesthetized. The carotid bodies were enzymatically dissociated as previously described (Buckler 1997). Isolated cells were plated onto poly-D-lysine coated coverslips and kept at 37 °C, in Ham's F-12 medium (supplemented with 10% heat-inactivated foetal calf serum, 100 U/ml penicillin, 100 μg/ml streptomycin and 84 U/l insulin) equilibrated with %5 CO_2 in air until use (2-6 hrs).

Experiments were conducted using the inside-out configuration of the patch clamp technique. Single channels recordings were preformed using an Axopatch 200B amplifier. Membrane currents were filtered at 2 kHz and recorded at a sampling frequency of 20 kHz (using a CED Power 1401). Electrodes were fabricated from borosilicate glass capillaries and were Silgard coated and fire-polished before use. Electrode resistance was between 5-15 MΩ and seal resistances were ≥10 GΩ. Inside out single channel recordings were performed at a membrane potential of -70 mV.

Cells were bathed with standard HCO_3^--buffered saline prior to, and during, patch formation in the cell attached configuration. Cells were then transferred to an intracellular (inside out) solution prior to patch excision and recording in the inside-out configuration. Standard HCO_3^--buffered saline contained (in mM) 117 NaCl, 4.5 KCl, 23 $NaHCO_3$, 1.0 $MgCl_2$, 2.5 $CaCl_2$ and 11 glucose; and was bubbled with 5% CO_2 and 95% air (pH 7.4- 7.45). Intracellular (inside-out) solution contained (in mM) 130 KCl, 5 $MgCl_2$, 10 EGTA, 10 HEPES, 10 glucose, pH 7.2 (adjusted with KOH, final K^+ concentration = 152 mM). Pipette (extracellular) solutions contained (mM) 140 KCl, 4 $MgCl_2$, 1 EGTA, 10 HEPES, 10 tetraethylammonium (TEA)-Cl and 5 4-aminopyridine, pH 7.4 (adjusted with KOH, final K^+ concentration: 146 mM). K_2ATP, K_2ADP and AMP-PCP (β,γ-methylene ATP) were added directly to the saline solution before use. In experiments with mM K_2ATP, Mg^{2+} was increased in order to maintain a free Mg^{2+} level of approx 3-4 mM. Experiments were conducted at 28 to 32 °C.

Single channel recordings were analysed with Spike software (4.0, Cambridge Electronic Design). Open events were defined using the 50% threshold criteria with events of 150%, 250%, etc. of threshold counted as multiple channel openings. Channel activity is reported as the open probability times the number of channels in a given patch (NPo). Mean values are expressed as MEAN ± SEM.

Sigmaplot 8.0 (Chicago, IL) was used to calculate nonlinear regression of plotted data. Statistical significance of the dose dependent activation of the channel by ATP was assessed using one-way analysis of variance followed by post-hoc test (Bonferroni). When comparing the effects of ATP and AMP-PCP the two-tail paired Student's t-test was used.

3. RESULTS

3.1 Background K^+ Channel Activity in Inside-out Patches of Carotid Body Type I Cells

As previously reported, single channel activity ran down rapidly upon patch excision (the transition from the cell-attached to the inside-out configuration). After completion of rundown (~40 sec), the most commonly observed form of channel activity consisted of brief flickery openings with a main single current amplitude of ~1.3 pA (Figure 1, see also Williams & Buckler 2004). On occasion we have also observed other forms of channel activity but, due to their low frequency of occurrence, these other types of channels have not been studied further.

3.2 Effect of ATP on Background K⁺ Channel Activity

After completion of channel rundown addition of ATP to the cytosolic side of membrane patches produce a rapid and reversible increase in channel activity in the majority of patches tested. Figure 2 shows a recording in which there is a very rapid increase in channel activity upon switching from a control inside out solution to one containing 10 mM ATP. We investigated the effects of ATP over a wide range of different concentrations ranging from 0.1 to 20 mM. Stimulation of channel activity was detectable at concentrations of ATP as low as 0.5 mM ATP (basal NPo= 0.02 ± 0.01, 0.5 mM ATP= 0.03 ± 0.01, p <0.05) and increased in a dose dependent manner to reach a maximum at around 10 mM ATP, where a ~6 fold increase in channel activity was observed (NPo= 0.17 ± 0.01, P< 0.02). Figure 3 summarises this dose dependent activation of channel activity by ATP.

3.3 Effects of AMP-PCP and ADP on Background K⁺ Channel Activity

In order to investigate the mechanisms by which ATP regulates channel activity we have studied the effects of the non-hydrolysable ATP analogue AMP-PCP. In all seven patches studied, 10 mM AMP-PCP increased background K⁺ channel activity to a level comparable (~5 fold $P < 0.05$) to that obtained with 10 mM ATP (a comparison of the effects of 10 mM AMP-PCP with those of 10 mM ATP showed no statistically significant difference). Figure 4 shows recording from the same patch in the presence of 10 mM ATP (Fig 4a) and 10 mM AMP-PCP (Fig 4b).

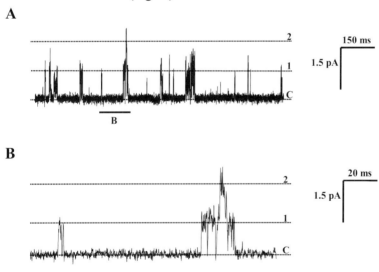

Figure 1. A, representative trace of an excised patch showing single channel activity at -70 mV membrane potential. This is a very abundant channel present in virtually all the patches tested. *B,* data from the same patch showing an expanded display of the trace underlined in A.. C, closed; 1 open level 1; 2, open level 2.

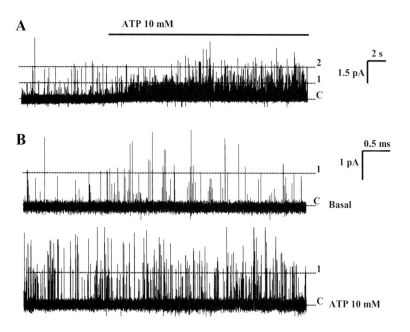

Figure 2. A, a single channel recording at -70 mV membrane potential showing that when ATP is applied to the cytosolic side of the membrane patch there is a rapid increase in channel activity. *B*, data from the same patch depicted in A showing increase in channel activity due to ATP at higher temporal resolution. C, closed; 1 open level 1; 2, open level 2.

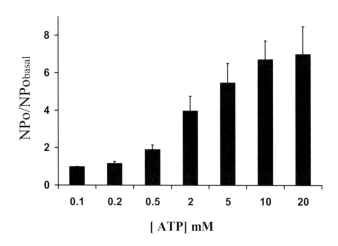

Figure 3. Dose dependent activation of background potassium channel by ATP. Plot of the normalised channel activity (NPo/NPobasal) as a function of the ATP concentration in the bathing solution. Bars indicate standard error of the mean.

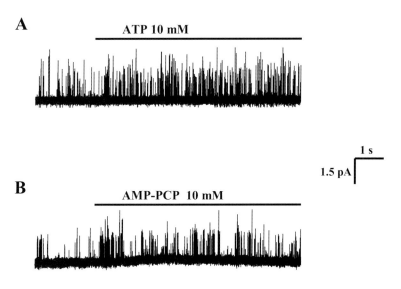

Figure 4. Channel activity recorded at -70 mV membrane potential showing increase in channel activity in the presence of a) a maximally effective concentration of ATP (10 mM) and b) an equal concentration of the non-hydrolysable ATP analogue AMP-PCP.

4. CONCLUSION

This report shows that background K^+ channels are very sensitive to changes in ATP levels within the physiological range (i.e. mid to low mM). In addition, the non-hydrolysable ATP analogue AMP-PCP also increases channel activity to a level comparable to the maximal stimulation seen with ATP, suggesting that ATP sensitivity is likely to be mediated through nucleotide binding rather than through protein phosphorylation. Thus direct ATP sensing by the channel, or associated regulatory protein, could account for the potent stimulatory effects of metabolic poisons on chemoreceptor activity. These data are also consistent with the metabolic theory of oxygen sensing wherein oxygen is indirectly sensed through changes in mitochondrial ATP synthesis.

ACKNOWLEDGEMENTS

This work was supported by the British Heart Foundation.

REFERENCES

Anichkov SV & Belen´kii ML (1963) Pharmacology of the carotid body chemoreceptors. Macmillan Publishing, New York

Buckler KJ (1997) A novel oxygen sensitive potassium current in rat carotid body type-I cells. J Physiol 498.3: 649-662.

Buckler KJ, Williams BA, Honore E (2000) An oxygen-, acid- and anaesthetic-sensitive TASK-like background potassium channel in rat arterial chemoreceptor cells. J Physiol 525.1: 135-42.

Gonzalez C, Almaraz L, Obeso A, Rigual R (1994) Carotid body chemoreceptors: from natural stimuli to sensory discharges. Physiol Rev; 74: 829-898.

Williams BA & Buckler KJ (2004) Biophysical properties and metabolic regulation of a TASK-like potassium channel in rat carotid body type-1 cells. Am J Physiol Lung Cell Mol Physiol 286 (1): L221-L230.

Wyatt CN & Buckler KJ (2004) The effect of mitochondrial inhibitors on membrane currents in isolated neonatal rat carotid body type I cells. J Physiol 556.1: 175-191.

Accumulation of Radiolabeled *N*-Oleoyl-Dopamine in the Rat Carotid Body

MIECZYSŁAW POKORSKI[1], DOMINIKA ZAJĄC[1], ANDRZEJ KAPUŚCIŃSKI[1], ZDZISŁAW MATYSIAK[2], AND ZBIGNIEW CZARNOCKI[3]

[1]*Medical Research Center, Polish Academy of Sciences, Warsaw,* [2]*Institute of Biochemistry and Biophysics, Polish Academy of Sciences, Warsaw, and* [3]*Faculty of Chemistry, Warsaw University, Warsaw, Poland*

1. INTRODUCTION

Exogenously administered dopamine (DA) is liable to penetrate into the carotid body, which, as opposed to the brain, has no endothelial barrier. DA is stored in the secretory vesicles of chemoreceptor cells. The vesicles are reminiscent of micelle-like entities and the hydrophilic properties of DA molecules make it dubious that DA could be packed and stay sustained in such an environment. The possible problems with the intravesicular arrangement of DA molecules may be one reason for thecomplex, often erratic, and as yet not full well understood DA action in the chemosensing process. DA displays a spate of varying effects, from stimulation to inhibition, on carotid chemosensory discharge and ventilation, depending on the species, the dose, and the presynaptic or postsynaptic dopamine D_2 receptor it interacts with (see for review Gonzales et al., 1994).

The present study was focused on extending the understanding of how the physicochemical properties of DA molecules could influence DA absorption by the carotid body. We investigated the hypothesis of whether administration of a substance in which the DA molecule is attached to a fatty acid chain motif could enhance carotid body uptake of DA. We chose *N*-oleoyl-dopamine (*N*-OL-DA) for the study, a product of bonding of the oleic acid chain to DA at its amino terminal, belonging to a class of novel biologically active lipid compounds, known under the summary name of dopamides (Pokorski and Matysiak, 1998). Recently, *N*-OL-DA has been identified as an endogenous participant of mammalian brain biochemistry with potent affinity to the central vanilloid VR1 receptor (Chu et al., 2003). *N*-OL-DA also is an active regulator of intracellular Ca^{2+} (Chu et, 2003) and injected intracarotidly causes a transient inhibition of respiratory drive in the cat (unpublished observation), which recalls the respiratory effects of exogenous DA (Ide et al., 1995). This study seeks to determine the uptake of a radiolabeled N-OL-DA by the carotid body and to compare it with that of DA alone. To this end, we started off by synthesizing N-OL-DA de novo and tagging it with a radioactive tritium.

2. MATERIALS AND METHODS

The animals from which the carotid body data outlined below were derived also were used for another study in which the uptake of radiolabeled N-oleoyl-dopamine by the brain was assessed. The results of that study were published elsewhere (Pokorski et al., 2003). The carotid body ramification of the experiment, perceived as a single unrelated object and a disconnected technical and research issue, was analyzed separately and presented herein. The study protocol was approved by a local Ethics Committee.

2.1 Synthesis of Unlabeled and Tritium Labeled N-Oleoyl-Dopamine

Details of the process of N-OL-DA synthesis can be found elsewhere (Czarnocki et al., 1998). For the unlabeled synthesis, briefly, N-OL-DA was prepared from dopamine hydrochloride and an equimolar amount of oleic acid mixed in the presence of benzotriazol-l-yloxy-tris (dimethylamino)-phosphonium hexafluorophosphate reagent in dry tetrahydrofuran. The purification of the product was performed with column chromatography. N-OL-DA purity was checked with thin-layer-chromatography. The radiolabeled synthesis was a two-step process (Pokorski et al., 2003). First, the isotope exchange procedure was optimized using deuterium oxide (D_2O). N-OL-DA was treated with D_2O in the presence of sodium hydroxide under argon atmosphere at 90°C for 8 h, and the product purified with column chromatography. The ^{1}HNMR spectrum taken on a sample of the product showed that deuterium combined only with the C2', C5', and C6' carbon atoms. The second step consisted of tritiation. Tritium labeling of N-OL-DA was accomplished by isotope exchange between the dopamide and tritiated water. The exchange was done in alkaline medium, elevated temperature, and the atmosphere free of oxygen. After a series of sequential steps to accomplish the isotope exchange, tritium was removed from the labile positions of the labeled product by use of the reverse isotope exchange technique. Then, the labeled product underwent a series of lyophilization procedures until its specific activity settled at a constant level. The finishing step was the product extraction with diethyl ether and evaporation to dryness. The end result was 1 mg of the labeled N-OL-DA of a total radioactivity of 7 µCi.

2.2 Animal Preparation and Experimental Paradigm

The data were collected from 17 anesthetized (chloral hydrate; 300 mg/kg intraperitoneally), tracheostomized, and spontaneously breathing female Wistar rats that received injections of [^{3}H]N-OL-DA, 1.33 µCi/ml, and [^{3}H]DA (NEN

Perkin Elmer, Life Science Products, Boston, MA), 10 µCi/ml. There were 7 injections of [^{3}H]N-OL-DA and 6 of [^{3}H]DA into the right common carotid artery just downstream the carotid body, after ligation of both common carotid and external carotid arteries, and 2 each injections into the right femoral artery. [^{3}H]N-OL-DA was dissolved in 0.3 ml DMSO and [^{3}H]DA in 1 ml physiological saline. Each animal was used for one injection once. The animals were sacrificed 15 min after the injections by inducing cardioplegia with a saturated KCl solution and 1 cm long regions of the carotid bifurcation containing the carotid body from the side of injection, the contralateral side, and a distal to the carotid body segment of the contralateral common carotid artery wall were excised. The ^{3}H radioactivity was measured in a Beckman LS5000TA liquid scintillation counter (Beckman Instruments, Fullerton, CA) after an overnight digestion by a tissue solubilizer (Soluene 350, Packard, Canberra Co., Meriden, CT). Standards consisted of 1:10 and 1:100 dilutions of the injected dose of radioactivity. Radioactivity was measured in disintegrations per minute (dpm).

Results are expressed as mean ±SE values of the percentage of the entire dose of radioactivity injected per gram of tissue. Differences in the level of radioactivity resulting from injections of either labeled compound across the three carotid artery fragments studied were analyzed with one-way ANOVA followed the Scheffe *post hoc* test. Other statistical comparisons were made with a two-tailed paired or unpaired *t*-test as required.

3. RESULTS

The ability of carotid bodies to accumulate both labeled derivatives of DA after intracarotid injections is demonstrated in Fig. 1. The carotid body at the side of injection accumulated, on average, over 30% of the entire radioactivity tagged to [^{3}H]N-OL-DA. This accumulation was about 3-fold greater than that for the ipsilateral [^{3}H]DA. At the contralateral side, carotid body uptake of [^{3}H]N-OL-DA and [^{3}H]DA was about 4.5-fold and 2-fold smaller, respectively. The contralateral carotid body uptake of the labeled compounds was similar to the background uptake by a segment of the common carotid artery wall distant to the carotid body location (data shown in Table 1).

After intravenous injections, the uptake of both labeled compounds was but a fraction of the afore mentioned for the intracarotid route and the distribution of radioactivity for either compound was about even in the samples studied, irrespective of the injection side (Table 1). The uptake of the intravenous [^{3}H]N-OL-DA remained, however, 4-5 times greater than for [^{3}H]DA. It should be pointed out that the calculated differences between the uptake of [^{3}H]N-OL-DA and [^{3}H]DA, given both intravenously and intracarotidly, did not take into consideration that the load of radioactivity carried by the intravenous [^{3}H]DA was several-fold greater. Although the intravenous injection is liable to dilute the compound, the ability of the carotid body to accumulate [^{3}H] N-OL-DA could

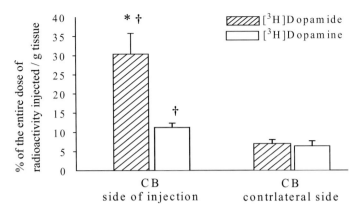

Figure 1. Accumulation of the labeled N-oleoyl-dopamine (dopamide) and dopamine in the carotid bodies (CB), ipsilateral and contralateral to intracarotid injection side. *Significant difference between the level of [^{3}H]dopamide and [^{3}H]dopamine at the side of injection at P<0.006; †Significant differences in the level of corresponding compounds between the ipsilateral and contralateral sides at P<0.007 for [^{3}H]dopamide and P<0.02 for [^{3}H]dopamine.

Table 1. Radioactivity level in the extracts of carotid artery fragments studied after intracarotid and intravenous injections of labeled N-oleoyl-dopamine ([^{3}H]N-OL-DA) and dopamine ([^{3}H]DA).

	Carotid body; contralateral	Carotid body; side of injection	Contralateral carotid artery segment; distal to carotid body
INTRACAROTID			
[^{3}H]N-OL-DA	7.03 ±1.14	31.22 ±5.01*	6.60 ±1.13
[^{3}H]DA	6.47 ±1.41	11.26 ±1.24*	5.42 ±0.73
INTRAVENOUS			
[^{3}H]N-OL-DA	7.59	6.73	5.29
[^{3}H]DA	1.48	1.77	1.15

Data are means ±SE of a percent of the entire radioactivity injected, expressed per gram of tissue. No SE values are provided for the intravenous route, as there were only two injections of each compound in this group. *Significantly different from the other two at P<0.01, one-way ANOVA.

plausibly have been much more than it was measured had the radioactive load been equal. The mean values of the measured level of radioactivity in the three carotid artery fragments for both intracarotid and intravenous injections of [^{3}H]N-OL-DA and [^{3}H]DA are presented in Table 1.

4. DISCUSSION

Here we report that the carotid body absorbs, to a significant extent, N-OL-DA loaded into the incoming blood. Moreover, this absorption is several times greater than that for DA. N-OL-DA is thus the preferred form of DA transport into the organ.

DA has a wide range of physiological functions. One of such function is its role as a major putative neurotransmitter in the sensory organ of the carotid body that generates hypoxic hyperventilation. The importance of DA in carotid body function swings back and forth over the years. The results vary diametrically; from ascribing parallelism between the rate of DA synthesis and release and the level of carotid body neural discharge (Gonzalez et al, 1992) to disavowing the determinative role of DA in chemosensing by showing that its release from the organ lags behind the hypoxic stimulation of chemoreceptor discharge and that the pool of DA is exhausted during the repetitive hypoxic stimulation, as opposed to the sustained ability of the chemoreceptor discharge to rise (Donnelly, 1995).

The contentious results of DA studies may, in part, stem from the properties of DA molecules. The hydrophilicity of DA makes it inaccessible to the membrane signaling target sites and may erratically hamper the stability of its enclosure in secretory vesicles, which all may be the source of a blur on DA effects. Such problems are circumvented by N-OL-DA, in which DA is functionalized with a long chain polyunsaturated fatty acid. N-OL-DA has recently been recognized as an endovanilloid with a high affinity to the brain VR1 receptor and a participant of essential for the cell functioning processes, such as Ca^{2+} trafficking (Chu et al., 2003). Carotid body predilection to accumulate N-OL-DA demonstrated in the present study gives rise to the following assumptions. The DA moiety of the compound, carried along, could interact with the D2 receptors present on the chemoreceptor cell membrane (Gonzalez et al., 1994). This interaction could help unleash the signaling cascade, e.g., phosphoinosities that seem operative in shaping cellular responses in the carotid body (Pokorski et al., 2000), which might help streamline the hypoxia-sensing process. In this regard, it is of interest to note that another flag dopamide, N-arachidonoyl-dopamine, shows little interaction with the D2 receptor (Bisogno et al., 2000). That, however, hardly precludes the potential interaction of N-OL-DA with DA receptors, by far an unstudied aspect of N-OL-DA metabolism, since dopamides do not act biologically in like manner. Another possibility might be that N-OL-DA exercises its high affinity to the VR1 receptor after being taken up by the tissue. That, in turn, would incriminate the VR1 receptor in carotid body function, but the presence of this receptor beyond the brain has not yet been substantiated.

The present study does not resolve the enigma of the exact determinants of DA role in the carotid body. We believe, however, we have shown that making DA lipophilic offers an attractive alternative to circumvent the drawbacks linked to DA inability to penetrate and stabilize in the signaling target areas of carotid body cells. The demand for DA in such areas is apparently substantial, as evidenced by a markedly higher uptake of exogenous [^{3}H]N-OL-DA compared with [^{3}H]DA. N-OL-DA seems the right compound to take part in the elaborate signaling process at the carotid body. The possibility arises that dopamides may integrate the use of lipid signals in DA-mediated signal transduction. We conclude that the recent progress in unraveling the biological role of dopamides opens up new leads of researching the DA role in chemosensing.

ACKNOWLEDGEMENTS

The study was supported by statutory budgets of the Polish Academy of Sciences Medical Research Center and Warsaw University. M. Pokorski was a visiting scientist at the Division of Histology, Department of Cell Biology, Tohoku University Graduate School of Medicine in Sendai, supported by the Japan Society for the Promotion of Science, at the time of the writing of this article.

REFERENCES

Bisogno T., Melck D., Yu M., Bobrov M.Y., Gretskaya N.M., Bezuglov V.V., De Petrocellis L., Di Marzo V. N-acyl-dopamines: novel synthetic CB_1 cannabinoid-receptor ligands and inhibitors of anandamide inactivation with cannabimimetic activity *in vitro* and *in vivo*. Biochem. J. 351: 817-824, 2000.

Chu C.J., Huang S.M., De Petrocellis L., Bisogno T., Ewing S.A., Miller J.D., Zipkin R.E., Daddario N., Appendino G., Di Marzo V., Walker J.M. N-oleoyl-dopamine, a novel endogenous capsaicin-like lipid that produces hyperalgesia. J. Biol. Chem. 278: 13633-13639, 2003.

Czarnocki Z., Matuszewska M.P., Matuszewska I. Highly efficient synthesis of fatty acid dopamides. Org. Prep. Proced. Int. 30: 699-702, 1998.

Donnelly D.F. Does catecholamine secretion mediate the hypoxia-induced increase in nerve activity? Biol. Signals 4: 304-309, 1995.

Gonzalez C., Almaraz L., Obeso A., Rigual R. Oxygen and acid chemoreception in the carotid body chemoreceptors. Trends Neurosci. 15: 146–153, 1992.

Gonzalez C., Almaraz L., Obeso A., Rigual R. Carotid body chemoreceptors: from natural stimuli to sensory discharges. Physiol. Rev. 74: 829–898, 1994.

Ide T., Shirahata M., Chou C.L., Fitzgerald R.S. Effects of a continuous infusion of dopamine on the ventilatory and carotid body responses to hypoxia in cats. Clin. Exp. Pharmacol. Physiol. 22: 658-664, 1995.

Pokorski M., Matysiak Z. Fatty acid acylation of dopamine in the carotid body. Med. Hypoth. 50: 131-133, 1998.

Pokorski M., Matysiak Z., Marczak M. Ostrowski R.P., Kapuściński A., Matuszewska I., Kańska M., Czarnocki Z. Brain uptake of radiolabeled N-oleoyl-dopamine in the rat. Drug Dev. Res. 60: 217-224, 2003.

Pokorski M., Sakagami H., Kondo H. Classical protein kinase C and its hypoxic stimulus-induced translocation in the cat and rat carotid body. Eur. Respir. J. 16:459-463, 2000.

Profiles for ATP and Adenosine Release at the Carotid Body in Response to O_2 Concentrations

SÍLVIA V. CONDE AND EMÍLIA C. MONTEIRO

Department of Pharmacology, Faculty of Medical Sciences, New University of Lisbon, Campo Mártires da Pátria, 130, 1169-056 Lisbon, Portugal

1. INTRODUCTION

Excitatory effects on carotid body (CB) chemotransduction have been described for both adenosine and ATP. Adenosine when applied exogenously increases carotid sinus nerve (CSN) discharges in the cat, *in vivo* (McQueen and Ribeiro, 1983) and *in vitro* (Runold et al., 1990). Administration of adenosine and drugs that increase its endogenous levels stimulate ventilation in rats, an effect abolished by the section of CSN and mediated by A_2 receptors (Monteiro and Ribeiro, 1987, 1989; Ribeiro and Monteiro, 1991). In humans, the intravenous infusion of adenosine causes hyperventilation and dyspnoea, an effect attributed to the activation of CB (Watt and Routledge, 1985, Watt et al., 1987; Maxwell et al., 1986; 1987, Uematsu et al., 2000). The excitatory effect of ATP at the CB described by Zhang et al. (2000) in co-cultures of type I cells with petrosal neurons was further supported by the finding that mice deficient in $P2X_2$ showed a markedly attenuated ventilatory response to hypoxia (Rong et al., 2003) and by the detection of hypoxia- evoked ATP release from chemoreceptor cells of the rat carotid body (Buttigieg and Nurse, 2004).

Two metabolic sources of extracellular adenosine, catabolism of ATP by ecto-5'-nucleotidase and adenosine transport by equilibrative nucleoside transporters were demonstrated for the rat CB (Conde and Monteiro, 2004). The hypothesis that both adenosine and ATP contribute to chemosensory activity and that the balance between their extracellular concentrations is dependent on the intensity of the hypoxic stimuli was tested in the present work. We also investigated the role of external calcium mobilization in the release of both mediators at the CB.

2. METHODS

The present work was carried out in *Wistar* rats (250-350g), anaesthetized with sodium pentobarbital (60 mg/kg ip., Sigma). The rats were tracheostomized and breathed spontaneously during surgical procedure. Carotid bodies, superior cervical ganglion (SCG) and common carotid arterial tissue were removed *in situ*, under a Nikon SMZ-2B dissection scope and placed in 500 µl of ice-cold 95% O_2+ 5% CO_2 equilibrated medium containing different drugs in accordance with the protocol used. The incubation medium composition was (mM): NaCl 116; $NaHCO_3$ 24; KCl 5; $CaCl_2$ 2; $MgCl_2$ 1.1; HEPES 10; glucose 5.5; pH 7.42. After removal of the tissues, the animals were sacrificed by an intracardiac injection of a lethal dose of pentobarbital in agreement with the directives of the European Union (Portuguese law n° 1005/92 and 1131/97). After 30 min of pre-incubation in hyperoxia (95% O_2 + 5% CO_2) at 37°C, the CBs, SCG and arterial tissue were incubated during 10 min in normoxia (20% O_2 + 5% CO_2), hypoxia (2%, 5% and 10%O_2 + 5%CO_2) or hyperoxia (95% O_2 + 5% CO_2) in a medium containing erythro-9-(2-hydroxy-3-nonyl)adenine (EHNA, 2.5 µM, Sigma), inhibitor of adenosine deamination. The effect of extracellular Ca^{2+} on the release of ATP and adenosine from CBs was assessed in normoxia and 10% O_2 in incubation medium with 0 Ca^{2+} and with 10 mM of EDTA. In some experiments the inhibitor of 5'-ectonucleotidase, α,β-methylene-ADP (AOPCP, 100 µM, Sigma) was also added to the incubation medium. After the incubation period, nucleotides were extracted from the incubation medium and aliquots of the neutralized supernatants were kept at –20°C until subsequent analysis. Adenosine was quantified by reverse-phase HPLC with UV detection at 254nm (Conde and Monteiro, 2004). For ATP quantification 100 µL of the samples were added to 100 µL of luciferine-luciferase (FLE50, Sigma) and to 4 mL of buffer (in mM: HEPES 20; MgCl2 25; Na_2HPO_4 5). The reaction starts when the enzyme is added to the mixture and the samples were analyzed in triplicate, for 1 minute, by bioluminescence using a luminescence counter (Beckham). Data were evaluated using Graph Pad Prism (Graph Pad Software Inc., San Diego, CA, USA) software and were presented as mean ± SEM values. The significance of the differences between the groups' means was calculated by One-Way or Two-Way ANOVA with Dunnett's and Bonferroni multiple comparison post tests, respectively. Values of $P<0.05$ were considered as representing significant differences.

3. RESULTS

Adenosine and ATP extracellular concentrations obtained in basal conditions (20% O_2) were respectively 67.23 ± 5.36 pmol/CB (n=5) and 6.05 ± 0.70 pmol/CB (n=8). The effect of different O_2 concentrations on the release of adenosine and ATP from CBs is shown in Fig. 1. Hyperoxia did not modify the basal release of adenosine and ATP. Both moderate (10% O_2) and intense hypoxia (2% O_2) increased adenosine extracellular concentrations with the maximal effect (60.8 ± 17.7 %) being achieved with moderate hypoxia (Fig. 1). The effect of hypoxia on ATP extracellular concentrations was more pronounced and maximal concentrations of 16.09 ± 2.64 pmol/CB (n=6) were obtained during intense hypoxic conditions.

ATP and Adenosine Release at the CB to O_2

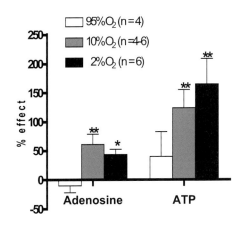

Figure 1. Effect of different oxygen concentrations (2, 10, and 95% O_2) on the release of adenosine and ATP from rat carotid bodies. 0% effect corresponds to concentrations obtained during normoxia (20% O_2). Experiments were performed in the presence of EHNA and 5% CO_2. * $P<0.05$ and ** $P<0.01$, One-Way ANOVA with Dunnett's multiple comparison test corresponding to the differences in adenosine or ATP levels between different O_2 concentrations and control conditions (20% O_2).

The effect of different O_2 concentrations (2, 5, 10, 20 and 95%) on the release of ATP from SCG and arterial tissue is shown in Fig. 2. ATP extracellular concentrations obtained during normoxic conditions in SCG and arterial tissue were respectively 16.23 ± 3.61 pmol/mg (n=6) and 24.68 ± 7.42 pmol/mg. No changes in ATP concentrations in response to O_2 were found in arterial tissue (Fig. 2). In contrast, increases in ATP concentrations of about 154% and 298% were observed in SCG during intense (respectively 5 and 2% O_2) hypoxic conditions. Moderate hypoxia did not cause significant increases in ATP concentrations in SCG.

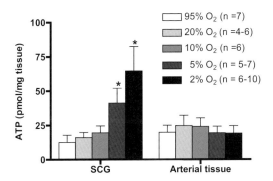

Figure 2. Effect of different oxygen concentrations (2, 5, 10, 20 and 95% O_2) on the release of ATP from superior cervical ganglion (SCG) and carotid arterial tissue Experiments were performed in the presence of EHNA and 5% CO_2. * $P<0.05$ One-Way ANOVA with Dunnett's multiple comparison test corresponding to the differences in ATP values between different O_2 concentrations and normoxic conditions (20% O_2).

The contribution of extracellular calcium to the mechanisms involved in the release of adenosine and ATP from the CB was assessed in experiments in the presence of EDTA and the absence of Ca^{2+} in the incubation medium. The effects of external Ca^{2+} mobilization in normoxia and moderate hypoxia

(10% O_2) are summarized in Table 1. Removal of extracellular calcium did not modify the basal release of either adenosine or ATP in normoxic conditions but completely abolished the release of those transmitters evoked by moderate hypoxia (Table 1). The reduction in the release of adenosine and ATP in moderate hypoxia caused by the absence of extracellular calcium was of the same magnitude ($\approx$ 50%) for both transmitters.

Table 1. – Influence of external calcium mobilization on adenosine and ATP release from rat carotid body.

	Adenosine (pmol/CB)		ATP (pmol/CB)	
	20% O_2	10% O_2	20% O_2	10% O_2
Control	67.56 ± 6.32 (6)	107.1 ± 11.55[++] (6)	6.05 ± 0.70 (6)	13.21 ± 1.97[++] (6)
$0Ca^{2+}$ + EDTA	70.73 ±7.30 (4)	53.14 ± 6.43 *** (5)	6.65 ± 1.02 (11)	6.59 ± 0.63*** (13)

Values represent mean ± SEM (n). [++] P<0.01, compared with 20% O_2 in control conditions; *** P<0.001 compared with 10% O_2 in control conditions (Two-Way ANOVA with Bonferroni multiple comparison test)

In order to investigate whether the amount of adenosine released in hypoxia by the CB through an extracellular Ca^{2+}-dependent mechanism came from primary ATP release and its further catabolism, experiments were performed in the presence of the inhibitor of 5'-ectonucleotidases, AOPCP. In the absence of extracellular calcium, AOPCP reduced adenosine concentrations from 70.73 ± 7.3 pmol/CB to 45.90 ± 5.11 pmol/CB (n=6) in normoxia (*P<0.05). In contrast, no statistically significant differences were found between adenosine concentrations measured in the absence of extracellular calcium in moderate hypoxia, before (53.14 ± 6.43 pmol/CB) and after (42.76 ± 4.37 pmol/CB, n=8) the addition of AOPCP to the medium.

4. DISCUSSION

In response to hypoxia, the CB releases both adenosine and ATP and moderate hypoxia (10% O_2) was a strong enough stimulus to trigger the release of both transmitters increasing the ATP/adenosine ratio with the intensity of the hypoxic conditions.

The magnitude of the effect of moderate hypoxia on adenosine extracellular concentrations at the CB is in agreement with that previously described (Conde and Monteiro, 2004) in the rat but the effect of more intense hypoxic stimulations have never been reported. The release of ATP from the CB was first shown by Buttigieg and Nurse (2004) in the rat in vitro in response to intense hypoxic stimulations (15 - 20 mmHg $\approx$ 2% O_2). In the present work extracellular concentrations of adenosine and ATP were measured in the same CB and in response to different O_2 concentrations in order to understand the relative contribution of both transmitters to chemosensory activity.

Maximal effect on adenosine extracellular concentrations was achieved with moderate hypoxia suggesting that the contribution of adenosine receptor activation at the CB is probably particularly relevant in these circumstances. Extracellular concentrations of adenosine depend on extracellular catabolism of ATP but are also regulated by bi-directional equilibrative nucleoside transporters. An enhanced uptake activity of these transporters induced by extracellular accumulation of adenosine can explain how adenosine concentrations are higher in moderate hypoxia than in strong acute conditions. In turn, the profile of ATP release showing a direct linear correlation with the intensity of the hypoxic stimulus is in agreement with the CSN discharge pattern in response to oxygen concentrations obtained *in vivo* in mice deficient in $P2X_2/P2X_3$ receptors (Rong et al., 2003). Further comparisons based on the absolute values of adenosine/ATP extracellular concentrations are not possible because all the experiments were performed in the presence of an inhibitor of adenosine deamination. ATP extracellular concentrations quantified in the CB were similar to those found by Cunha et al., (2001) in rat hippocampus slices but are about 10 times lower than those described in CB homogenates in cats (Acker and Starlinger, 1984; Obeso et al., 1986) and in rabbits (Verna et al., 1990).

Since the present experiments were performed in whole CB preparations the cell origin of ATP and adenosine cannot be advanced and the release of ATP from SCG terminals in response to hypoxia can of course contribute to the values quantified in the CB. However, 10% O_2 does not seem to be a strong enough stimulus to induce the release of adenosine (Conde and Monteiro, 2004) and ATP (present) in SCG and arterial tissue.

The results of the present work confirmed that in normoxia, approximately 35% of the adenosine released from the CB comes from ATP extracellular catabolism (Conde and Monteiro, 2004) but also provided evidence that the amount of ATP further catalyzed in adenosine by 5'-ectonucleotidases in normoxia originates from an extracellular Ca^{2+}-independent mechanism.

The increase in the release of both adenosine and ATP triggered by moderate hypoxia was completely prevented by removal of extracellular calcium. From this finding we first advanced with the hypothesis that in moderate hypoxia ATP is released from vesicles and extracellular adenosine comes from its catabolism. Further inhibition of 5'-ectonucleotidases did not reduce adenosine concentrations quantified in the absence of extracellular calcium supporting the hypothesis that actually all adenosine originating in moderate hypoxia from ATP is dependent on vesicular release of the nucleotide.

Independently of the cellular and molecular origin of extracellular adenosine in the CB, the nucleoside accumulation in response to hypoxia supports its excitatory effects on ventilation and together with the activation of ATP-P2X receptors can contribute to CSN responses to hypoxia.

ACKNOWLEDGMENTS

The authors are grateful to Ms Elizabeth Halkon for reviewing the English. This research was supported by CEPR/FCT. SV Conde is funded by a PhD grant from FCT.

REFERENCES

Acker, H. and Starlinger, H., 1984. Adenosine triphosphate content in the cat carotid body under different arterial O_2 and CO_2 conditions. *Neurosci Lett.*, **50**: 175-179.

Buttigieg, J. and Nurse, C.A., 2004. Detection of hypoxia-evoked ATP release from chemoreceptor cells of the rat carotid body. *Biochem. Biophy. Res. Com.*, **322**: 82-87.

Conde, S.V. and Monteiro, E.C., 2004. Hypoxia induces adenosine release from the rat carotid body. *J. Neurochem.*, **89**: 1148-1156.

Cunha RA, Almeida T and Ribeiro JA., 2001, Parallel modification of adenosine extracellular metabolism and modulatory action in the hippocampus of aged rats. J. Neurochem., **76**: 372-382.

Maxwell D. L., Fuller R. W., Nolop K. B., Dixon C. M. S. and Hughes M. B. (1986) effects of adenosine on ventilatory responses to hypoxia and hypercapnia in humans. *J. Appl. Physiol.* **61**: 1762-1766.

Maxwell D. L., Fuller R. W., Conradson T-B., Dixon C. M. S., Aber V., Hughes M. B. and Barnes P. J. (1987) Contrasting effects of two xanthines, theophyline and enprofylline, on the cardio-respiratory stimulation of infused adenosine in man. *Acta Physiol. Scand.* **131**: 459-465.

McQueen, D.S. and Ribeiro, J.A., 1983. On the specificity and type of receptor involved in carotid body chemoreceptor activation by adenosine in the cat. *Br. J. Pharmacol.*, **80**: 347-354.

Monteiro, E.C. and Ribeiro, J.A., 1987. Ventilatory effects of adenosine mediated by carotid chemoreceptors in the rat. *Naunyn-Schmiedeberg's Arch. Pharmacol.*, **335**:143–148.

Monteiro, E.C. and Ribeiro, J.A., 1989. Adenosine deaminase and adenosine uptake inhibitors facilitate ventilation in rats. *Naunyn-Schmiedeberg's Arch. Pharmacol.*, **340**:230-238.

Obeso, A., Almaraz, L. and Gonzalez, C., 1986. Effects of 2-Deoxy-D-Glucose on In Vitro Cat Carotid Body. Brain Res., **371**:25-36

Ribeiro, J.A. and Monteiro, E.C., 1991. On the adenosine receptor involved in the excitatory action of adenosine on respiration: antagonist profile. *Nucleosides Nucleotides*, **10**: 945-953.

Rong, W., Gourine, A.V., Cockayne, D.A., Xiang, Z., Ford, A.P.D.W., Spyer, M. and Burnstock, G., 2003. Pivotal role of Nucleotide P_2X_2 receptor subunit of the ATP-gated ion channel mediating ventilatory responses to hypoxia, *J. Neurosci.*, **23**: 11315-11321.

Runold, M., Cherniak, N.S. and Prabhakar, N.R., 1990. Effect of adenosine on isolated and superfused cat carotid body activity, *Neurosci. Lett.*, **113**: 111-114.

Uematsu T., Kozawa O., Matsuno H., Yoshikoshi H., Oh-uchi M., Kohno K., Nagashima S. and Kanamaru M. (2000) Pharmacokinetics and tolerability of intravenous infusion of adenosine (SUNY4001) in healthy volunteers. *Br. J. Clin. Pharmacol.* **50**: 177-181

Verna, A., Talib, N., Roumy, M. and Pradet, A., 1990. Effects of metabolic inhibitors and hypoxia on the ATP, ADP and AMP content of the rabbit carotid body in vitro: the metabolic hypothesis in question. *Neurosci Lett.*, **116**: 156-161.

Watt A. H., Reid P. G., Stephens M. R. and Routledge P. A. (1987) Adenosine –induced respiratory stimulation in man depends on site of infusion. Evidence for an action on the carotid body? *Br. J. Clin. Pharmacol.* **23**: 486-490.

Watt A. H. and Routledge P. A. (1985) adenosine stimulates respiration in man. *Br. J. Clin. Pharmacol.* **20**: 503-506.

Zhang, M., Zhong, H., Vollmer, C. and Nurse, C.A., 2000, Co-release of ATP and ACh mediates hypoxic signaling at rat carotid body chemoreceptors. *J. Physiol.,* **525**: 143-158

Hypoxic Regulation of Ca^{2+} Signalling in Astrocytes and Endothelial Cells

CHRIS PEERS[1], PARVINDER K. ALEY[1], JOHN P. BOYLE[1], KAREN E. PORTER[1], HUGH A. PEARSON[2], IAN F. SMITH[3] AND PAUL J. KEMP[4]

Schools of Medicine[1] and Biological Sciences[2], University of Leeds, UK, [3]Department of Neurobiology & Behavior, University of California, Irvine, CA, USA, [4]School of Biosciences, Cardiff University, Cardiff UK

Hypoxic modulation of K^+ channels is now firmly established in a variety of tissue types (Lopez-Barneo et al., 1988; Weir and Archer, 1998; Franco-Obregon et al., 1995; Youngson et al., 1993; Hool, 2001; Jiang and Haddad, 1994; Rychkov et al., 1998), and hypoxic modulation of specific channel types can also be reproduced in recombinant expression systems (Fearon et al., 2000; Lewis et al., 2001; Lewis et al., 2002; Williams et al., 2004), providing an opportunity to probe the molecular mechanism(s) of O_2 sensing by ion channels (see e.g. Kemp et al., this volume). In addition, the consequences for cell function of hypoxic ion channel modulation are fairly well established. In most cases, an appropriate response to hypoxia (such as systemic vasodilation, pulmonary vasoconstriction or carotid body glomus cell transmitter release – see (Lopez-Barneo et al., 2001) for review) involves modulation of $[Ca^{2+}]_i$ and this occurs primarily via modulation of Ca^{2+} influx (but in the lung vasculature this is contentious - see e.g. (Evans and Dipp, 2002)). Ca^{2+} influx can be regulated either through control of membrane potential via modulation of K^+ channel activity (Buckler and Vaughan-Jones, 1994; Wyatt et al., 1995; Osipenko et al., 1997; Weir and Archer, 1998), or via a direct effect on Ca^{2+} channels (Franco-Obregon et al., 1995; Hool, 2001).

Currently, scant attention is paid to the effects of hypoxia on non-excitable cells, which express voltage-gated Ca^{2+} channels only at very low levels, if at all. Such cells (e.g. astrocytes and endothelial cells), are worthy of study with respect to hypoxic responses, as they influence the physiological activity of neighbouring excitable cells (e.g. neurones and vascular smooth muscle). Furthermore, many of their vital roles depend on the precise control of $[Ca^{2+}]_i$. For these reasons, we have explored the ability of acute hypoxia to modulate $[Ca^{2+}]_i$ in both primary cultures of rat cortical astrocytes and human saphenous vein endothelial cells. $[Ca^{2+}]_i$ was monitored in Fura-2 loaded cells following incubation with the acetoxymethylester form of the dye. Cells were excited alternately at 340 and 380nm and emitted light collected at 510nm whilst cells were perfused with a HEPES-buffered physiological saline. For Ca^{2+}-free solutions, Ca^{2+} was omitted and replaced with 1mM EGTA. All protocols and procedures have been detailed elsewhere e.g (Smith et al., 2003; Budd et al., 1991).

In the absence of extracellular Ca^{2+}, application of agonists (ATP for endothelial cells, bradykinin (BK) for astrocytes), caused transient, receptor-dependent dependent rises of $[Ca^{2+}]_i$ (Figure 1A,C). Following agonist washout, exposure of either cell type to hypoxia (pO_2 20-30mmHg) was without effect on $[Ca^{2+}]_i$ (Figure 1A,C). However, when this protocol was reversed and cells were exposed firstly to hypoxia, small transient rises of $[Ca^{2+}]_i$ were apparent in the majority of recordings (>80%) from both cell types (Figure 1B,D). Following restoration of normoxia, applications of agonists also evoked transient rises of $[Ca^{2+}]_i$ in both cell types (Figure 1B,D). These were attenuated when compared to responses evoked before exposure to hypoxia, suggesting that hypoxia mobilized Ca^{2+} from an agonist-sensitive pool in both cell types.

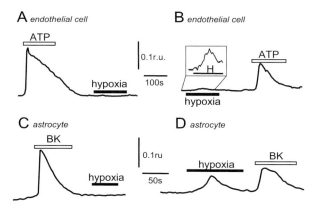

Figure 1. Acute hypoxia mobilizes Ca^{2+} from an intracellular pool. Example recordings of $[Ca^{2+}]_i$ in endothelial cells (A,B) and astrocytes (C,D). In each case, cells were exposed to an agonist (either 10μM ATP; or 100nM bradykinin (BK)) for the periods indicated by the open horizontal bars, and to hypoxia (pO_2 20-30mmHg) for the periods indicated by the solid horizontal bars (in B, this has been enlarged in the boxed section for clarity). Throughout these experiments, Ca^{2+} was omitted from the extracellular solution and replaced with 1mM EGTA.

Further support for this idea came from the observations that in both cell types neither hypoxia nor the relevant agonist could evoke changes in $[Ca^{2+}]_i$ following store depletion by pre-treatment of cells with the endoplasmic reticulum Ca^{2+}-ATPase inhibitor, thapsigargin (not shown).

Some effects of hypoxia have been attributed to a seemingly paradoxical increase in reactive oxygen species (ROS) generation (Chandel et al., 1998; Chandel and Schumacker, 2000; Duranteau et al., 1998; Pearlstein et al., 2002; Waypa et al., 2001; Leach et al., 2001). We therefore tested the ability of two distinct antioxidants to interfere with such hypoxic signalling; trolox (McClain et al., 1995) and TEMPO (with catalase (Abramov et al., 2004)). Both manoeuvres fully prevented rises in $[Ca^{2+}]_i$ evoked by hypoxia in both cell types (not shown), indicating that hypoxic mobilization of Ca^{2+} from an intracellular pool(s) requires ROS formation. There are numerous sites where ROS can be generated including mitochondria; hypoxia can lead to a rise of ROS derived from site(s) within the electron transport chain (Chandel et al., 2000; Chandel et al., 1998; Chandel and Schumacker, 2000). We therefore examined responses to hypoxia during mitochondrial uncoupling with FCCP (10μM, applied with 2.5μg/ml oligomycin). In endothelial cells, FCCP and oligomycin caused a transient rise of $[Ca^{2+}]_i$ (Figure 2A) and, although hypoxic rises of $[Ca^{2+}]_i$ could still be detected, they were attenuated as compared with responses seen in cells with functional mitochondria (Figure 2A,B). Responses in astrocytes were very different (Figure 2C,D). FCCP and oligomycin caused small rises of $[Ca^{2+}]_i$ (see also (Smith et al., 2003)) but, in its presence, responses to hypoxia were greater than those seen in the absence of inhibitors (Figure 2C,D).

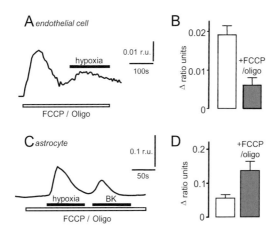

Figure 2. Modulation of hypoxic responses by mitochondrial uncoupling. (A) Example recording of $[Ca^{2+}]_i$ in an endothelial cell which was exposed to 10μM FCCP together with 2.5μg/ml oligomycin for the period indicated by the open bar. For the period indicated by the solid bar, the cell was also exposed to hypoxic solution (pO$_2$ 20-30mmHg). (B) Bar graph shows mean (± s.e.m.) peak rise of $[Ca^{2+}]_i$ evoked by hypoxia alone (open bar) or in the presence of FCCP and oligomycin (hatched bar). (C) As (A),(D) as (B) except that data were acquired in astrocytes. Note the opposite effect on hypoxic changes of $[Ca^{2+}]_i$ caused by mitochondrial inhibition.

Given that hypoxia was capable of partially depleting intracellular stores in both cell types, it was possible that it may initiate capacitative Ca^{2+} entry (CCE), an influx pathway important for various cell functions (Putney, 2001; Putney, Jr. et al., 2001). To investigate this, we first exposed both endothelial cells and astrocytes to Ca^{2+}-free perfusate (under normoxic conditions), then re-admitted Ca^{2+} to the perfusate. As exemplified in Figure 3 (A, C), this caused no marked change in $[Ca^{2+}]_i$ in either cell type. However, when cells were exposed to hypoxia during perfusion with Ca^{2+}-free solution, subsequent re-addition of Ca^{2+} caused marked rises of $[Ca^{2+}]_i$ Figure 3B, D). These rises could be blocked fully by co-application of Gd^{3+} (1mM; not shown). These data indicate that acute hypoxia, by stimulating Ca^{2+} release from intracellular pools, can activate CCE.

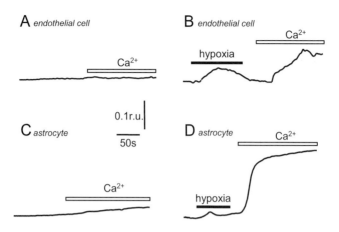

Figure 3. Acute hypoxia stimulates capacitative Ca^{2+} entry. (A, C) Example recordings of $[Ca^{2+}]_i$ in an endothelial cell (A) and an astrocyte (C) during perfusion with a Ca^{2+}-free perfusate in normoxia. For the period represented by the open horizontal bar in each case, Ca^{2+} (2.5mM) was readmitted to the perfusate. (B, D) as (A, C), except that before re-admission of Ca^{2+} to the perfusate, cells were exposed to hypoxia (pO$_2$ 20-30mmHg) for the period indicated by the horizontal bar. Note the marked rise of $[Ca^{2+}]_i$ when Ca^{2+} is added to the perfusate. Scale bars apply to all traces.

A number of important issues concerning cellular responses to hypoxia arise from the present work. Most importantly, we have demonstrated in two cell types that hypoxia can mobilize Ca^{2+} from intracellular pools that are also susceptible to depletion by agonists; presumably, this pool is the endoplasmic reticulum. Whilst the rises of cytosolic $[Ca^{2+}]$ were modest when compared to those evoked by agonists, they were sufficient to trigger CCE, an important Ca^{2+} influx pathway for various cellular functions (Putney, 2001; Putney, Jr. et al., 2001). Clearly, these responses to hypoxia were dependent on cellular production of ROS, since they were abolished in both cell types by antioxidants. However, cell-specific responses to hypoxia were revealed during mitochondrial uncoupling, indicating that potential roles for mitochondria in hypoxic Ca^{2+} signalling differed between the two cell types. In endothelial cells hypoxic responses were suppressed by FCCP and oligomycin, a finding consistent with the idea that mitochondria are the primary source of ROS in these cells during hypoxia. By contrast, hypoxic responses in astrocytes were strikingly potentiated by FCCP and oligomycin. This finding suggests that ROS are generated in astrocytes during hypoxia from a non-mitochondrial source and also that mitochondria serve an additional function, since their inhibition caused potentiation of the hypoxic response. In this regard, our previous studies have shown that mitochondria can be an important sink for Ca^{2+} in these cells (Smith et al., 2003), raising the possibility that mitochondria are effective as buffers for hypoxia-evoked rises of cytosolic $[Ca^{2+}]$. Thus, their inhibition uncovers a much greater effect of hypoxia to mobilize Ca^{2+}.

Both endothelial cells and astrocytes have a strict requirement for closely controlled regulation of $[Ca^{2+}]_i$ in order to perform many of their specific functions (Adams and Hill, 2004; Verkhratsky et al., 1998). Clearly, episodes of hypoxia are likely to interfere with such functions through disruption of Ca^{2+} signalling. However, it should be borne in mind that both cell types exist under physiological conditions in environments of relative hypoxia, and so the effects of hypoxia described here may reflect physiological rather than potentially pathophysiological effects of hypoxia.

ACKNOWLEDGEMENTS

This work was supported by The British Heart Foundation, The Wellcome Trust, The Medical Research Council, and Pfizer Central Research.

REFERENCES

1. Abramov AY, Canevari L, Duchen MR (2004) Beta-amyloid peptides induce mitochondrial dysfunction and oxidative stress in astrocytes and death of neurons through activation of NADPH oxidase. J Neurosci 24: 565-575.
2. Adams DJ, Hill MA (2004) Potassium channels and membrane potential in the modulation of intracellular calcium in vascular endothelial cells. J Cardiovasc Electrophysiol 15: 598-610.
3. Buckler KJ, Vaughan-Jones RD (1994) Effects of hypoxia on membrane potential and intracellular calcium in rat neonatal carotid body type I cells. J Physiol 476: 423-428.
4. Budd JS, Allen KE, Bell PR (1991) Effects of two methods of endothelial cell seeding on cell retention during blood flow. Br J Surg 78: 878-882.
5. Chandel NS, Maltepe E, Goldwasser E, Mathieu CE, Simon MC, Schumacker PT (1998) Mitochondrial reactive oxygen species trigger hypoxia-induced transcription. Proc Natl Acad Sci USA 95: 11715-11720.
6. Chandel NS, McClintock DS, Feliciano CE, Wood TM, Melendez JA, Rodriguez AM, Schumacker PT (2000) Reactive oxygen species generated at mitochondrial complex III stabilize hypoxia-inducible factor-1alpha during hypoxia: a mechanism of O2 sensing. J Biol Chem 275: 25130-25138.
7. Chandel NS, Schumacker PT (2000) Cellular oxygen sensing by mitochondria: old questions, new insight. J Appl Physiol 88: 1880-1889.
8. Duranteau J, Chandel NS, Kulisz A, Shao Z, Schumacker PT (1998) Intracellular signaling by reactive oxygen species during hypoxia in cardiomyocytes. J Biol Chem 273: 11619-11624.
9. Evans AM, Dipp M (2002) Hypoxic pulmonary vasoconstriction: cyclic adenosine diphosphate-ribose, smooth muscle Ca2+ stores and the endothelium. Respir Physiol Neurobiol 132: 3-15.
10. Fearon IM, Varadi G, Koch S, Isaacsohn I, Ball SG, Peers C (2000) Splice variants reveal the region involved in oxygen sensing by recombinant human L-type Ca2+ channels. Circ Res 87: 537-539.
11. Franco-Obregon A, Urena J, Lopez-Barneo, J. (1995) Oxygen-sensitive calcium channels in vascular smooth muscle and their possible role in hypoxic arterial relaxation. Proc Natl Acad Sci USA 92: 4715-4719.
12. Hool LC (2001) Hypoxia alters the sensitivity of the L-type Ca2+ channel to alpha-adrenergic receptor stimulation in the presence of beta-adrenergic receptor stimulation. Circ Res 88: 1036-1043.
13. Jiang C, Haddad GG (1994) A direct mechanism for sensing low oxygen levels by central neurons. Proc Natl Acad Sci USA 91: 7198-7201.

14. Leach RM, Hill HM, Snetkov VA, Robertson TP, Ward JP (2001) Divergent roles of glycolysis and the mitochondrial electron transport chain in hypoxic pulmonary vasoconstriction of the rat: identity of the hypoxic sensor. J Physiol 536: 211-224.
15. Lewis A, Hartness ME, Chapman CG, Fearon IM, Meadows HJ, Peers C, Kemp PJ (2001) Recombinant hTASK1 is an O2-sensitive K+ channel. Biochem Biophys Res Comm 285: 1290-1294.
16. Lewis A, Peers C, Ashford MLJ, Kemp PJ (2002) Hypoxia inhibits human recombinant maxi K+ channels by a mechanism which is membrane delimited and Ca2+ sensitive. J Physiol 540: 771-780.
17. Lopez-Barneo J, Lopez-Lopez JR, Urena J, Gonzalez C (1988) Chemotransduction in the carotid body: K+ current modulated by PO2 in type I chemoreceptor cells. Science 241: 580-582.
18. Lopez-Barneo J, Pardal R, Ortega-Saenz P (2001) Cellular mechanism of oxygen sensing. Annu Rev Physiol 63: 259-287.
19. McClain DE, Kalinich JF, Ramakrishnan N (1995) Trolox inhibits apoptosis in irradiated MOLT-4 lymphocytes. FASEB J 9: 1345-1354.
20. Osipenko ON, Evans AM, Gurney AM (1997) Regulation of the resting potential of rabbit pulmonary artery myocytes by a low threshold, O2-sensing potassium current. Brit J Pharmacol 120: 1461-1470.
21. Pearlstein DP, Ali MH, Mungai PT, Hynes KL, Gewertz BL, Schumacker PT (2002) Role of mitochondrial oxidant generation in endothelial cell responses to hypoxia. Arterioscler Thromb Vasc Biol 22: 566-573.
22. Putney JW (2001) The pharmacology of capacitative calcium entry. Molecular Interventions 1: 84-94.
23. Putney JW, Jr., Broad LM, Braun FJ, Lievremont JP, Bird GS (2001) Mechanisms of capacitative calcium entry. J Cell Sci 114: 2223-2229.
24. Rychkov GY, Adams MB, McMillen IC, Roberts ML (1998) Oxygen-sensing mechanisms are present in the chromaffin cells of the sheep adrenal medulla before birth. J Physiol 509: 887-893.
25. Smith IF, Boyle JP, Plant LD, Pearson HA, Peers C (2003) Hypoxic remodeling of Ca2+ stores in type I cortical astrocytes. J Biol Chem 278: 4875-4881.
26. Verkhratsky A, Orkand RK, Kettenmann H (1998) Glial calcium: homeostasis and signaling function. Physiol Rev 78: 99-141.
27. Waypa GB, Chandel NS, Schumacker PT (2001) Model for hypoxic pulmonary vasoconstriction involving mitochondrial oxygen sensing. Circ Res 88: 1259-1266.
28. Weir EK, Archer SL (1998) The mechanism of acute hypoxic pulmonary vasoconstriction: the tale of two channels. FASEB J 9: 183-189.
29. Williams SE, Wootton P, Mason HS, Bould J, Iles DE, Riccardi D, Peers C, Kemp PJ (2004) Hemoxygenase-2 is an Oxygen Sensor for a Calcium-Sensitive Potassium Channel. Science 306: 2093-2097.
30. Wyatt CN, Wright C, Bee D, Peers C (1995) O2-sensitive K+ currents in carotid-body chemoreceptor cells from normoxic and chronically hypoxic rats and their roles in hypoxic chemotransduction. Proc Natl Acad Sci USA 92: 295-299.
31. Youngson C, Nurse C, Yeger H, Cutz E (1993) Oxygen sensing in airway chemoreceptors. Nature 365: 153-155.

Does AMP-activated Protein Kinase Couple Hypoxic Inhibition of Oxidative Phosphorylation to Carotid Body Excitation?

CN WYATT[1], P KUMAR[2], P ALEY[3], C PEERS[3], DG HARDIE[4] AND AM EVANS[1]

[1] *School of Biology, Bute Building, St Andrews, Fife, Scotland.* [2] *Department of Physiology, The Medical School, University of Birmingham, UK.* [3] *Institute of Cardiovascular Research, Worsley Building, University of Leeds, UK.* [4] *Division of Molecular Physiology, School of Life Sciences, University of Dundee, Scotland.*

1. INTRODUCTION

The carotid bodies are the primary peripheral chemoreceptors. They respond to a fall in blood pO_2, a rise in blood pCO_2 and consequent fall in pH by releasing neurotransmitters. These increase the firing frequency of the carotid sinus nerves which then correct the pattern of breathing via an action at the brainstem. It is now generally accepted that the type 1 or glomus cells are the chemosensory element within the carotid body. However, the precise mechanism by which a fall in pO_2 excites the neurotransmitter rich type 1 cells has been the subject of hearty debate for decades now.

It has been known for many years that agents that inhibit mitochondrial function excite the carotid body (Heymans et al., 1931; Krylov and Anichkov, 1968). These observations led to work which suggested that O_2-sensing in the carotid body was mediated by an aspect of mitochondrial function (Mills and Jobsis, 1972). More recently it has been demonstrated that hypoxia and mitochondrial inhibitors excite carotid body type 1 cells via inhibition of membrane K^+ currents, causing depolarization and voltage-gated calcium entry (Peers, 1990; Buckler and Vaughan-Jones, 1994; Barbé et al., 2002; Wyatt and Buckler, 2004). However, the mechanism by which inhibition of oxidative phosphorylation couples to K^+ channel closure remains unknown.

In this article we present our preliminary findings indicating that the 'metabolic fuel gauge', AMP-activated protein kinase (AMPK), may be the missing link in the hypoxic chemotransduction pathway. It is known that any small decrease in the cellular ATP/ADP ratio, such as would be seen with hypoxic inhibition of oxidative phosphorylation, is translated into an increase in the AMP/ATP ratio via the adenylate kinase reaction. Adenylate kinase converts 2 molecules of ADP to ATP + AMP in an attempt to maintain ATP levels. The increased AMP/ATP ratio leads to subsequent activation of the enzyme AMP-kinase (Hardie, 2004). Whilst the majority of work on AMPK has focused on its role in energy metabolism, recent data has indicated that AMP-kinase can affect membrane ion channels (Hallows et al., 2003; Light et al., 2003). We therefore considered the proposal (Evans, 2004) that in carotid body type 1 cells, hypoxic inhibition of oxidative phosphorylation may activate AMPK and that AMPK may then mediate cell depolarization, voltage-gated calcium entry, transmitter release and hence an increase in the firing frequency of the carotid sinus nerve (CSN; Evans et al., 2005; Wyatt et al., 2004).

2. METHODS

Neonatal rat carotid body type 1 cells were isolated as previously described (Wyatt and Peers, 1993). Cells were allowed to adhere to poly-d-lysine coated coverslips before being used for immunocytochemistry, calcium imaging or electrophysiology.

2.1 Immunocytochemistry

Cells were fixed in ice cold methanol for 15 min, permeabilised and incubated overnight at 4°C with antibodies against the AMPK α1 subunit (1:500). Coverslips were washed with blocking solution and incubated (1hr, 22°C, dark) with FITC-conjugated secondary antibodies. Images were acquired using a Deltavision microscope system (Applied Precision), on an Olympus IX70 microscope, x60 objective (1.4 n.a.). Images were deconvolved and analysed off-line via Softworx (Applied Precision).

2.2 Electrophysiology

Cells were recorded using the amphotericin perforated-patch technique at 35°C. Solutions were as follows. Pipette solution (in mM): K_2SO_4, 55; KCl, 30; $MgCl_2$, 5; EGTA, 1; glucose, 10; HEPES, 20, adjusted to pH 7.2 with NaOH. Extracellular solution (in mM): KCl, 4.5; NaCl, 140; $CaCl_2$, 2.5; $MgCl_2$, 1; glucose, 11; HEPES, 20, adjusted to pH 7.4. Records were obtained in the current-clamp ($I=0$) configuration with a sampling frequency of 0.1 kHz, filtered at 2 kHz. Access resistance was typically 15MΩ and was not compensated for.

2.3 Ca^{2+} Imaging

Intracellular Ca^{2+} concentration was reported by Fura-2 fluorescence ratio (F340 / F380 excitation; emission 510 nm). Emitted fluorescence was recorded at 22°C with a sampling frequency of 0.02 Hz, using a Hamamatsu 4880 CCD camera via a Nikon Fluor 40x, 1.3 n.a. oil immersion lens and an inverted Nikon microscope. Image analysis was via Openlab (Improvision, UK). Background subtraction was performed on-line.

2.4 Carotid Sinus Nerve Recording

Rats were anaesthetized with 1-4% halothane in O_2 (Pepper et al., 1995), killed and exsanguinated. Left and right carotid bifurcations were identified and removed, pinned on Sylgard (184, Farnell, U.K.) in a 0.2 ml chamber and perfused (3 ml min^{-1}) with PSS-C (mM): 125 NaCl, 3 KCl, 1.25 NaH_2PO_4, 5

Na_2SO_4, 1.3 $MgSO_4$, 24 $NaHCO_2$, 2.4 $CaCl_2$, 10 glucose, pH 7.4, 37°C. Perfusate was equilibrated to 40 Torr PCO_2 and 400 Torr PO_2 via precision flow valves (Cole-Parmer, USA). The sinus nerve was sectioned at the junction with the glossopharyngeal nerve, extracellular recordings of afferent fibre spike activity were made with glass suction electrodes, recorded on video tape, and action potentials sampled digitally via LabVIEW 2 (National Instruments Co).

3. RESULTS

3.1 Immunocytochemistry

AMPKα1 staining was observed in type 1 cells. It was predominantly expressed at the plasma membrane although there was some cytoplasmic staining (Fig. 1).

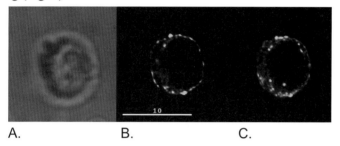

A. B. C.

Figure 1. AMPKα1 staining in a typical carotid body type 1 cell. A. Transmission. B. z-section. C. 3d reconstruction. Scale bar 10μm.

3.2 Electrophysiology and Ca^{2+} Imaging

At 37°C the AMP mimetic 5-aminoimidizole-4-carboxamide riboside (AICAR, 1 mM), a compound commonly used to activate AMPK, caused 6 of 8 type 1 cells to depolarize by 12.6 ± 0.9mV (n=6, current clamp I=0, see Fig 2A) the other 2 cells failed to respond.

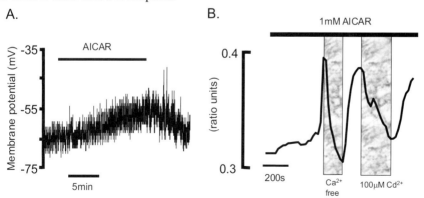

Figure 2. A, Example current-clamp recording showing effect of AICAR on carotid body type 1 cell resting membrane potential. B, Effect of AICAR on type 1 cell intracellular Ca^{2+} concentration. The effects of removal of extracellular Ca^{2+} and block of voltage-gated Ca^{2+} entry with Cd^{2+} are also shown.

Membrane depolarization by AICAR (1 mM) was associated with an increase in intracellular Ca^{2+} concentration in acutely isolated carotid body type 1 cells, the Fura-2 fluorescence ratio increasing by 0.07 ± 0.02 (n = 11). This increase in intracellular Ca^{2+} concentration was abolished by removal of extracellular Ca^{2+} (n = 6) and attenuated by blockade of voltage-gated Ca^{2+} influx with Cd^{2+} (100 µM; 0.009 ± 0.006, n = 6; Fig. 2B).

3.3 *In Vitro* Carotid Body Preparation

In the *in vitro* carotid body preparation hypoxia causes a rise in carotid sinus nerve activity which is reversed upon removal of extracellular Ca^{2+}. Consistent with the effects of hypoxia the application of AICAR (1mM) to this preparation induced a relatively rapid and reversible, ca. 10-fold increase in single fibre sensory afferent discharge from 0.22 ± 0.03 to 2.8 ± 0.56 spikes s^{-1} (n = 19, Fig. 3). This too was abolished by removal of extracellular Ca^{2+} (0.11 ± 0.03 spikes s^{-1}, n = 5) and was attenuated by blockade of voltage-gated Ca^{2+} influx with Cd^{2+} (100 µM; 0.91 ± 0.18 spikes s^{-1}, n = 5).

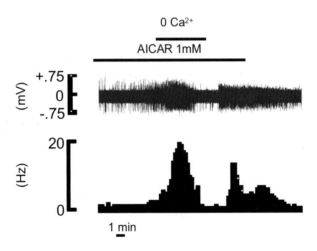

Figure 3. Example trace showing effect of AICAR on multiple fibre sensory afferent discharge and the effect of removal of extracellular Ca^{2+}.

4. DISCUSSION

Our findings demonstrate that AMPK is predominantly targeted to the plasma membrane of carotid body type 1 cells. This localization is ideal if AMPK is to interact with membrane ion channels as has been demonstrated in other tissues. AMPK activation by AICAR mirrored the effect of hypoxia on the carotid body. In isolated type 1 cells AICAR induced membrane depolarization and caused transmembrane Ca^{2+} influx resulting in an increase in sensory afferent discharge in the *in vitro* carotid body.

These results are consistent with our proposal (Evans, 2004; Evans et al., 2005; Wyatt et al., 2004) that AMPK may mediate carotid body excitation in response to hypoxia. Indeed, they present compelling evidence that AMPK may be the missing link between hypoxic inhibition of mitochondrial oxidative phosphorylation and excitation of carotid body type 1 cells.

Therefore, we propose that hypoxia inhibits mitochondrial oxidative phosphorylation and thus ATP production in type 1 cells. A consequent rise in the ADP/ATP ratio may then be converted to a rise in the AMP/ATP ratio by adenylate kinase in an effort to maintain ATP supply. This rise in the AMP/ATP ratio leads to AMP-kinase activation resulting in depolarisation, voltage-gated Ca^{2+} entry, neurotransmitter release and hence increased carotid sinus nerve activity (Fig 4).

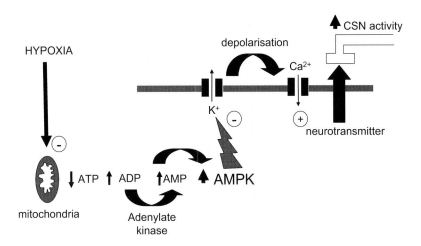

Figure 4. Schematic showing the proposed mechanism by which hypoxic inhibition of mitochondrial oxidative phosphorylation couples to type 1 cell depolarization, voltage-gated Ca^{2+} entry, neurotransmitter release and a consequent increase in carotid sinus nerve activity.

These data taken together with our results investigating the role of AMPK in hypoxicpulmonary vasoconstriction (Evans et al., 2005) strongly suggest that AMPK may couple a fall in pO_2 to Ca^{2+} signalling mechanisms in O_2-sensing cells.

ACKNOWLEDGEMENTS

This work was supported by the Wellcome Trust (Grant Ref: 070772)

REFERENCES

Barbé C, Al-Hashem F, Conway AF, Dubuis E, Vandier C and Kumar P. A possible dual site of action for carbon monoxide-mediated chemoexcitation in the rat carotid body. *J. Physiol-Lond.* 2003; 543, 933-945.

Buckler KJ and Vaughan-Jones. Effects of hypoxia on membrane potential and intracellular calcium in rat neonatal carotid body type 1 cells. J. Physiol-Lond. 1994; 476, 423-428.

Evans AM. Hypoxia cell metabolism and cADPR accumulation. In: Yuan X-J (ed). Hypoxic pulmonary vasoconstriction: cellular and molecular mechanisms. Kluwer Academic Publications. p 313-338

Evans AM, Hardie DG, Galione A, Peers C, Kumar P and Wyatt CN. AMP-activated protein kinase couples mitochondrial inhibition by hypoxia to cADPR dependent Calcium mobilization from the sarcoplasmic reticulum and/or transmembrane Calcium influx in Oxygen-sensing cells. In: Signalling pathways in acute Oxygen sensing. 2005. Novartis open meeting 272.

Hallows KR, Kobinger GP, Wilson JM, Witters LA and Foskett JK. Physiological modulation of CFTR activity by AMP-activated protein kinase in polarized T84 cells. Am. J Physiol 2003; 284, 1297-1308.

Hardie DG. The AMP-activated protein kinase pathway—new players upstream and downstream. *J Cell Sci*. 2004; 117, 5479-5487.

Heymans C, Bouckaert JJ and Dautreband L. Sinus carodidien et reflexes respiratoires; sensibilite des sines carotidiens aux substances chimiques. Acion stimulante respiratoire reflex du sulfre de sodium, du cyanure de potassium, de la nicotine et de la lobeline. *Arch. Int. Pharmacodyn. Ther*. 1931; 40, 54-91.

Krylov SS and Anichkov SV. The effect of metabolic inhibitors on carotid chemoreceptors. In Torrance RW (ed). Arterial Chemoreceptors. Blackwell, Oxford. P 103-109.

Light PE, Wallace CH and Dyck JR. Constitutively active adenosine monophosphate-activated protein kinase regulates voltage-gated Sodium currents in ventricular myocytes. *Circulation*. 2003; 1962-1965.

Mills E and Jobsis FF. Mitochondrial respiratory chain of carotid body and chemoreceptor response to changes in Oxygen tension. J. Neurophysiol. 1972; 35, 405-428.

Peers C. Hypoxic suppression of K^+ currents in type 1 carotid body cells: selective effect on the Ca^{2+}-activated K^+ current. Neurosci. Lett. 1990; 119, 253-256.

Pepper DR, Landauer RC, Kumar P. Post-natal development of CO_2-O_2 interaction in the rat carotid body *in-vitro*. *J. Physiol-lond*. 1995;485, 531-541.

Wyatt CN and Peers C. Nicotinic acetylcholine receptors in isolated type 1 cells of the neonatal rat carotid body. *Neuroscience* 1993; 54, 275-281.

Wyatt CN and Buckler KJ. The effect of mitochondrial inhibitors on membrane currents in isolated neonatal rat carotid body type I cells. *J. Physiol-Lond*. 2004; 556, 175-191.

Wyatt CN, Kumar P, Peers C, Kang P, Hardie DG and Evans AM. The potential role for AMP-kinase in hypoxic chemotransduction of rat carotid body. *J. Physiol-Lond*. 2004; 560P, C44.

Mitochondrial ROS Production Initiates Aβ$_{1-40}$-Mediated Up-Regulation of L-Type Ca^{2+} Channels during Chronic Hypoxia

IAN M. FEARON[1], STEPHEN T. BROWN[2], KRISTIN HUDASEK[2], JASON L. SCRAGG[3], JOHN P. BOYLE[3], AND CHRIS PEERS[3]

[1]*Faculty of Life Sciences, The University of Manchester, Manchester M13 9PT, U.K.*
[2]*Department of Biology, McMaster University, Hamilton, ON L8S 4K1, Canada,* [3]*School of Medicine, The University of Leeds, Leeds LS2 9JT, U.K.*

1. INTRODUCTION

Exposure to chronic hypoxia (CH) initiates cellular responses designed to counteract this deleterious stimulus, providing a physiological response to low oxygen. However, long-term exposure to CH, such as that which occurs in cardiorespiratory diseases such as ischaemic stroke, can also have pathological consequences. In many cases, CH alters the transcription of genes encoding numerous proteins, secondary to accumulation of the transcriptional activator hypoxia inducible factor-1 (HIF-1) (Schofield and Ratcliffe, 2004). In contrast, we recently reported that hypoxic regulation of the plasma membrane expression of L-type Ca^{2+} channel α_{1C} subunits occurred in a post-transcriptional manner due to the trafficking of these subunits towards, and / or their retention within, the plasma membrane (Scragg et al., 2004). This process involved the altered production of amyloid β peptides (AβPs), since it was inhibited by selective inhibitors of the secretases involved in the production of these peptides, and mimicked by exogenous AβP. This regulation of the functional membrane expression of a voltage-gated Ca^{2+} channel may contribute to the Ca^{2+} dyshomeostasis seen in Alzheimer's disease, a prevalent disorder in which hypoxia / ischaemia is a predisposing factor (Moroney et al., 1996).

While CH enhances Ca^{2+} currents, the full pathway connecting lowered O$_2$ levels and α_{1C} subunit expression has yet to be elucidated. To address this, in the present study we investigated the mechanisms underlying CH and AβP-mediated regulation of the α_{1C} subunit. We demonstrate that during CH reactive oxygen species (ROS) production by complex I of the mitochondrial electron transport chain (ETC) precedes the production of AβPs and Ca^{2+} current enhancement. These studies provide compelling evidence for the temporal contributions of mitochondrial ROS and AβPs in the O$_2$ sensing pathway.

2. METHODS

2.1 Electrophysiology

Ca^{2+} currents were recorded in HEK293 cells stably transfected with the α_{1C} subunit of the L-type Ca^{2+} channel (Hudasek et al., 2004) using the whole-cell configuration of the patch-clamp technique. Cells were perfused with a solution composed of (mM): NaCl, 95; CsCl, 5; $MgCl_2$, 0.6; $BaCl_2$ 20; Hepes, 5; D-glucose, 10; TEA-Cl, 20 (pH7.4 with NaOH). Patch electrodes were filled with (mM): CsCl, 120; TEA-Cl, 20; $MgCl_2$, 2; EGTA, 10; Hepes, 10; ATP, 2 (pH7.2 with CsOH). Cells were voltage-clamped at -80mV, and whole-cell currents evoked by step depolarising the membrane to various test potentials for 100ms at a frequency of 0.1Hz. All recordings were made at room temperature (22±2°C).

2.2 Tissue Culture

The culture media used was described previously (Hudasek et al., 2004). Cells were incubated in a humidified atmosphere of air/CO_2 (95%:5%). In experiments examining the effects of CH, cells were incubated for 24 h in an humidified environment of 6% O_2 / 5% CO_2 / 89% N_2 or 2.5% O_2 / 5% CO_2 / 92.5% N_2.

2.3 Creation of ρ^0 HEK293 Cells

Cells depleted of a functional mitochondrial ETC (ρ^0 cells) were created by incubating cells for >2 months in 2 μg/ml ethidium bromide. Culture media was supplemented with 50μg/ml uridine and 1 mM pyruvate, to stimulate cell growth (King and Attardi, 1989). ρ^0 cells were deficient in the mitochondrially-encoded gene for cytochrome C oxidase and failed to take up reduced mitotracker red (CM- H_2XROS; Invitrogen; Figure 1). This probe does not fluoresce until it enters an actively respiring cell, where it is oxidized to the corresponding fluorescent mitochondrion-selective probe and sequestered into the mitochondria.

3. RESULTS

Exposure of HEK293 cells to CH enhanced the functional expression of α_{1C} subunits. At a test potential of +10mV (the peak of the I-V relationship), currents in cells incubated in normoxia were -7.8±0.8pA/pF (n=20), while following

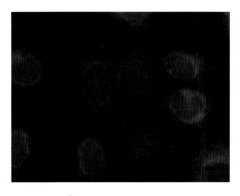 

Figure 1. ρ^0 cells do not take up the mitochondria-selective dye, reduced mitotracker red (CM-H$_2$XROS). Wild-type (left) and ρ^0 (right) cells were incubated in 500nM CM-H$_2$XROS for 15min at 37 °C in normal culture media. Cells were then fixed in 4% paraformaldehyde in PBS and mounted onto slides using Vectashield hardset mounting media. Slides were visualised on a Zeiss Axioplan 2 upright microscope equipped with epifluorescence and a rhodamine filter set. Images were taken with a QImaging QICAM 12 bit monochrome digital camera and QCapture Pro 5.0 software.

exposure to CH (6% O$_2$ for 24 h) currents were -13.1±2.5pA/pF (n=18; P<0.05, unpaired Students' t-test; Figure 2). Enhancement of Ca^{2+} currents was due to the production of oxidant molecules, since it was abolished by the antioxidants ascorbic acid (200μM) and TROLOX (500μM; Figure 2). This enhancement was also abolished by selective inhibitors of β and γ secretases (Scragg et al., 2004) which are essential for amyloid β peptide formation (Mattson, 1997). This suggests that CH enhanced Ca^{2+} currents in a manner involving ROS and ABP production.

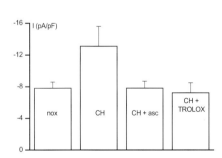

Figure 2. Hypoxic enhancement of Ca^{2+} currents involves cellular oxidant production. Each bar shows the mean (± s.e.m.) current density evoked when step depolarising cells to +10mV (holding potential, -80mV). Currents were evoked in cells incubated under normoxic and hypoxic conditions, as indicated. The antioxidants ascorbic acid (200μM) and TROLOX (500μM) abolished hypoxic enhancement of Ca^{2+} currents.

To test the hypothesis that altered function of the mitochondrial ETC mediates the response to CH, we examined hypoxic regulation of α$_{1C}$ subunits in cells depleted of a functional ETC. In these ρ^0 cells, CH failed to enhance Ca^{2+} current amplitudes. Thus, mean (±s.e.m.) currents were -6.5±0.6pA/pF (n=30) in cells incubated in normoxia and -5.6 ± 0.7pA/pF (n=16) in cells incubated under CH conditions (P>0.05, unpaired Students' t-test). Hypoxic enhancement was similarly abolished when cells were co-incubated with the complex I inhibitor rotenone (1μM) under hypoxic conditions. In the ρ^0 cells, Ca^{2+} currents were enhanced (to -10.8 ± 1.3pA/pF) following a 6h incubation in the superoxide generation system, xanthine/xanthine oxidase (X/XO; 100μM/5mUml^{-1}), a response abolished by secretase inhibitors (data not shown). Taken together, these data suggest that hypoxic enhancement of Ca^{2+} currents involves superoxide production in the mitochondrial ETC.

To investigate whether AβP production occurs upstream or downstream of altered mitochondrial function during CH, we applied AβP$_{1-40}$ (50nM) to ρ^0 cells expressing α_{1C} subunits. In these cells, Ca^{2+} currents were increased from -6.5 ± 0.6pA/pF (n=30) in control cells to -9.3 ± 1.0pA/pF (n=12) following incubation in AβP$_{1-40}$ (P<0.05, unpaired Student's t-test). This effect was not altered by co-incubating cells with AβP$_{1-40}$ and ascorbic acid (200μM). Thus, altered mitochondrial function during CH lies upstream of AβP production in this O_2-sensing pathway.

4. DISCUSSION

Previously, we demonstrated the role of AβP production in the enhancement of Ca^{2+} currents due to CH (Scragg et al., 2004). Here, we have extended this work to demonstrate that during CH, ROS production in the mitochondrial ETC provides a stimulus for altered synthesis of AβPs. These data provide convincing evidence for the temporal sequence of events linking CH to Ca^{2+} channel expression.

AβPs themselves have been shown to produce ROS *in vitro* [Hensley et al., 1994]. Furthermore, a recent study demonstrated that AβPs caused mitochondrial dysfunction and ROS production, although these effects were observed in isolated astrocytes but not neurones, suggestive of cell specificity of this process [Abramov et al., 2004]. Contrastingly, there is evidence to suggest that compromised mitochondrial function and the ensuing ROS production are responsible for elevating AβP levels, as part of a physiological neuroprotective mechanism with the ability to utilise AβPs as antioxidants [Smith et al., 2002]. Here, AβP$_{1-40}$ enhanced Ca^{2+} currents in HEK293 cells depleted of a functional mitochondrial ETC. This is in stark contrast to the lack of effect of CH in these ρ^0 cells. Thus, CH causes altered mitochondrial function, which is a trigger for AβP production and subsequently for the trafficking / membrane retention of α_{1C} subunits [Scragg et al., 2004]. Thus, these data provide strong evidence that, in terms of Ca^{2+} dyshomeostasis due to CH, altered mitochondrial function temporally precedes enhanced AβP production.

ACKNOWLEDGEMENTS

IMF was supported by a Grant-in-Aid (Grant # NA 5230) from the Heart and Stroke Foundation of Ontario. Equipment was provided by New Opportunities grants to IMF from the Canada Foundation for Innovation (# 7400) and the Ontario Innovation Trust. CP was supported by the British Heart Foundation, The Medical Research Council, The Alzheimer's Research Trust and the Alzheimer's Society. JLS was supported by the British Heart Foundation. We thank Dr. Gyula Varadi (University of Cincinnati) for providing us with the pCDNA3.1-α_{1C} construct.

REFERENCES

Abramov, A.Y., Canevari, L. and Duchen, M.R. (2004). Beta-amyloid peptides induce mitochondrial dysfunction and oxidative stress in astrocytes and death of neurons through activation of NADPH oxidase. *J. Neurosci.* **24**, 565-575.

Hensley, K., Carney, J.M., Mattson, M.P., Aksenova, M., Harris, M., Wu, J.F., Floyd, R.A. and Butterfield, D.A. (1994). A model for beta-amyloid aggregation and neurotoxicity based on free radical generation by the peptide: relevance to Alzheimer disease. *Proc. Natl. Acad. Sci. U.S.A.* **91**, 3270-3274.

Hudasek, K., Brown, S.T. and Fearon, I.M. (2004). H_2O_2 regulates recombinant Ca^{2+} channel a_{1C} subunits but does not mediate their sensitivity to acute hypoxia. *Biochem. Biophys. Res. Commun.* **318**, 135-141.

King, M.P. and Attardi, G. (1989). Human cells lacking mtDNA: repopulation with exogenous mitochondria by complementation. *Science* **246**, 500-503.

Mattson, M.P. (1997). Cellular actions of β-amyloid precursor protein and its soluble and fibrillogenic derivatives, *Physiol. Rev.* **77**, 1081-1132.

Moroney, J.T., Bagiella, E., Desmond, D.W., Paik, M.C., Stern, Y. and Tatemichi, T.K. (1996). Risk factors for incident dementia after stroke. Role of hypoxic and ischemic disorders. *Stroke* **27**, 1283-1289.

Schofield, C.J. and Ratcliffe, P.J. (2004). Oxygen sensing by HIF hydroxylases. *Nat. Rev. Mol. Cell Biol.* **5**, 343-354.

Scragg, J.L., Fearon, I.M., Boyle, J., Ball, S.G., Varadi, G. and Peers, C. (2004). Alzheimer's amyloid peptides mediate hypoxic up-regulation of L-type Ca^{2+} channels. *FASEB J.* **19**, 150-152.

Smith, M.A., Drew, K.L., Nunomura, A., Takeda, A., Hirai, K., Zhu, X., Atwood, C.S., Raina, A.K., Rottkamp, C.A., Sayre, L.M., Friedland, R.P. and Perry, G. (2002). Amyloid-β, tau alterations and mitochondrial dysfunction in Alzheimer disease: the chickens or the eggs? *Neurochem. Int.* **40**, 527-531.

Acute Hypoxic Regulation of Recombinant THIK-1 Stably Expressed in HEK293 Cells

IAN M. FEARON[1], VERONICA A. CAMPANUCCI[2], STEPHEN T. BROWN, KRISTIN HUDASEK, ITA M. O'KELLY[1] AND COLIN A. NURSE

Department of Biology, McMaster University, Hamilton, ON L8S 4K1, Canada [1]Faculty of Life Sciences, The University of Manchester, Manchester M13 9PT, U.K. [2]Department of Physiology, McGill University, Montréal, Québec, H3G 1Y6, Canada.

1. INTRODUCTION

Hypoxic inhibition of O_2-sensitive K^+ channels plays a key role in mediating numerous cellular responses which counteract the deleterious effects of hypoxia. In type I cells of the carotid body (CB), a neurosecretory organ that responds to hypoxia by releasing neurotransmitters from specialized O_2-sensing type I cells onto sensory nerve endings, hypoxic inhibition of K^+ channels underlies the membrane depolarisation (Lopez-Barneo et al., 1988) that stimulates Ca^{2+} entry and neurotransmitter release (Urena et al., 1994). In other neurosecretory cells, such as those located in the neuroepithelial cell bodies of the lung (Youngson et al., 1993) and the adrenal medulla (Thompson and Nurse, 1998), hypoxic inhibition of K^+ channels provides a critical link between O_2 levels and the appropriate cellular responses.

In rat CB type I cells, acute hypoxia inhibits both voltage-dependent Ca^{2+}-activated K^+ (BK) channels (Peers, 1990) and a member of the tandem pore domain family of background K^+ channels (TASK-like; Buckler et al., 2000). Background K^+ channels are constitutively active ionic channels responsible for regulating cell resting membrane potential, and play a dominant role in cell firing and excitability (Goldstein et al., 2001). Since the initial demonstration of the O_2-sensitivity of a TASK-1-like conductance in the rat CB, hypoxic inhibition of this and other background K^+ channels has been reported in cerebellar granule neurons (Plant et al., 2002), glossopharyngeal neurons (Campanucci et al., 2003), and the lung neuroepithelial cell line H146 (Kemp et al., 2002). These findings are supported by the O_2-sensitivity of these channels in recombinant expression systems (Lewis et al., 2001; Miller et al., 2003).

Nitric oxide synthase containing neurons of the petrosal ganglion and glossopharyngeal nerve (GPN) mediate efferent inhibition of the CB chemoreceptors (Wang et al., 1994; 1995). GPN neurons themselves possess intrinsic O_2 sensitivity via the hypoxic inhibition of a background K^+

conductance (Campanucci et al., 2003), which leads to increased excitability (Campanucci & Nurse, 2005). The properties of this conductance strongly resembled those of the recently-described TWIK-related halothane-inhibitable K^+ channel THIK-1 (Kcnk12; Rajan et al., 2001) channel. Here, we directly examined the O_2-sensitivity of THIK-1, which we stably expressed in HEK293 cells. Hypoxia inhibited THIK-1 currents and depolarised cells in which the channel was expressed. Thus THIK-1 is an O_2-sensitive K^+ channel. We further demonstrate that the mechanism underlying its O_2 sensitivity differs from that proposed for a further background K^+ channel family member, TASK-1 (Wyatt & Buckler, 2004).

2. METHODS

In transient transfection studies, wild-type HEK293 cells were split prior to the day of transfection such that cells were 60-70% confluent in 35mm dishes when transfected. Cells were transfected with 5μg of either pCDNA3.1-THIK-1 or pCDNA3.1-THIK-2 using ExGen 500 (Fermentas, Burlington, ON, Canada), according to the manufacturer's instructions. The following day, the medium was removed and cells were washed with PBS and fresh medium added. To allow visual selection of transfected cells, they were co-transfected with 0.5μg of pEGFP-C1 (Clontech, Mississauga, ON, Canada), causing the co-expression of the enhanced green fluorescent protein in successfully transfected cells. Cells were used in electrophysiological studies 48h post-transfection. To create a cell line stably expressing THIK-1, cells were initially transfected as described above. Three days after transfection, the medium was supplemented with 400μg/ml G418 (Invitrogen; Mississauga, ON, Canada). Selection was applied for 2 weeks, after which time individual colonies were picked and transferred to 35mm dishes for further culture and examination of K^+ currents. A single clone was identified for further study based on expression levels within this clone.

K^+ currents were recorded using the perforated-patch (nystatin, 500μg/ml) and whole-cell configurations of the patch-clamp technique. Cells were perfused with (mM): NaCl, 135; KCl, 5; $MgCl_2$, 2; $CaCl_2$, 2; Hepes, 10 and D-glucose, 10 (pH 7.4 with NaOH). Patch electrodes were filled with (mM): KCl, 135; NaCl, 5; $CaCl_2$, 2; EGTA, 11; Hepes, 10 and MgATP, 2 (pH 7.2 with KOH.

Hypoxia was produced by bubbling the extracellular perfusate with 100% N_2 gas for >30minutes prior to experimentation. Bath Po_2 was measured using a polarised (-600mV) carbon fibre electrode, and was always stable at ~20mmHg within 30-45s of exchanging solution. Bubbling with N_2 caused no change in the either the pH or the osmolarity of the extracellular perfusate. In control studies, the perfusate was bubbled with compressed air.

3. RESULTS

The majority of untransfected (wild-type; WT) HEK293 cells displayed little or no outward K^+ current in asymmetrical K^+ solutions. However, in ~5% of cells, ramp depolarisations evoked small, outwardly-rectifying currents (see e.g. Figure 1). These currents were small, being around 5 pA/pF in magnitude at a highly depolarised test potential (+50mV). The O_2-sensitive K^+ current we describe below cannot be attributed to these endogenous currents, since O_2-sensitive difference currents in transfected cells were much larger, even at less depolarised test potentials.

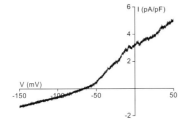

Figure 1. Endogenous K^+ current in a WT HEK293 cell. These currents were seen in ~5% of cells examined. The current was evoked by ramp depolarising cells between -150 and +50mV over a period of 1s.

In cells co-transfected with pCDNA3.1-THIK-1 and pEGFP-C1 constructs and exhibiting positive GFP fluorescence, step depolarisations evoked large, moderately outwardly-rectifying K^+ currents in each of 7 cells examined (e.g. Figure 2). These currents reversed close to the estimated equilibrium potential for K^+ ions (E_k). The mean (±s.e.m.) membrane potential (V_m) of these cells was -71.3±2.1 mV. In contrast, when transiently expressing pCDNA3.1-THIK-2 and pEGFP-C1, GFP-positive cells displayed no discernible K^+ currents (n=10). The mean V_m of these cells was not significantly different to that seen in WT cells.

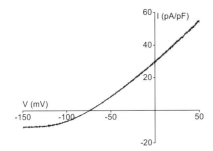

Figure 2. K^+ current in a HEK293 cell transiently expressing THIK-1. The cell was identified by positive GFP fluorescence following co-transfection of pCDNA3.1-THIK-1 and pEGFP-C1. The current was evoked by ramp depolarising cells between -150 and +50mV over a period of 1s.

In cells stably expressing THIK-1, currents were reversibly inhibited (by ~40%) by the inhalational anaesthetic, halothane (5mM). Currents were enhanced by ~40% in the presence of 5μM arachidonic acid. Thus, the recombinant channel displayed pharmacological properties of THIK-1.

Exposure to acute hypoxia caused a reversible reduction in THIK-1 current amplitudes (see e.g. Figure 3). In 6 cells examined, currents were reduced to ~87% of control values. The O_2-sensitive difference current (IKO$_2$) reversed at the estimated E_k. In control experiments, THIK-1 current amplitudes were unchanged when cells were perfused with a solution that had been continuously bubbled with compressed air for 1h. Hypoxic inhibition of K^+ currents was completely occluded in the presence of 5mM halothane, demonstrating that the O_2-sensitive channel was attributable to THIK-1. Hypoxic inhibition was unaltered in the presence of the mitochondrial complex I inhibitor rotenone, at a concentration (1μM) which ablated the hypoxic response of a TASK-1-like channel in rat carotid body type I cells (Wyatt & Buckler, 2004).

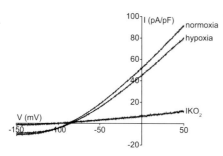

Figure 3. Hypoxic inhibition of THIK-1.. Currents were evoked by ramp depolarising cells between -150 and +50mV over a period of 1s. Recordings were made under normoxic and hypoxic (P_{O_2}, 20mmHg) as indicated. IKO_2, calculated O_2-sensitive difference current.

4. DISCUSSION

In specialised chemosensing cells such as CB type I and lung neuroepithelial cells, and also in non-specialised cells such as vascular smooth muscle, acute hypoxia regulates K^+ channels in the cell membrane. This function mediates homeostatic, physiological responses designed to maintain an O_2 supply commensurate with demand. Recently, attention has moved towards a role for tandem-pore (2P) domain background K^+ channels (Goldstein et al., 2001) in the acute O_2 sensitivity of CB type I cells (Buckler et al., 2000), cerebellar granule neurones (Plant et al., 2002), and H146 cells (O'Kelly et al., 1999). These findings are supported by the O_2 sensitivity of these channels in recombinant systems (Lewis et al., 2001; Miller et al., 2003).

Similar to these studies, the data presented here expressing THIK-1 in a recombinant system support our previous finding of an O_2-sensitive THIK-like background K^+ conductance in neurones of the glossopharyngeal nerve (GPN; Campanucci et al., 2003). Like the IKO_2 in the GPN neurones, the recombinant O_2-sensitive current was inhibited by halothane, a defining characteristic of the THIK-1 channel (Rajan et al., 2001).

The O_2-sensitivity of a TASK-1-like background K^+ channel in carotid body type I cells was occluded by 1μM rotenone, an inhibitor of complex I of the mitochondrial electron transport chain (ETC). In direct contrast, hypoxia robustly inhibited recombinant THIK-1 in the presence of the inhibitor, suggesting that hypoxic regulation of different members of the tandem-pore background K^+ channel family does not share a common molecular sensor. The exact mechanism by which THIK-1 senses low O_2 has yet to be elucidated. However this recombinant system will likely form an excellent paradigm in which to uncover the mechanisms of O_2 sensing by this channel.

ACKNOWLEDGEMENTS

Supported by a Grant-in-Aid (Grant # NA 5230) to IMF from the Heart and Stroke Foundation of Ontario, by an Operating Grant (Grant # MOP-57909) to CAN from the Canadian Institutes for Health Research, and by a CIHR scholarship to VAC. Equipment was provided by New Opportunities grants to IMF from the Canada Foundation for Innovation (#7400) and the Ontario Innovation Trust. We gratefully acknowledge Regina Preisig-Müller and Jürgen Daut for providing us with the pCDNA3.1-THIK-1 and pCDNA3.1-THIK-2 constructs.

REFERENCES

Buckler KJ, Williams BA and Honore E (2000). An oxygen-, acid- and anaesthetic-sensitive TASK-like background potassium channel in rat arterial chemoreceptor cells. *J Physiol* **525**, 135–142.

Campanucci VA, Fearon IM and Nurse CA (2003). A novel O_2-sensing mechanism in rat glossopharyngeal neurones mediated by a halothane-inhibitable background K^+ conductance. *J Physiol* **548**, 731–743.

Campanucci VA and Nurse CA (2005). Biophysical characterization of whole-cell currents in O2-sensitive neurons from the rat glossopharyngeal nerve. *Neuroscience* **132**, 437–51.

Goldstein SA, Bockenhauer D, O'Kelly I and Zilberberg N (2001). Potassium leak channels and the KCNK family of two-P-domain subunits. *Nat Rev Neurosci* **2**: 175–184, 2001.

Kemp PJ, Lewis A, Hartness ME, Searle GJ, Miller P, O'Kelly I and Peers C (2002). Airway chemotransduction: from oxygen sensor to cellular effector. *Am J Respir Crit Care Med* **166**, S17–S24.

Lewis A, Hartness ME, Chapman CG, Fearon IM, Meadows HJ, Peers C and Kemp PJ (2001). Recombinant hTASK1 is an O_2-sensitive K^+ channel. *Biochem Biophys Res Commun* **285**, 1290–1294.

Lopez-Barneo J, Lopez-Lopez JR, Urena J and Gonzalez C (1988). Chemotransduction in the carotid body: K^+ current modulated by PO_2 in type I chemoreceptor cells. *Science* **241**, 580–582.

Miller P, Kemp PJ, Lewis A, Chapman CG, Meadows HJ and Peers C (2003). Acute hypoxia occludes hTREK-1 modulation: re-evaluation of the potential role of tandem P domain K^+ channels in central neuroprotection. *J Physiol* **548**, 31–37.

O'Kelly I, Stephens RH, Peers C, Kemp PJ (1999). Potential identification of the O_2-sensitive K^+ current in a human neuroepithelial body-derived cell line. *Am J Physiol* **276**, L96-L104.

Peers C (1990). Hypoxic suppression of K^+ currents in type I carotid body cells: selective effect on the Ca^{2+}-activated K^+ current. *Neurosci Lett* **119**, 253–256.

Plant LD, Kemp PJ, Peers C, Henderson Z and Pearson HA (2002). Hypoxic depolarization of cerebellar granule neurons by specific inhibition of TASK-1. *Stroke* **33**, 2324–2328.

Rajan S, Wischmeyer E, Karschin C, Preisig-Muller R, Grzeschik KH, Daut J, Karschin A and Derst C (2001). THIK-1 and THIK-2, a novel subfamily of tandem pore domain K^+ channels. *J Biol Chem* **276**, 7302-7311.

Thompson RJ and Nurse CA (1998). Anoxia differentially modulates multiple K^+ currents and depolarizes neonatal rat adrenal chromaffin cells. *J Physiol* **512**, 421–434.

Urena J, Fernandez-Chacon R, Benot AR, Alvarez de Toledo GA and Lopez-Barneo J (1994). Hypoxia induces voltage-dependent Ca^{2+} entry and quantal dopamine secretion in carotid body glomus cells. *Proc Natl Acad Sci USA* **91**, 10208–10211.

Wang ZZ, Stensaas LJ, Bredt DS, Dinger B and Fidone SJ (1994) Localization and actions of nitric oxide in the cat carotid body. *Neurosci* **60**, 275–286.

Wang ZZ, Stensaas LJ, Dinger BG and Fidone SJ (1995). Nitric oxide mediates chemoreceptor inhibition in the cat carotid body. *Neurosci* **65**, 217–229.

Wyatt CN and Buckler KJ (2004). The effect of mitochondrial inhibitors on membrane currents in isolated neonatal rat carotid body type I cells. *J Physiol* **556**, 175–191.

Youngson C, Nurse C, Yeger H and Cutz E (1993). Oxygen sensing in airway chemoreceptors. *Nature* **365**, 153–155.

Differential Expression of Oxygen Sensitivity in Voltage-Dependent K Channels in Inbred Strains of Mice

TOSHIKI OTSUBO, SHIGEKI YAMAGUCHI, MARIKO OKUMURA, MACHIKO SHIRAHATA

Department of Environmental Health Science, Johns Hopkins Bloomberg School of Public Health, Baltimore, USA

1. INTRODUCTION

Oxygen sensitivity of voltage-dependent K channels (Kv channels) in chemosensory glomus cells is responsible for hypoxic chemotransduction processes in the carotid body. Human studies in twins and in individuals over time suggest that hypoxic sensitivity of the carotid body is genetically controlled (Collins et al., 1978; Kawakami et al., 1982; Nishimura et al., 1991; Thomas et al., 1993). The concept is further confirmed in the studies using inbred strains of mice (Tankersley et al., 1994; Campen et al., 2004) and rats (Weil et al., 1998) which are genetically almost identical within a strain. In these studies, respiratory or cardiovascular responses to hypoxia vary among several strains, but are similar within a strain. Thus, some proteins which are differentially expressed in individuals due to genetic differences likely cause variable carotid body responses. We have hypothesized that differential expression of oxygen-sensitive Kv channels contributes to the differences in hypoxic sensitivity of DBA/2J and A/J strains of mice.

2. MATERIALS AND METHODS

2.1 Patch Clamp Experiments

Male DBA/2J and A/J mice (3-6 weeks old) were used. They were deeply anesthetized with 50 mg/kg ketamine and 100 mg/kg sodium pentobarbital (i.p.). After the heart was removed to avoid bleeding in the neck, the carotid body was harvested together with the carotid bifurcation, cleaned, and then placed in a recording chamber which was attached to a stage of an upright microscope (Axioskop 2, Zeiss). After the tissue was treated with 0.0375% collagenase (Type IX; Sigma) in Krebs solution (in mM : NaCl, 118; KCl, 4.7; $CaCl_2$, 1.8; KH_2PO_4, 1.2; $MgSO_4·7H_2O$, 1.2; $NaHCO_3$, 25; glucose, 11.1; EDTA, 0.0016; pH, 7.4 with 5% CO_2/air), the carotid body was visualized using a water immersion lens (ACHROPLAN, 40X) combined with an infrared differential interference video camera (DAGE-MTI Inc.). All experiments were performed approximately at 37.0 °C. A conventional tight-seal whole-cell recording was applied using the patch electrodes with resistance of 4~6MΩ (the internal solution in mM: K gluconate, 90; KCl, 33; NaCl, 10; $CaCl_2$, 1; EGTA, 10; MgATP, 5; HEPES, 10; pH 7.2 with KOH). Voltage-dependent whole cell current was evoked by a voltage clamp pulse (from 80 mV of holding potential to +20 mV) for 100 ms. The current was processed using an Axopatch 200B patch-clamp amplifier with the combination of Digidata 1320A, and pCLAMP 8.1 (Axon Instrument). Iberiotoxin (Alomon Labs) was included in Krebs solution when necessary. Hypoxia was applied by switching Krebs solutions from the one saturated with 5% CO_2/air to another saturated with 5% CO_2/0% O_2 for 5 minutes. PO_2 in the chamber was measured with a small oxygen electrode (MI-730 Oxygen Electrode, OM-4 Oxygen Meter; Microelectrodes, Inc) in separate experiments.

2.2 RT-PCR Analysis

The carotid bodies were harvested as described above. Total RNA of the carotid body was isolated in TRIzol, and genomic DNA was digested with RQI DNase I (Invitrogen). First strand cDNA was obtained using SuperScript III™ (Invitrogen), and multiplex PCR was performed following the manufacture's instruction (Quiagen). Previous studies have shown that Kv1.2, Kv4.3 and voltage-dependent Ca^{2+}-activatec K (BK) channels are inhibited by hypoxia. Hence, the primer pairs were designed for β-actin, Kv1.2α, Kv4.3α, BKα, BKβ2, and BKβ4 subunit (final concentrations: 1 μM for β-actin and 5 μM for other primers). The conditions for PCR amplification were as follows. The template was denatured at 95 °C for 15 min followed by 35 cycles of denaturation at 94 °C for 30 sec, annealing of primers at 60 °C for 1.5 min, extension at 72 °C for 1.5 min. Final cycle was 72 °C for 10 min. Positive tissue control was cDNA from the SCG. The negative reaction controls were total RNA and water as templates for PCR.

3. RESULTS

Hypoxic sensitivity of Kv current in glomus cells was tested by applying Krebs solution which was saturated with 5% CO_2/0% O_2. Changes in Kv current were recorded at 1 minute, 3 minutes, and 5 minutes after switching the Krebs solution (Fig. 1). Kv current was significantly reduced along with the decrease in PO_2 in the chamber in DBA/2J mice. The inhibitory effect was reversible. In A/J mice, Kv current in glomus cells was not significantly affected. To evaluate the contribution of BK channels to hypoxic sensitivity, the effect of hypoxia on Kv current was monitored with a presence of iberiotoxin (200 nM). Mild hypoxia did not significantly influence Kv current with a presence of iberiotoxin in either strain of mice (Fig. 2).

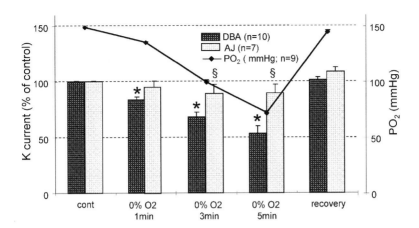

Figure 1. The effect of mild hypoxia on Kv current in glomus cells of DBA/2J and A/J mice. Kv current was almost lineally inhibited with decreasing PO_2 in glomus cells of DBA/2J mice, but not of A/J mice. *, significantly different from control (cont) and recovery (p<0.01). §, significantly different between DBA/2J and A/J mice. There is no statistical difference between control and recovery.

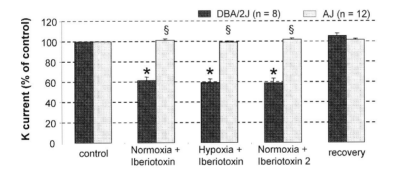

Figure 2. The effect of mild hypoxia on Kv current of glomus cells in DBA/2J and A/J mice with iberiotoxin. Initially the carotid body was superfused with Krebs equilibrated with 5% CO_2/air. With the presence of iberiotoxin K current was not further affected by mild hypoxia in either strain. The effects of iberiotoxin and hypoxia were reversible. *, significantly different from control and recovery (p<0.01). §, significantly different from DBA/2J mice. There is no statistical difference between control and recovery.

PCR products from whole carotid bodies showed clear bands for β-actin, Kv1.2α, and BKα subunit in both DBA/2J and A/J mice (Fig. 3). The band for Kv4.3α was very weak in both strains and not always observed. Clear differences between two strains were observed in BK subunits. BKα and β2 subunits were more expressed in the carotid body of DBA/2J mice than those of A/J mice. The expression of BKβ4 subunit mRNA in relative to BKβ2 subunit was low in the carotid body of DBA/2J mice, but high in that of A/J mice.

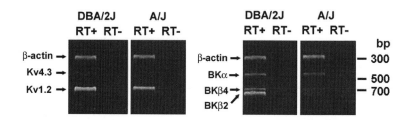

Figure 3. Agarose gel electrophoresis of mRNAs for Kv channel subunits in the carotid body of 4-week-old mice. Carotid bodies from five mice in each strain were separately harvested, RNA was extracted, reverse transcribed, and multiplex PCR was performed for β-actin, Kv1.2 α, Kv4.3α, BKα, BKβ2, and BKβ4 subunits. For negative reaction controls, total RNA (RT-) and water (data not shown; no bands) was used as templates in PCR. Similar data were obtained in other four mice in each strain.

4. DISCUSSION

Major findings of this study are threefold. First, Kv channels in glomus cells of DBA/2J mice are more sensitive to hypoxia than those of A/J mice. Second, BK channels, which are sensitive to iberiotoxin, are major O_2-sensitive Kv channels in glomus cells of DBA/2J mice. Third, differential expression of BK channel subunits was seen between the two strains and this difference may result in different hypoxic sensitivity of glomus cells in these mice.

The different sensitivity of the carotid body to hypoxia between DBA/2J and A/J mice could be, at least in part, based on the morphological differences in the carotid bodies. That is, the carotid body of DBA/2J mice is larger and contains more glomus cells than that of A/J mice (Yamaguchi et al., 2003). However, the current study has clearly shown that the function of glomus cells differs between these strains of mice. A decrease in O_2 tension significantly reduced Kv current in glomus cells of DBA/2J mice, but not in those of A/J mice.

A type of O_2-sensitive Kv channels in glomus cells differs among the rabbit, the rat, and the cat (Shirahata and Sham, 1999). Molecular biological techniques have been extensively applied to rabbit glomus cells, and Kv channels including Kv4.1α and Kv4.3α subunits are suggested to be O_2-sensitive components (Sanchez et al., 2002). These channels are fast inactivating channels. However, in glomus cells of DBA/2J and A/J mice, a major part of Kv current was not fast-inactivating current (data not shown). Further, the expression of mRNA for Kv4.3α subunits was very low (Fig. 3). Hence, it is unlikely that the Kv4.3 channel is a major O_2-sensitive Kv channel in these mice. With application of iberiotoxin, hypoxia did not further influence Kv current. The data suggest that O_2-sensitive Kv channels in glomus cells of DBA/2J mice are most likely BK channels. A question remains whether Kv1.2 channels in these mice are sensitive to hypoxia. mRNA for Kv1.2α subunits was strongly expressed in the carotid bodies of both strains. Hypoxic inhibition of Kv1.2 channels has been shown in PC12 cells (Conforti et al., 2000). Our current data clearly indicate that mild hypoxia mainly inhibits BK channels. However, we cannot dismiss a possibility that severe hypoxia inhibits Kv1.2 channels in these mice.

Pharmacological studies using iberiotoxin indicate that Kv channels in glomus cells of A/J mice are mostly insensitive to this blocker. However, RT-PCR analysis suggests that α subunits of BK channels exist in glomus cells in A/J mice as well. BK channels consist of the four α subunits, which form the pore of the channel, and the auxiliary β subunits (β1-4). Although α subunits alone can form functional channels when expressed in some cell lines, in native tissues they are likely associated with β subunits. Distribution of β subunits has been recently investigated, and β1 mRNA is high in smooth muscle; β2, in chromaffin cells and brain; β3, in testis, pancreas, and spleen; and β4, in brain (Gribkoff et al., 2001;Orio et al., 2002). Because the carotid body is originated from the neural crest, we focused on α, β2 and β4 mRNA levels in the carotid body of the two strains of mice. It appears that mRNAs for α and β2 subunits were significantly more expressed in the carotid body of DBA/2J than those of A/J mice (Fig. 3). Further, relative expression of β4 subunits to α and β2 subunits was higher in glomus cells of A/J mice, suggesting that β4 subunits contribute more as an associated protein in BK channels in glomus cells of A/J mice. When α subunits are expressed together with β4 subunits in HEK293 or CHO cells, I-V curve of BK current is shifted to the right by 50 mV. Further, neither iberiotoxin nor charybdotoxin inhibits the channel activity (Weiger et al., 2000b; Meera et al., 2000). It is likely that BK channels in glomus cells of A/J mice are insensitive to iberiotoxin, because of close association of α subunits with β4 subunits. The functional presence of oxygen-sensitive BK channels may result in differential hypoxic sensitivity of carotid bodies between DBA/2J and A/J mice.

ACKNOWLEDGEMENTS

This work was supported by AHA0255358N, NHLBI HL61596, and NHLBI HL72293.

REFERENCES

Campen M.J., Tagaito Y., Li J., Balbir A., Tankersley C.G., Smith P., Schwartz A., O'Donnell C.P., 2004. Phenotypic variation in cardiovascular responses to acute hypoxic and hypercapnic exposure in mice. *Physiol. Genomics* 20: 15-20.

Collins D.D., Scoggin C.H., Zwillich C.W., Weil J.V., 1978. Hereditary aspects of decreased hypoxic response. *J. Clin. Invest.* 62: 105-110.

Conforti L., Bodi I., Nisbet J.W., Millhorn D.E., 2000. O_2-sensitive K^+ channels: role of the Kv1.2 -subunit in mediating the hypoxic response. *J. Physiol.* 524: 783-793.

Gribkoff V.K., Starrett J.E., Dworetzky S.I., 2001. Maxi-K potassium channels: Form, function, and modulation of a class of endogenous regulators of intracellular calcium. *Neuroscientist* 7: 166-177.

Kawakami Y., Yoshikawa T., Shida A., Asanuma Y., Murao M., 1982. Control of breathing in young twins. *J. Appl. Physiol.* 52: 537-542.

Meera P., Wallner M., Toro L., 2000. A neuronal beta subunit (KCNMB4) makes the large conductance, voltage- and Ca^{2+}-activated K^+ channel resistant to charybdotoxin and iberiotoxin. *Proc. Natl. Acad. Sci. U S A* 97: 5562-5567.

Nishimura M., Yamamoto M., Yoshioka A., Akiyama Y., Kishi F., Kawakami Y., 1991. Longitudinal analyses of respiratory chemosensitivity in normal subjects. *Am. Rev. Respir. Dis.* 143: 1278-1281.

Orio P., Rojas P., Ferreira G., Latorre R., 2002. New disguises for an old channel: MaxiK channel beta-subunits. *News Physiol. Sci.* 17: 156-161.

Rubin A.E., Polotsky V.Y., Balbir A., Krishnan J.A., Schwartz A.R., Smith P.L., Fitzgerald R.S., Tankersley C.G., Shirahata M., O'Donnell C.P., 2003. Differences in sleep-induced hypoxia between A/J and DBA/2J mouse strains. *Am. J. Resp. Crit. Care Med.* 168: 1520-1527.

Sanchez D., Lopez-Lopez J.R., Perez-Garcia M.T., Sanz-Alfayate G., Obeso A, Ganfornina MD, Gonzalez C, 2002. Molecular identification of Kvalpha subunits that contribute to the oxygen-sensitive K^+ current of chemoreceptor cells of the rabbit carotid body. *J Physiol.* 542: 369-382.

Shirahata M., Sham J.S., 1999. Roles of ion channels in carotid body chemotransmission of acute hypoxia. Jpn. J. Physiol. 49: 213-228.

Tankersley C.G., Fitzgerald R.S., Kleeberger S.R. (1994) Differential control of ventilation among inbred strains of mice. *Am. J. Physiol. 267: R1371-7.*

Thomas D.A., Swaminathan S., Beardsmore C.S., McArdle E.K., MacFadyen U.M., Goodenough PC, Carpenter R, Simpson H (1993) Comparison of peripheral chemoreceptor responses in monozygotic and dizygotic twin infants. Am. Rev. Respir. Dis. 148: 1605-1609.

Weiger T.M., Holmqvist M.H., Levitan I.B., Clark F.T., Sprague S., Huang W.J., Ge P., Wang C., Lawson D., Jurman M.E., Glucksmann M.A., Silos-Santiago I, DiStefano PS, Curtis R, 2000. A novel nervous system beta subunit that downregulates human large conductance calcium-dependent potassium channels. *J. Neurosci.* 20: 3563-3570.

Weil J.V., Stevens T., Pickett C.K., Tatsumi K., Dickinson M.G., Jacoby C.R., Rodman D.M 1998. Strain-associated differences in hypoxic chemosensitivity of the carotid body in rats. *Am. J. Physiol.* 274: L767-L774.

Yamaguchi S., Balbir A., Schofield B., Coram J., Tankersley C.G., Fitzgerald R.S., O'Donnell CP, Shirahata M. 2003, Structural and functional differences of the carotid body between DBA/2J and A/J strains of mice. *J. Appl. Physiol.* 94: 1536-1542.

An Overview on the Homeostasis of Ca^{2+} in Chemoreceptor Cells of the Rabbit and Rat Carotid Bodies

S. V. CONDE, A.I. CACERES, I. VICARIO, A. ROCHER, A. OBESO AND C. GONZALEZ

Departamento de Bioquimica y Biología Molecular y Fisiología/IBGM Facultad de Medicina. Universidad de Valladolid/CSIC. 47005Valladolid. Spain

1. INTRODUCTION

Carotid body (CB) chemoreceptors sense arterial PO_2 and PCO_2/pH becoming activated in hypoxic hypoxia and in all types of acidosis. The sensing structures in the CB are chemoreceptor cells (CBCC), which are connected synaptically with the sensory nerve endings of the carotid sinus nerve (CSN). In situations of hypoxia and acidosis, CBCC are activated and their rate of release of neurotransmitters (NT) increase, promoting an increase in the activity of the CSN and subsequent ventilatory and cardiovascular reflexes (5).

The release of NT induced by natural stimuli is dependent on the presence of Ca^{2+} in the extracellular milieu (4, 10). Ca^{2+} enters the cells via voltage operated-Ca^{2+} channels (VOCC; 10, 11) and triggers the exocytotic release of NT. However, there are several aspects of the homeostasis of Ca^{2+} by CBCC during stimulation and the recovery of basal Ca^{2+} levels in the post stimulus period that have nor been studied.

Figure 1 summarizes the main aspects of Ca^{2+} metabolism in excitable mammalian cells. Labeled as 1 are represented the main pathways for Ca^{2+} entry, represented by VOCC. In rabbit CBCC it is known that basically all members of the VOCC, except for the T-channels, are expressed but only the L and P/Q subtypes participate in the exocytotic responses (11); in the rat it is known that L-type are expressed and support around 70% of increase in $[Ca^{2+}]i$ elicited by hypoxia (1). Labeled 2 are represented the ryanodine- and IP_3-receptors/channels located in the endoplasmic reticulum, which represents the main Ca^{2+}-storing organelle in most cells. In rabbit CBCC the apparent capacity of this store is very limited as activation or inhibition of the ryanodine/IP_3 receptors produced very small alterations in the $[Ca^{2+}]i$ and did not alter the exocytotic responses (14). In rat CBCC there are no data regarding the potential contribution of the Ca^{2+} stores to the exocytosis. Label as 3 are the endoplasmic reticulum ATPases that load the stores with Ca^{2+}. Rabbit CBCC inhibition of this pump did not alter the $[Ca^{2+}]i$ nor the exocytotic responses (14), but there are not comparable data for the rat CBCC. Label as 4 and 5 are the plasma membrane Ca^{2+}-ATPase and the Na^+/Ca^{2+} exchanger, which represent the two unique pathways of efflux of Ca^{2+} of the cells against its electrochemical gradient; they are responsible for returning the $[Ca^{2+}]i$ levels to basal in the post stimulus periods. Label as 6 are other potential intracellular Ca^{2+} stores that have not been explored in the CBCC. Labeled with a 7 are mitochondria, which usually contribute to buffer Ca^{2+} in the stimulus and immediate post stimulus periods; later they dispose Ca^{2+} to cell cytoplasm slowly so that the $[Ca^{2+}]i$ does not reach levels high enough to trigger Ca^{2+}-dependent responses (8). Finally, label 8 is a family of Ca^{2+} channels (cationic channels), which represent additional pathways for Ca^{2+} entry and include, among others, store- and receptor-operated Ca^{2+} channels (13). There are not data on these channels in CBCC.

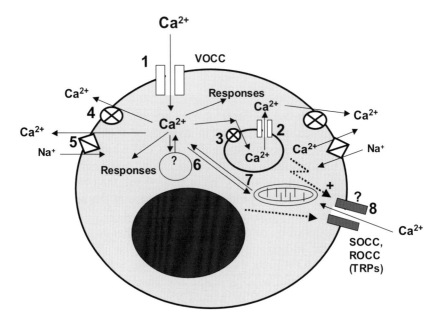

Figure 1. Schema showing the main pathways involved in the homeostasis of Ca^{2+} in mammalian excitable cells.

In the present study we have performed experiments to characterize the significance of mechanisms 1, 2, 3, 4 and 5 in the homeostasis of Ca^{2+} assessed by their participation in the exocytotic release of catecholamines (CA), which are the most abundant NT in CBCC.

2. METHODS

Surgical procedures. The experiments were performed in intact CB of adult New Zealand rabbits and adult Wistar rats anaesthetized with sodium pentobarbital, 40 (i.v.) and 60 (i.p.) mg/Kg, respectively. After tracheostomy, a block of tissue containing the carotid artery bifurcation was removed and placed in a lucite chamber filled with ice-cold 100% O_2 Hepes-buffered Tyrode and the CBs were cleaned of surrounding tissues under a dissecting microscope.

Measurement of the release of CA. To label the CA stores, the CBs (6-8/experiment) were incubated during 2 h in small vials containing 0.5ml of 100% O_2 Hepes-buffered Tyrode (in mM: NaCl, 140; KCl, 5; $CaCl_2$, 2; $MgCl_2$, 1.1; glucose, 5.5; HEPES, 10) and placed in a metabolic shaker at 37 °C. The incubating solution contained ^{3}H-tyrosine (20-30 µM) with high specific activity (30 and 50 Ci/mmol for rabbit and rat CBs, respectively), 100 µM 6-methyl-tetrahydropterine and 1 mM ascorbic acid as cofactors for tyrosine hydroxylase and dopamine-β-hydroxylase, respectively. At the end of the labeling period individual CB were transferred to a glass vial containing 4 ml of precursor-free Tyrode bicarbonate solution (24 mM NaCl was substituted by 24 mM $NaHCO_3$), and kept at 37 °C for the rest of the experiment. Solutions were continuously bubbled with a gas mixture saturated with water vapor of composition 20% O_2/5% CO_2/75% N_2, except when hypoxia was applied (see Results). When high external K^+-containing solutions were used as stimulus, equimolar amounts of Na^+ were removed. The incubating solutions were renewed every 20 min. for 2 hours (rabbit) and 1 h (rat) and discarded. Thereafter, incubating solutions were collected and saved for analyses in ^{3}H-CA content. Specific protocols for stimulus and drug application and for incubating solutions are shown in the Results.

Analytical procedures. The analysis of ^{3}H-catechols present in the collected solutions included: adsorption to alumina (100 mg) at alkaline pH (obtained by the addition under shaking of 5 ml of 2.5 M TRIS-buffer, pH=8.6), extensive washing of the alumina with distilled water, bulk elution of all ^{3}H-catechols with 1 ml of 1 N HCl and liquid scintillation counting of the eluates. No further analysis of the radioactivity present in the eluates was performed because previous studies from our laboratory have shown that most of the tritium in the eluates corresponds to ^{3}H-DA plus is catabolite ^{3}H-dihydroxy phenyl acetic acid. (14, 15). At the end of the experiments, the CBs were transferred to cold eppendorf tubes containing 200 µl of 0.4 N perchloric acid. Thereafter the tissues were weighed in an electrobalance (Supermicro, Sartorius), homogenized at 0-4 °C and the homogenates analyzed for their ^{3}H-CA content.

3. RESULTS

Figure 2 shows experiments directed to explore the identity of VOCC supporting the exocytotic release of ^{3}H-CA. Part A (left) shows single experiments in the rabbit CB in which the release of ^{3}H-CA induced by hypoxia (7% O_2-equilibtaed solutions) in control conditions, in the presence 2 µM nisoldipine (a blocker of L-type Ca^{2+} channels) and in the presence of 2 µM nisoldipine + 3 µM MVIIC (a toxin blocker of N and P/Q Ca^{2+} channels).

Similar experiments were conducted for 30 mM K^+ as stimulus. In the right panel of part A are shown mean results of 6-10 individual values.

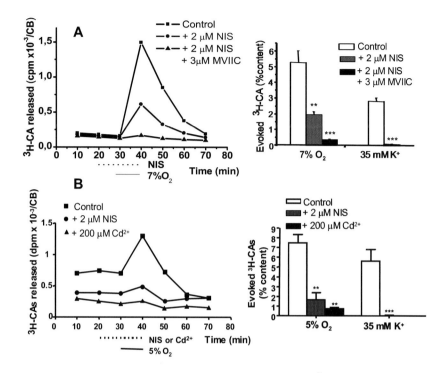

Figure 2. Effects of different blockers of VOCC on the release of ^{3}H-CA induced by hypoxia and high external K^+ (A, rabbit CB; B; rat CB).

Note that nisoldipine inhibited the release by about 60% and in combination with MVIIC fully inhibited the release. In additional experiments (not-shown) GVIA (a toxin blocker of N-type channels) did not affect the release. Figure 2 part B shows alike experiments carried out with rat CB and the results are comparable. L-type channels supported around 75% of the release response elicited by hypoxic stimuli, but it remains to be identified the subtypes of Ca^{2+} responsible for the 25% of the release that being in part sensitive to Cd^{2+} is not blocked by nisoldipine. The release induced by high K^+ was fully sensitive to the blocker of L-type channels in both species.

Figure 3 shows the effects of inhibition of the Na^+/Ca^{2+} exchanger on the release response elicited by high external K^+ in the rabbit (left) and rat (right) CBs. Inhibition of the Na^+/Ca^{2+} exchanger was achieved by removing Na^+ from the incubation solution. The Figure shows mean results of the time course of 6 experiments. Note that in both species the short period of incubation in Na^+-free solutions did not alter basal release of ^{3}H-CA, but in the Na^+-free solutions the release induced by high external K^+ was dramatically potentiated to nearly double the release induced by the same $[K^+]$ in normal Na^+-containing solutions.

An Overview on the Homeostasis of Ca^{2+}

Attempts to block the plasma membrane Ca^{2+}-ATPase with 300 µM La^{3+} in CBs incubated in Na^+-free solutions did not produce further enhancement of the release response elicited by high external K^+ in neither species (Figure 4).

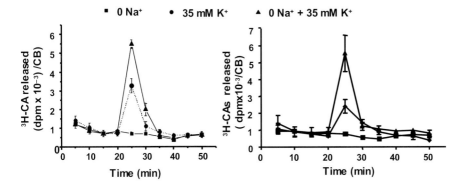

Figure 3. Effects of inhibition of the Na^+/Ca^{2+} exchanger by removal of extracellular Na^+ on the release of ^3H-CA induced by high external K^+.

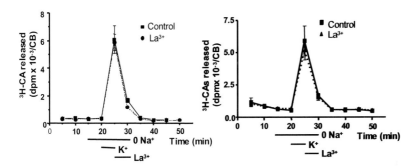

Figure 4. Effects of inhibition of the Na^+/Ca^{2+} exchanger and subsequently of the plasma membrane Ca^{2+}-ATPase on the release of ^3H-CA induced by high external K^+. Note that application of La^{3+} was immediately after K^+ stimulation because in addition to the Ca^{2+} pump it also inhibits VOCC.

Manipulation of the intracellular Ca^{2+} stores with ryanodine at concentrations that activate the ryanodine receptors, or the IP_3 receptors by incubating with muscarinic IP_3-generating agonists, and the endoplasmic reticulum Ca^{2+}-ATPases by incubating with thapsigargine did not produce changes in the release of ^3H-CA elicited by high external K^+ in either species (Table I). Only terbutyl-hydroquinone, that in addition of a blocker of the reticulum ATPase, is also a powerful inhibitor of VOCC (9), altered the evoked release response.

4. DISCUSSION

The data related to the role of VOCC and to drugs affecting intracellular Ca^{2+}-storing organelles in the rabbit CB were already known (11, 14), but the rest of the data in the rabbit CB and all data related to rat CB are reported here by the first time. However, we have wanted to present the data for both species side to side to emphasize that there are minimal differences in both species regarding the homeostasis of Ca^{2+} in CBCC, evaluated by the Ca^{2+}-dependent exocytotic release of ^{3}H-CA, in spite of marked differences between them in the density of Na^+ and Ca^{2+} currents and in the nature of K^+ currents (6, 7), and marked differences in the turnover rates of their CA content (3, 15).

Table 1. Effects of different agents capable of altering the intracellular Ca^{2+} stores on the basal and high external K^+-induced release of ^{3}H-CA.

Experimental maneuver	Rabbit CB		Rat CB	
	Basal release	High K^+-induced release	Basal release	High K^+-induced release
Ryanodine (0.5-2 µM)	No effect	No effect	No effect	No effect
Bethanechol (50-300 µM)	No effect	No effect	No effect	No effect
Thapsigargine (0.1-1 µM)	No effect	No effect	No effect	No effect
T-B-hydroquinone (10 µM)	No effect	≈ –50%	No effect	≈ –80%

In both species the release of ^{3}H-CA by chemoreceptor cells in response to hypoxic and high external K^+ stimuli is reduced by around 95-98% in nominally Ca^{2+}-free solutions (10, 15). In both species, Ca^{2+} enters during hypoxic and high K^+ stimulation via VOCC, mainly trough L-type channels, although other types of VOCC also provide Ca^{2+} for the exocytotic release of NT (Figure 2). The present data using the release of NT as an index of Ca^{2+} dynamics in rat CB are nearly identical to the findings of Buckler and Vaughan Jones (1) showing that about 70% of the increase in $[Ca^{2+}]i$ was suppressed by blocker of L-type Ca^{2+} channels. In the rat we have not yet characterized the VOCC subtypes providing the remaining Ca^{2+} supporting the dihydropyrine unsensitive secretion, but in the rabbit CBCC, P/Q channels are complementary to L-channels (Figure 2), being their contribution greater as the intensity of the hypoxic stimulation increases (11).

Ca^{2+} that has entered the cells during stimulation must efflux in the post stimulus period to restore the cells to resting state. Our data of Figures 3 and 4 are the first ones addressing the pathways for Ca^{2+} efflux in CBCC and evidence the prime importance of Na^+/Ca^{2+} exchanger. The significance of plasma membrane Ca^{2+} pump seems to be small. With the same protocol used in our experiments, Sasaki et al. (12) have evidenced a very important role for the plasma membrane Ca^{2+} pump in neurohypophysial terminals, where La^{3+} nearly doubled the secretory response, but has not effect in our preparation. This finding suggest that the role of the plasma membrane Ca^{2+} pump in CBCC is small, or alternatively, that La^{3+} is not efficacious to block the isoform of the plasma membrane Ca^{2+} pump expressed in rat and rabbit CBCC.

The negligible significance of intracellular Ca^{2+} stores in the homeostasis of Ca^{2+} was evidenced in rabbit CBCC by Vicario et al. (14) using the tools shown in Table I, and additional ones. In rat CBCC, we expected a greater role for the intracellular stores owed to the smaller amplitude of VOCC (7), but results indicate that also in this species the intracellular Ca^{2+} stores have a small capacity, or alternatively, that the Ca^{2+} they can provide or buffer is not coupled to the exocytotic machinery. In this regard is worth mentioning that Dasso et al. (2) showed that muscarinic agonists produced a significant increase in $[Ca^{2+}]i$ in a subpopulation of rat CBCC, yet the same agonists did not modify the exocytosis of ^{3}H-CA either in the rat or in the rabbit CBCC.

As depicted in Figure 1, to complete the picture of the homeostasis of Ca^{2+} in CBCC it remains to be explored the potential role of mitochondria, of unidentified intracellular Ca^{2+} stores and of cationic channels of the TRP family in different experimental situations.

ACKNOWLEDGEMENTS

Supported by Spanish DGICYT grant BFU2004-06394 and by grants from the ICiii/Fiss, Red Respira-Separ and PI042462.

REFERENCES

1. Buckler KJ, Vaughan-Jones RD. Effects of hypoxia on membrane potential and intracellular calcium in rat neonatal carotid body type I cells. *J Physiol* 476: 423-8, 1994.
2. Dasso LL, Buckler KJ and Vaughan-Jones RD. Muscarinic and nicotinic receptors raise intracellular Ca^{2+} levels in rat carotid body type I cells. *J Physiol*. 498:327-38, 1997.
3. Fidone S and Gonzalez C. Catecholamine synthesis in rabbit carotid body in vitro. *J Physiol* 333: 69-79, 1982.
4. Fidone S, Gonzalez C and Yoshizaki K. Effects of low oxygen on the release of dopamine from the rabbit carotid body in vitro. *J Physiol* 333: 93-110, 1982.
5. Gonzalez C, Almaraz L, Obeso A and Rigual R. Carotid body chemoreceptors: From natural stimuli to sensory discharges. *Physiol Rev* 74: 829-898, 1994.
6. Gonzalez C, Rocher A and Zapata P. Quimiorreceptores arteriales: mecanismos celulares y moleculares de las funciones adaptativa y homeostática del cuerpo carotídeo. *Rev. Neurol.* 36: 239-254, 2003.
7. Lopez-Lopez J.R and Peers, C. Electrical properties of chemoreceptor cells, In: *The carotid body Chemoreceptors* (Ed. C. González) Springer, NY. pp 65-77, 1997.
8. Montero M, Alonso MT, Albillos A, Garcia-Sancho J and Alvarez J. Mitochondrial Ca^{2+}-induced Ca^{2+} release mediated by the Ca(2+) uniporter. *Mol Biol Cell.* 12:63-71, 2001.
9. Nelson EJ, Li CC, Bangalore R, Benson T, Kass RS and PM Hinkle. Inhibition of L-type calcium-channel activity by thapsigargin and 2,5-t-butylhydroquinone, but not by cyclopiazonic acid. *Biochem J.* 302:147-54, 1994.
10. Obeso A, Rocher A, Fidone S and Gonzalez C. The role of dihydropyridine-sensitive Ca^{2+} channels in stimulus-evoked catecholamine release from chemoreceptor cells of the carotid body. *Neuroscience* 47: 463-472, 1992.
11. Rocher A, Geijo-Barrientos E, Caceres AI, Rigual R, Gonzalez and Almaraz L. *Role of voltage-dependent calcium channels in stimulus-secretion coupling in rabbit carotid body chemoreceptor cells.* J Physiol. 562:407-20, 2005.

12. Sasaki N, Dayanithi G and Shibuya I. Ca^{2+} clearance mechanisms in neurohypophysial terminals of the rat. *Cell Calcium.* 37:45-56, 2005.
13. Vazquez G, Wedel BJ, Aziz O, Trebak M and Putney JW Jr. The mammalian TRPC cation channels. Biochim Biophys Acta. 1742:21-36, 2004.
14. Vicario I, Obeso A, Rocher A, Lopez-Lopez JR and Gonzalez C. Intracellular Ca^{2+} stores in chemoreceptor cells of the rabbit carotid body: significance for chemoreception. Am J Physiol Cell Physiol. 279:C51-61, 2000a.
15. Vicario I, Rigual R, Obeso A and Gonzalez C. Characterization of the synthesis and release of catecholamine in the rat carotid body *in vitro*. *Am J Physiol Cell Physiol.* 278: C490-C499, 2000b.

Midbrain Neurotransmitters in Acute Hypoxic Ventilatory Response

HOMAYOUN KAZEMI

Pulmonary and Critical Care Unit, Massachusetts General Hospital, Harvard Medical School 55 Fruit Street, BUL 148, Boston, MA 02114 USA

1. INTRODUCTION

In control ventilation, chemical stimuli are paramount in setting the level of ventilation. These are primarily changes in concentration of hydrogen ions, as well as changes in PO2 and PCO2. A number of receptors, both in the periphery and in the central nervous system, respond to these changes. Of the three so-called "chemical" stimuli, the response to CO2 is most prominent and for any 1 mm change in PCO2, ventilation changes by about 2 to 2.5 L/min. The ventilatory response to hypoxia becomes quite prominent once arterial PO2 has reached values of about 60 mm Hg. It has been well documented that the primary effect of hypoxia is stimulation of the carotid chemoreceptors with transmission of signal to the NTS. With acute hypoxia, there is a biphasic ventilatory response with an initial hyperventilation followed by a fall in ventilation, the so-called "roll-off", to values above those in the pre-hypoxic level. This biphasic response is present in man as well as in a large number of other mammals tested and central neurotransmitters are essential in this response (4,5,7-10). This presentation will concentrate on the effects of acute hypoxia on the ventilatory response in anesthetized dog and rat, and the relationship between the ventilatory response and the release of neurotransmitters in the central nervous system, but in particular, in the medial chemosensitive area on the ventral surface of the medulla and summarizes the work from our laboratory from the past decade. The amino acids of interest are those that excite ventilation, which are primarily glutamate and aspartate; and those that depress ventilation, which include GABA, taurine, and glycine (3). Earlier work from this laboratory showed that inhibition of glutamate by intravenous administration of the specific NMDA receptor antagonist MK801 in anesthetized dog leads to a significant reduction in the hyperventilatory response to hypoxia (1). The subsequent studies in anesthetized rat showed that ventricular cisternal perfusion of MK801 abolished hyperventilatory response of acute hypoxia and infusion of GABA antagonist bicuculline caused augmentation of the hyperventilatory response to acute hypoxia and the "roll-off" was no longer observed (10).

These observations led to a series of experiments in our laboratory with microinjections of both agonists and antagonists at specific sites on the ventral surface of the medulla to identify areas of interest for subsequent microdialysis during acute hypoxia (2).

2. METHODS

The experimental model utilized in our studies was the anesthetized rat, mechanically ventilated to maintain normal arterial pH as well as normal $PaCO_2$. They underwent bilateral vagotomy in the neck and subsequently microdialysis catheters were placed 1.5 mm below the ventral surface of the medulla in the intermediate chemosensitive area. The microdialysis with artificial SCF was continuous during experiments, and samples collected during 20 minutes of normoxia every 1-3 minutes. Subsequently, the animals were exposed to hypoxia (10% oxygen), and microdialysis continued during 20 minutes of hypoxia and 10 minutes of recovery on room air. Continuous phrenic nerve recordings were made during the entire experimental procedure, and the data from the phrenic nerve tracings were divided into frequency and amplitude burst. In another group of rats the same experimental procedure was carried out after bilateral chemodenervation of the carotid bodies at the bifurcation of the carotid arteries.

More recently we have conducted a limited number of experiments with placement of the dialysate catheters in the caudal region of the NTS.

3. RESULTS

During acute hypoxia, the classic biphasic response in phrenic nerve output was observed with a rise in output during the first three minutes of hypoxia and a subsequent reduction over the next 5-7 minutes to a new steady stage for the remainder of the hypoxic period. The biphasic response was seen in both frequency and amplitude burst of the phrenic nerve discharge. After bilateral peripheral chemodenervation, there was no rise in phrenic output during hypoxia and instead there was a progressive fall in output during the period of hypoxia. There was a rise in concentration of glutamate from the dialysate during acute hypoxia in the intact animals, but not in those who had undergone bilateral chemodenervation. GABA increased in the dialysate after the first three minutes of hypoxia, and was as prominent in chemodenervated animals as intact animals.

Taurine showed a biphasic response, initially falling and subsequently increasing after about 4 minutes of hypoxia. There were no significant changes in aspartate or glycine in any of these experiments.

4. DISCUSSION

These results demonstrate that simulation of peripheral chemoreception by hypoxia leads to release of glutamate in the midbrain, which causes the observed hyperventilatory response. In chemodenervated animal there was no rise in midbrain glutamate and no increased phrenic output during acute hypoxia.

The rise in GABA, a ventilatory depressant, occurred in the midbrain during hypoxia with or without chemodenervation, suggesting the direct effect of hypoxia on the brain. In order to relate the changes in amino acids to the phrenic output during hypoxia in intact or chemodenervated animals we utilized a mathematical model. In this model we calculated the expected change in phrenic output for any change in the concentration of any single amino acid during the experimental procedures. Then, we plotted arithmetic sum of changes in ventilation expected for the changes in amino acid concentration, be it excitatory or inhibitory, and plotted the theoretical output of the phrenic nerve based on these calculations and compared that to the measured changes in phrenic nerve output at any given moment during hypoxia with or without chemodenervation. The measured values very closely matched the calculated values for phrenic output during the entire period of hypoxia in both intact and chemodenervated animals (2, 6). In summary, stimulation of peripheral chemoreceptors leads to the release of glutamate centrally and the effect of hypoxia on the brain itself leads to change in concentration of amino acids in respiratory related nuclei, with release of GABA, and biphasic change in concentration of taurine. Level of ventilation at any instant depends on interaction between amino acid neurotransmitters in the midbrain, producing the final integrated ventilatory response.

ACKNOWLEDGEMENTS

The studies reported here have been supported by grants from the NIH and the Shoolman Fund of the Pulmonary Unit. I am grateful to my many collaborating colleagues in conduct of these studies and to Lynn Wilcott for her help with this manuscript.

REFERENCES

1. Ang RC, Hoop B, Kazemi H. Role of glutamate as the central transmitter in the hypoxic ventilatory response. J Appl Physiol 1992; 72:1480-1487.
2. Beagle JL, Hoop B, Kazemi H. Phrenic Nerve response to glutamate antagonist microinjection in the ventral medulla. In: Hughson RL, Cunningham DA, Duffin J (Eds), Advances in Modeling and Control of Ventilation. Plenum Press, New York 1998:61-65.
3. Burton MD, Kazemi H. Neurotransmitters in central respiratory control. Respiration Physiology, 2000: 122(2-3): 111-122.
4. Easton PA, Slykerman LJ, Anthonisen NR. Ventilatory response to sustained hypoxia in normal adults. J Appl Physiol 1999; 61:906-911.
5. Gozal D, Gozal E, Torres JE, Gozal YM, Nuckton TJ, Hornby PH. Nitric oxide modulates ventilatory responses to hypoxia in the developing rat. Am J Respir Crit Care Med 1997; 155:1755-1762.
6. Hoop B, Beagle JL, Maher TJ, Kazemi H. Brainstem amino acid neurotransmitters and hypoxic ventilatory response. Respiratory Physiology 1999; 118:117-129.
7. Housley GD, Sinclair JD. Localization by kainic acid lesions of neurones transmitting the carotid chemoreceptor stimulus for respiration in rat. J Physiol London 1988; 106:99-114.

8. Kazemi H, Hoop B. Glutamic acid and γ-aminobutyric acid neurotransmitters in central control of breathing. J appl Physiol 1991; 70:1-7.
9. Powell FL, Milson WK, Mitchell GS. Time domains of the hypoxic ventilatory response. Respir Physiol 1998; 112:123-134.
10. Soto-Arape I, Burton M, Kazemi H. Central amino acid neurotransmittersand the hypoxic ventilatory response. Am J Respir Crit Care Med 1995; 151:1112-1120.

Chronic Intermittent Hypoxia Enhances Carotid Body Chemosensory Responses to Acute Hypoxia

RODRIGO ITURRIAGA, SERGIO REY, *JULIO ALCAYAGA, RODRIGO DEL RIO

*Oratorio de Neurobiología, Facultad de Ciencias Biologicas; Universidad Católica de Chile y *Laboratorio de Fisiologia Celular, Facultad de Ciencias, Universidad de Chile. Santiago, Chile*

1. INTRODUCTION

Chronic intermittent hypoxia (CIH), characterized by short episodes of hypoxia followed by normoxia, is a common feature of obstructive sleep apnea (OSA). It has been proposed that CIH enhances the hypoxic ventilatory response (HVR) leading to hypertension, upregulation of catecholaminergic and renin-angiotensin systems (Fletcher, 2000; Prabhakar and Peng, 2004). Most of the information of the effects of CIH on peripheral chemoreflex control of cardiovascular and respiratory systems has been obtained from studies performed on OSA patients. However, conclusions from these studies are conflictive because of comorbidities associated with OSA (Narkiewicz et al., 1999). Experiments performed in rats showed that CIH enhances HVR (Ling et al., 2001) and produces long-term facilitation of respiratory motor activity (McGuire et al., 2003; Peng & Prabhakar, 2003). The facilitator effect of CIH on HVR has been attributed to a potentiation of the carotid body (CB) chemosensory responses to acute hypoxia. However, it is a matter of debate if the ventilatory potentiation induced by CIH is due to a CB enhanced activity or secondary to central facilitation of chemosensory input. Peng et al., (2001) found that basal CB discharges and chemosensory responses to acute hypoxia were enhanced in rats exposed to a pattern of 5% O_2 for 15s followed by normoxia for 5 min, repeated 8 hours/day for 10 days. However, this observation has not been confirmed in other animal models of CIH. Using a protocol of short hypoxic episodes, we studied the effects of CIH on cat cardiorespiratory reflexes and CB chemosensory responses induced by hypoxia.

2. METHODS

Male cats were placed in a chamber (35.2 l volume) flushed alternatively with 100% N_2 (13 l/min for 90s) or compressed air (20 l/min for 270s). This gas alternation produced a see-saw pattern of PO_2, which dropped to a minimum of~ 75 Torr in 100s and then returned to near 150 Torr, reducing the PO_2 below 100 Torr for 60s. This pattern was repeated 10 times/hour during 8 hours for 4 days. Acute experiments were performed the morning of the day after the end of hypoxic exposure. As a control group, male cats were kept in normoxic conditions. Cats were anaesthetized with pentobarbital (40 mg/Kg, I.P.) followed by additional doses (8-12 mg, I.V.) to maintain a level of surgical anaesthesia. End tidal PO_2 ($P_{ET}CO_2$) was measured, and minute inspiratory volume (V_I), tidal volume (V_T) and respiratory frequency (f_R) were obtained from the airflow signal. Heart rate was obtained from the ECG and mean arterial pressure (BP) from the arterial pressure recording. Physiological variables were recorded with an analog-digital Power Lab system connected to a computer. Under anesthesia and prior to any maneuver the ECG was recorded for 5 min at 2 KHz. The heart rate variability was analyzed with software produced by The Biomedical Signal Analysis Group, University of Kuopio Finland (Tarvainen et al., 2001). To assess the effects of CIH on peripheral chemoreceptor reflexes, we studied the ventilatory and cardiovascular responses to several isocapnic levels of PO_2 maintained during 1 min. After finishing the study of cardiorespiratory reflexes, the CB chemosensory discharge was recorded as previously described (Iturriaga et al., 1998). One carotid sinus nerve (CSN) and the ipsilateral ganglioglomerular nerves were dissected and cut. The CSN was placed on platinum electrodes and the neural signal was preamplified, amplified and fed to a spike-amplitude discriminator to select action potentials of given amplitude above the noise. Barosensory fibers were eliminated by crushing the common carotid artery wall between the carotid sinus and the CB. Chemosensory discharge was counted with a frequency meter and the frequency of chemosensory discharge (f_x) was expressed in Hz. f_x was measured at several isocapnic levels of PO_2 maintained for 1 min. Results are expressed as mean ± SEM. Statistical differences were analyzed by a two-way ANOVA and post hoc analyses with the Bonferroni test. The level of significance was $P < 0.05$.

3. RESULTS

Basal V_T, V_I, $P_{ET}CO_2$, BP, and f_H in CIH cats were not significantly different from those observed in control cats ($P > 0.05$) and were similar to those previously reported in cats anesthetized with pentobarbital (Iturriaga et al., 1998). Exposure of cats to CIH for 4 days enhanced V_I responses to acute hypoxia due to an increase in V_T over f_R in response to hypoxia. The effect of CIH on V_I measured in response to several PO_2 levels (f_{IO_2}= 21 to 0.3%) is shown in Fig. 1. The V_I was significantly higher in CIH cats as compared with control cats ($P<0.001$, two way ANOVA). The Bonferroni test showed that V_I was statistically higher at a PO_2 of about 20 Torr. Hypoxia (f_{IO_2}= 3%) increased BP and heart rate in both groups, but CIH did not enhance the responses with respect to the control group ($P > 0.05$, two-way ANOVA). However, the spectral analysis of heart rate variability showed that CIH modified the distribution of low (LF = 0.02-0.15 Hz) and high (HF = 0.15-0.60 Hz) components of heart rate variability without changing the peak frequencies.

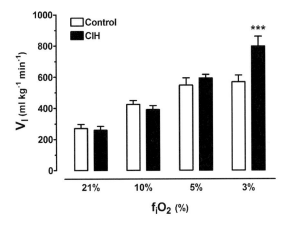

Figure 1. Effects of CIH on V_T induced by hypoxia in 4 cats compared with 6 control cats (***, P < 0.001, Bonferroni test after two way ANOVA).

Fig. 2 shows the distribution of the spectral components of heart rate variability normalized to the total power in one CIH and one control cat.

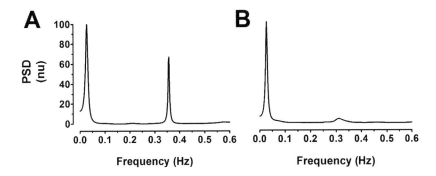

Figure 2. Effect of CIH on power spectral density of heart rate variability in one control (A) and one CIH cat (B). PSD, power spectral density in normalized units (n.u.).

In CIH cats, the power spectral density analysis showed a prevalence of low frequency over the high frequency components, while the opposite situation is observed in control animals. In fact, CIH cats showed a significantly higher LF/HF ratio of 2.4 ± 0.1 vs. 0.6 ± 0.1 in control cats ($P < 0.001$, n=6). The basal f_x measured at the beginning of the recordings was significantly higher ($P < 0.05$) in cats exposed to CIH (63.2 ± 7.3 Hz, n= 6) than basal f_x measured in control cats (36.3 ± 5.1 Hz, n = 6). In addition, chemosensory responses to acute hypoxia were enhanced in CIH cats as is shown in Fig. 3. The statistical analysis showed that the chemosensory responses for PO_2 were different in CIH cats ($P < 0.001$), while the Bonferroni test showed that the f_x was higher ($P < 0.001$) in the lowest hypoxic range in CIH cats.

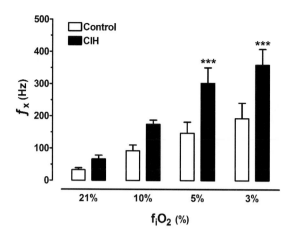

Figure 3. Effects of CIH on f_x induced by hypoxia in 4 cats compared with 6 control cats (***, P < 0.001, Bonferroni test after two way ANOVA).

4. DISCUSSION

We found that cats exposed to short hypoxic episodes ($PO_2 \sim 75$ Torr) followed by normoxia, during 8 hours for 4 days showed an enhanced ventilatory and CB chemosensory responses to acute hypoxia. Present results confirm and extend previous observations of Peng et al. (2001; 2003) who found that CIH enhances CB chemosensory response to acute hypoxia in rats. In addition, we found that CIH did not increase BP and heart rate responses to acute hypoxia, but the spectral analysis of basal heart rate variability shows an increase of the LF/HF ratio. Thus, our results support the idea that few days of CIH are enough to enhance the CB reactivity to hypoxia, which may contribute to the augmented ventilatory response, and to early alterations in the autonomic balance of heart rate variability.

Chronic intermittent hypoxia has been related with augmented HVR in OSA patients and animal models. Narkiewicz et al. (1999) found that ventilatory, sympathetic and cardiovascular reflex responses to acute hypoxia were enhanced in patients with recently diagnosed OSA. Studies performed in rats have shown different effects of CIH on HVR. Ling et al. (2001) found that phrenic HVR was enhanced in rats exposed to 5 min of 10% O_2, followed by 5 min of air, 12 hours/night during 7 nights. However, Greenberg et al. (1999) did not find any differences in HVR in rats exposed for 30s to 7% O_2 followed by 30s of normoxia, for 8 hours/day during 10 days. Thus, it is likely that effects of CIH on HVR are dependent on the pattern and severity of CIH. Our results show that CIH enhances cat CB chemosensory and ventilatory responses to hypoxia after few days. In addition, long-term CIH may lead to hypertension, which has been attributed to an increased sympathetic outflow due to repetitive CB stimulation (Greenberg et al., 1999; Fletcher, 2000). Fletcher et al. (1992) found an elevated BP in rats after 35 days of CIH exposure, but not after 10 days. On the contrary, Sica et al. (2000) and Peng & Prabhakar (2003) found that systolic BP increases after 7 days of CIH. Our results indicate that the time required for increasing BP and heart rate is longer than the time required to enhance HVR. Four days of intermittent hypoxia seem to be insufficient to increase basal BP and cardiovascular responses to hypoxia. However, the spectral analysis of the RR interval data suggests that autonomic modulation of heart rate variability is early modified by CIH. Indeed, the LF/HF ratio was significantly higher in CIH cats. The spectral analysis shows an effect of CIH that resembles what is observed in OSA patients. In fact, patients with OSA show an increased LF/HF ratio, with a relative predominance of the LF component (Narkiewicz et al., 1998a). The LF and HF components are related to autonomic influences on heart rate (Task Force, 1996). The HF band has been related to cardiac parasympathetic efferent activity, and the LF band is modulated by sympathetic output to the heart (Task Force, 1996).

ACKNOWLEDGEMENTS

We would like to thank Mrs. Carmen Gloria Leon and Paulina Arias for their assistance. Work supported by FONDECYT 1030330

REFERENCES

Fletcher EC (2000). Effect of episodic hypoxia on sympathetic activity and blood pressure. *Respir Physiol* **119**, 189-197.
Fletcher EC, Lesske J, Behm R, Miller CC, Stauss H, & Unger T (1992). Carotid chemoreceptors, systemic blood pressure, and chronic episodic hypoxia mimicking sleep apnoea. *J Appl Physiol* **72**, 1978-1984.
Greenberg HE, Sica A, Batson D & Scharf SM (1999). Chronic intermittent hypoxia increases sympathetic responsiveness to hypoxia and hypercapnia. *J Appl Physiol* **86**, 298-305.
Iturriaga R, Alcayaga J & Rey S (1998). Sodium nitroprusside blocks the cat carotid chemosensory inhibition induced by dopamine, but not that by hyperoxia. *Brain Res* **799**, 26-34.
Ling L, Fuller DD, Bach KB, Kinkead R, Olson EB & Mitchell GS (2001). Chronic intermittent hypoxia elicits serotonin-dependent plasticity in the central neural control of breathing. *J Neurosci* **21**, 5381-5388.
McGuire M, Zhang Y, White DP & Ling L (2003). Chronic intermittent hypoxia enhances ventilatory long-term facilitation in awake rats. *J Appl Physiol* **95**, 1499-1508.
Narkiewicz K, Montano N, Cogliati C, Van de Borne PJ, Dyken ME & Somers VK (1998). Altered cardiovascular variability in obstructive sleep apnea. *Circulation* **98**, 1071-1077.
Narkiewicz K, Van De Borne PJ, Pesek CA, Dyken ME, Montano N & Somers VK (1999). Selective potentiation of peripheral chemoreflex sensitivity in obstructive sleep apnea. *Circulation* **99**, 1183-1189.
Peng YJ, Kline D, Dick TE & Prabhakar NR (2001). Chronic intermittent hypoxia enhances carotid body chemoreceptor response to low oxygen. *Adv Exp Med Biol* **499**, 33-38.
Peng YJ & Prabhakar NR (2003). Reactive oxygen species in the plasticity of breathing elicited by chronic intermittent hypoxia. *J Appl Physiol* **94**, 2342-2349.
Prabhakar NR & Peng YJ (2004). Peripheral chemoreceptors in health and disease. *J Appl Physiol* **96**, 359 - 366.
Sica AL, Greenberg HE, Ruggiero DA & Scharf SM (2000). Chronic-intermittent hypoxia: a model of sympathetic activation in the rat. *Respir Physiol* **121**, 173-184.

Tarvainen MP, Ranta-Aho PO & Karjalainen PA (2002). An advanced detrending method with application to HRV analysis. *IEEE Trans Biomed Eng* **49**, 172-175.

Task Force of the European Society of Cardiology and the North American Society of Pacing and Electrophysiology (1996). Heart rate variability. Standards of measurement, physiological interpretation, and clinical use. *Eur Heart J* **17**, 354-381.

The Cell-Vessel Architecture Model for the Central Respiratory Chemoreceptor

MASUMASA OKADA[1], SHUN-ICHI KUWANA[2], YOSHITAKA OYAMADA[3] AND ZIBIN CHEN[4]

1.Department of Medicine, Keio University Tsukigase Rehabilitation Center, Izu City, Shizuoka 410-3215 Japan; 2.Department of Physiology, Teikyo University, Tokyo 173-8605 Japan; 3.Department of Pulmonary Medicine, School of Medicine, Keio University, Tokyo 160-8582 Japan; 4.Department of Biochemical and Analytical Pharmacology, GlaxoSmithKline, Research Triangle Park, NC 27709 USA

1. INTRODUCTION

The group of Loeschcke established that the superficial ventrolateral medulla contains chemosensitive regions (see the reviews [1-3]). Later, the group of Nattie conducted experiments of acetazolamide microinjection that induced local tissue acidosis, and found that the midline region (raphe), nucleus tractus solitarii and locus coeruleus are also chemosensitive (see the review [4]). To map the medullary chemosensitive regions using a more physiological stimulation technique, we microinjected CO_2-enriched saline into various regions of the in vivo and in vitro rat medulla, and found that the superficial midline, parapyramidal and ventrolateral regions are chemosensitive [5]. These findings extend other microinjection studies [6,7] that showed only that the superficial ventrolateral medulla is chemosensitive.

Fukuda et al. [8] used a medullary slice preparation to analyze neuronal responses to pH and found chemosensitive neurons in the rat superficial ventrolateral medulla. Since then, there have been several reports on in vitro analyses of neuronal chemoresponsiveness, and it has been accepted that chemosensitive neurons are located in widespread brainstem sites including the nucleus tractus solitarii, ventrolateral medullary regions, raphe and locus coeruleus [9-13]. Sato et al. [14] first applied c-fos immunohistology to identify chemosensitive cells in the medulla and succeeded in mapping the chemosensitive cell distribution in the surface layer of the ventrolateral medulla. This technique also has been applied by other investigators [15-18], and it has been demonstrated that chemosensitive cells are located in widespread brainstem sites including the ventrolateral medulla, parapyramidal region, raphe, locus coeruleus and nucleus tractus solitarii. The technique of c-fos immunohistology has enabled us to analyze the anatomical structure of the central respiratory chemoreceptor [18], and we propose that chemosensitive cells in the thickened marginal glial layer are primary chemoreceptor cells. These cells are small and have morphological characteristics similar to glial cells. We, therefore, consider that glial cells may be primary chemoreceptor cells. This idea is compatible with the report by Fukuda et al. [19] who found silent cells that are depolarized by lowered pH in the superficial medullary layer. On the other hand, the group of Guyenet has reported that some surface neurons in the ventral medulla are glutamatergic, and they argue that these neurons are respiratory chemoreceptors [20,21].

Although the sites of chemoreception and the cellular mechanism for central chemoreception have been extensively studied [2-4,22], the anatomical structure of the central respiratory chemoreceptor has not been identified. Our morphological and functional studies have led us to believe that the central respiratory chemoreceptor primarily senses the perivascular chemical composition in the lower brainstem surface. After reviewing our experimental data and other previous reports, we are led to propose a morphological architecture model for the central respiratory chemoreceptor that consists of primary chemoreceptor cells and vessels near the medullary surface.

2. ANATOMICAL RELATIONSHIP OF MEDULLARY SURFACE LARGE VESSELS WITH CHEMO-SENSITIVE REGIONS

We analyzed the anatomical relationship of large vessels and CO_2-activated cells in the ventral medulla [18] and found a common anatomical architecture in the superficial ventral medullary midline, parapyramidal and ventrolateral regions where chemosensitive cells are concentrated; in each area the medullary surface shows an indentation covered by a large surface vessel, and the marginal glial layer is thickened. Because we carried out the histological processing by embedding medullary preparations in paraffin, detachment of surface vessels was prevented, and we could thus observe this intimate anatomical relationship between surface vessels and chemosensitive cells. We demonstrated that the midline region lies beneath the basilar artery, the parapyramidal region lies beneath the inferior petrosal sinus, and the rostrolateral and caudolateral medulla lies beneath rich anastomoses of the anterior and posterior inferior cerebellar and paraolivary arteries. In transverse brain sections, several ventral medullary surface areas consistently display concave shapes, apparently compressed by surface vessels (Fig. 1).

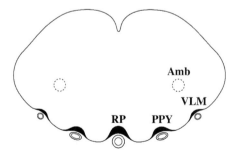

Figure 1. Schematic presentation of the ventral medullary chemosensitive regions shown as black areas. In the midline (RP), parapyramidal (PPY) and ventrolateral (VLM) chemosensitive regions, the medullary surface portions are covered and compressed by surface vessels, indented, and have the thickened marginal glial layer. Amb, nucleus ambiguus.

We consider that primary chemoreceptor cells respond to CO_2 that diffuse out of these surface vessels. This consideration is compatible with the observations by Eldridge et al. [23] and Kiley et al. [24] who demonstrated that not CSF pH but superficial ventral medullary tissue pH reflects arterial PCO_2. The concept that the central respiratory chemoreceptor responds not to CSF pH but to superficial ventral medullary tissue pH is further supported by our studies; selective injection of hypercapnic blood into various branches of cat brainstem arteries induced rapid respiratory augmentation, especially when injection was into the rostral ventrolateral medulla [25-26]. Bradley et al. [27] also reported that chemosensitive serotonergic neurons are closely associated with large medullary arteries, and claimed that these neurons are respiratory chemoreceptors, although they were criticized by the group of Guyenet who argued that these serotonergic neurons are not respiratory chemoreceptors but are innervating sympathetic efferent neurons [20,21].

Although the superficial ventral medulla is important in central chemosensitivity, deep medullary regions may also play a role in central chemosensitivity; Ichikawa et al. [28] demonstrated that injection of CO_2-saturated saline into the cat vertebral artery evokes acidic shift of local tissue pH with a time course analogous to ventilatory augmentation, not only in the superficial but also in various deep regions of the ventrolateral medulla.

3. ANATOMICAL RELATIONSHIP OF CHEMO-SENSITIVE CELLS WITH FINE PENETRATING VESSELS IN THE SUPERFICIAL MEDULLARY LAYER

It has been reported that the marginal glial layer is thickened in the classical chemosensitive area in the ventral medulla [29-31]. However, these were purely anatomical studies, and the physiological significance of the thickened marginal glial layer has not been elucidated. As noted above, we found a characteristic architecture in the thickened marginal glial layer; small cells, which are intrinsically chemosensitive, surround fine vessels that penetrate into the marginal glial layer from the ventral medullary surface [18]. Just beneath the marginal glial layer, large neurons are located [18]. Their chemosensitivity is lost with synaptic blockers [18], and we consider these as relay neurons (discussed later). Pan et al. [32] studied the morphological relationship between neurons and vessels in the rat superficial ventral medulla, and found a group of neurons closely associated with the walls of blood vessels. Hanson et al. [33] have demonstrated the existence of carbonic anhydrase in the classical chemosensitive areas of the cat medulla, and proposed that carbonic anhydrase accelerates CO_2/pH equilibration. These studies led us to consider that small chemosensitive cells surrounding fine vessels in the surface medullary layer sense perivascular PCO_2/pH acting as primary chemoreceptor cells. This idea is compatible with the theoretical study by Adams et al. [34], who showed that perivascular tissue PCO_2 around a small artery can easily be in equilibrium with blood PCO_2 of the small artery.

4. THE CELL-VESSEL ARCHITECTURE MODEL FOR THE CENTRAL RESPIRATORY CHEMO-RECEPTOR

On the basis of these findings, we propose a cell-vessel architecture model for the central respiratory chemoreceptor. Primary chemoreceptor cells are mainly located beneath large surface vessels in the superficial layer of the medulla, and surround fine penetrating vessels that branch from a large surface vessel (Fig. 2). These vessels are mainly arterial vessels, but can also be venous vessels.

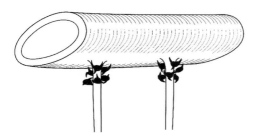

Figure 2. The cell-vessel architecture model for the central respiratory chemoreceptor. Primary chemoreceptor cells (shown in black) surround superficial parts of fine vessels that penetrate into the medulla. Primary chemoreceptor cells are covered by a large surface vessel from the surface side.

These primary chemoreceptor cells monitor the perivascular PCO_2/pH level. Responding to elevated PCO_2 in arterial and/or venous blood, primary chemoreceptor cells are depolarized, secrete neurotransmitter, and excite secondary relay neurons. Chemosensitive neurons in the ventrolateral medulla have been reported to have dendritic projections toward the medullary surface, and it has been proposed that surface-projecting dendrites may be responsible for directly sampling the chemical environment on the medullary surface [12]. We, however, consider that these surface-projecting dendrites are extending from secondary relay neurons and form synapses with primary chemoreceptor cells. The relay neurons convey the information to the deep respiratory neuronal network and work as amplifiers. Some respiratory neurons also send dendrites to the ventral medullary surface [12], and may directly form synapses with and receive information from primary chemoreceptor cells (Fig 3).

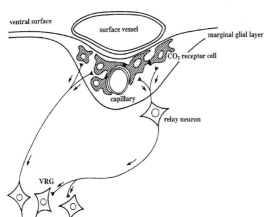

Figure 3. The anatomical architecture model for the chemoreceptor complex. Primary chemoreceptor cells that are located beneath a large surface vessel surround fine vessels or capillaries in the marginal glial layer and sense perivascular PCO_2/pH. Relay neurons beneath the marginal glila layer and some ventral respiratory group (VRG) neurons receive the chemical information from chemoreceptor cells and transmit the information to the VRG neuronal network.

In the present model proposal, we have focused on the superficial ventral medulla. However, some respiratory neurons in the medulla are intrinsically chemosensitive [12,35]. Also, central respiratory chemoreceptors appear to exist outside the medulla, e.g., in the pontine regions (locus coeruleus and A5 region) [4,13,16,36] and in the hypothalamus [37]. The anatomical structure of central respiratory chemoreceptors located elsewhere than at the ventral medullary surface may be different from that we propose here, and should be the subject of future studies.

REFERENCES

1. Loeschcke HH. Central chemosensitivity and the reaction theory. *J Physiol* 332: 1-24, 1982.
2. Kiwull-Schöne HF. The "Reaction Theory" of Hans Winterstein (1879-1963) in the light of today's research on the ventrolateral medulla. In *Ventral Brainstem Mechanisms and Control of Respiration and Blood Pressure.* Trouth CO, Millis RM, Kiwull-Schöne HF, Schläfke ME ed. Marcel Dekker, New York, pp1-39, 1995.
3. Okada Y, Chen Z, Kuwana S. Cytoarchitecture of central chemoreceptors in the mammalian ventral medulla. *Respir Physiol* 291: 13-23, 2001.
4. Nattie EE. Central chemosensitivity, sleep, and wakefulness. *Respir Physiol* 129: 257-268, 2001.
5. Okada Y, Chen Z, Jiang W, Kuwana S, Eldridge FL. Functional connection from the surface chemosensitive region to the respiratory neuronal network in the rat medulla. *Adv Exp Med Biol* 551: 45-51, 2004.
6. Issa FG, Remmers JE. Identification of a subsurface area in the ventral medulla sensitive to local changes in PCO_2. *J Appl Physiol* 72: 439-446, 1992.
7. Ribas-Salgueiro JL, Gaytán SP, Crego R, Pásaro R, Ribas J. Highly H^+-sensitive neurons in the caudal ventrolateral medulla of the rat. *J Physiol* 549: 181-194, 2003.
8. Fukuda Y, Honda Y. pH-sensitive cells at ventrolateral surface of rat medulla oblongata. *Nature* 256: 317-318, 1975.
9. Dean JB, Bayliss DA, Erikson JT, Lawing WL, Millhorn DE. Depolarization and stimulation of neurons in the nucleus tractus solitarii by carbon dioxide does not require chemical synaptic input. *Neuroscience* 36: 207-216, 1990.
10. Okada Y, Mückenhoff K, Scheid P. Hypercapnia and medullary neurons in the isolated brain stem-spinal cord of the rat. *Respir Physiol* 93: 327-336, 1993.
11. Richerson GB. Response to CO_2 of neurons in the rostral ventral medulla in vitro. *J Neurophysiol* 73: 933-944, 1995.
12. Kawai A, Ballantyne D, Mückenhoff K, Scheid P. Chemosensitive medullary neurones in the brainstem-spinal cord preparation of the neonatal rat. *J Physiol* 492: 277-292, 1996.
13. Oyamada, Y, Ballantyne D, Mückenhoff K, Schied P. Respiration-modulated membrane potential and chemosensitivity of locus ceruleus neurones in the in vitro brainstem-spinal cord of the neonatal rat. *J Physiol* 513: 381-398, 1998.
14. Sato M, Severinghous JW, Basbaum AI. Medullary CO_2 chemoreceptor neuron identification by c-fos immunocytochemistry. *J Appl Physiol* 73: 96-100, 1992.
15. Teppema LJ, Berkenbosch A, Veening JG, Olievier CN. Hypercapnia induces c-fos expression in neurons of retrotrapezoid nucleus in cats. *Brain Res* 635: 353-356, 1994.
16. Haxhiu MA, Yung K, Erokwu B, Cherniack NS. CO_2-induced c-fos expression in the CNS catecholaminergic neurons. *Respir Physiol* 105: 35-45, 1996.
17. Teppema LJ, Veening JG, Kranenburg A, Dahan A, Berkenbosch A, Olievier C. Expression of c-fos in the rat brainstem after exposure to hypoxia and to normoxic and hyperoxic hypercapnia. *J Comp Neurol* 388: 169-190, 1997.
18. Okada Y, Chen Z, Jiang W, Kuwana S, Eldridge FL. Anatomical arrangement of hypercapnia-activated cells in the superficial ventral medulla of rats. *J Appl Physiol* 93: 427-439, 2002.
19. Fukuda Y, Honda Y, Schlaefke ME, Loeschcke H. Effect of H^+ on the membrane potential of silent cells in the ventral and dorsal surface layers of the rat medulla in vitro. *Pflügers Arch* 376: 229-235, 1978.
20. Mulkey DK, Stornetta RL, Weston MC, Simmons JR, Parker A, Bayliss DA, Guyenet PG.. Respiratory control by ventral surface chemoreceptor neurons in rats. *Nat Neurosci* 7: 1360-1369, 2004.

21. Weston MC, Stornetta RL, Guyenet PG. Glutamatergic neuronal projections from the marginal layer of the rostral ventral medulla to the respiratory centers in rats. *J Comp Neurol* 473: 73-85, 2004.
22. Oyamada Y, Yamaguchi K, Murai M, Ishizaka A, Okada Y. Potassium channels in the central control of breathing. *Adv Exp Med Biol* (in press)
23. Eldridge FL, Kiley JP, Millhorn DE. Respiratory effects of carbon dioxide-induced changes of medullary extracellular fluid pH in cats. *J Physiol* 355: 177-189, 1984.
24. Kiley JP, Eldridge FL, Millhorn DE. The roles of medullary extracellular and cerebrospinal fluid pH in control of respiration. *Respir Physiol* 59: 117-130, 1985.
25. Kuwana S, Natsui T. Effect of hypercapnic blood injection into the vertebral artery on the phrenic nerve activity in cats. *Jpn J Physiol* 37: 155-159, 1987.
26. Kuwana S, Natsui T. Respiratory responses to occlusion or hypercapnic blood injection of the anterior inferior cerebellar artery in cats. *Jpn J Physiol* 40: 225-242, 1990.
27. Bradley SR, Pieribone VA, Wang W, Severson CA, Jacobs RA, Richerson GB. Chemosensitive serotonergic neurons are closely associated with large medullary arteries. *Nat Neurosci* 5: 401-402, 2002.
28. Ichikawa K, Kuwana S, Arita H. ECF pH dynamics within the ventrolateral medulla: a microelectrode study. *J Appl Physiol* 67:193-198, 1989.
29. Petrovicky P. Über die Glia marginalis und oberflächliche Nervenzellen im Hirnstamm der Katze. *Z Anat Entwick* 127: 221-231, 1968.
30. Trouth CO, Odek-Ogunde M, Holloway JA. Morphological observations on superficial medullary CO_2-chemosensitive areas. *Brain Res* 246: 35-45, 1982.
31. Filiano JJ, Choi C, Kinney HC. Candidate cell populations for respiratory chemosensitive fields in the human infant medulla. *J Comp Neurol* 293: 448-465, 1990.
32. Pan Y, Trouth CO, Douglas RM, Ting P. Age differences in choline acetyltrasferase, tyrsosine hydroxylase, and carbonic anhydrase neurons and opiate receptor binding at rat ventral brainstem. In *Ventral Brainstem Mechanisms and Control of Respiration and Blood Pressure.* Trouth CO, Millis RM, Kiwull-Schöne HF, Schläfke ME ed. Marcel Dekker, New York, pp563-588, 1995.
33. Hanson MA, Nye PC, Torrance RW. The location of carbonic anhydrase in relation to the blood-brain barrier at the medullary chemoreceptors of the cat. *J Physiol* 320: 113-125, 1981.
34. Adams JM, Banka C, Wojcicki WE, Roth AC. Carbon dioxide exchange across the walls of arterioles: implication for the location of the medullary chemoreceptors. *Ann Biomed Eng* 16: 311-322, 1988.
35. Solomon IC. Focal CO_2/H^+ alters phrenic motor output response to chemical stimulation of cat pre-Bötzinger complex in vivo. *J Appl Physiol* 94: 2151-2157, 2003.
36. Ito Y, Oyamada Y, Okada Y, Hakuno H, Aoyama R, Yamaguchi K. Optical mapping of pontine chemosensitive regions of neonatal rat. *Neurosci Lett* 366: 103-106, 2004.
37. Dillon GH, Waldrop TG. Responses of feline caudal hypothalamic cardiorespiratory neurons to hypoxia and hypercapnia. *Exp Brain Res* 96: 260-272, 1993.

Loop Gain of Respiratory Control upon Reduced Activity of Carbonic Anhydrase or Na^+/H^+ Exchange

HEIDRUN KIWULL-SCHÖNE[1], LUC TEPPEMA[2], MARTIN WIEMANN[3] AND PETER KIWULL[1]

[1]Dept. of Physiology, Ruhr-University D-44780 Bochum, Germany, [2]Dept. of Anesthesiology, Leiden University, The Netherlands; [3]Dept. of Physiology, University of Duisburg-Essen, Germany

1. INTRODUCTION

Considerations from control theory revealed that an elevated gain of the feedback loop may lead to instability of the respiratory system, e.g. during sleep [Longobardo et al., 1982; Honda et al., 1983; Khoo, 2000; Dempsey et al., 2004]. In respiratory medicine, the carbonic anhydrase (CA) inhibitor acetazolamide is known to reduce the incidence of apneas in mountain sickness [Swenson et al., 1991] or sleep disordered breathing [Tojima et al., 1988; Verbraecken et al., 1998]. Other clinical studies revealed that patients prone to sleep apnea showed an increased sodium/proton exchange activity in their lymphocytes [Tepel et al., 2000]. To predict possible protective effects of substances inhibiting either carbonic anhydrase activity or sodium/proton exchange, we evaluated steady state feedback loop characteristics of the respiratory control system from previous studies in anaesthetized rabbits [Kiwull-Schöne et al., 2001a,b]. Steady state loop gain (G) was assessed as ratio of the slope of the CO_2 response (S) and that of the metabolic hyperbola (S_L) [Honda et al., 1983; Khoo, 2000] at the intersection of both curves, by which also the arterial set point P_{CO_2} (Psp_{CO_2}) is determined.

2. METHODS

The experiments were performed in two groups (each N= 7) of adult male rabbits (average weight ±SEM: 3.6 ±0.1 kg). They were anaesthetized intravenously by pentobarbital sodium (initial dose: 58 ± 2 mg·kg^{-1}; continuous infusion: 7.6 ± 0.2 mg·kg^{-1}·h^{-1}).

Details of methods have been described previously [Kiwull-Schöne et al., 2001a,b]. Briefly, tidal volume (V_T) and inspiratory/expiratory durations (T_I, T_E) were continuously measured by pneumotachography, endtidal P_{CO_2} (Pet_{CO_2}) by infrared absorption. For artificial ventilation, animals were paralysed (Alloferin $^{(R)}$, Sandoz) and provided with a small animal respirator. Phrenic nerve compound potentials were processed by leakage integration. Throughout the experiment, arterial P_{O_2} values were maintained in the range of 150 mmHg. The arterial CO_2 pressure (Pa_{CO_2}) was determined indirectly from pH measurements by the two gas equilibration method at 38°C, the controlled body temperature.

Experimental protocol. In the first group (Fig. 1A), CO_2-responses of spontaneous ventilation were tested at different levels of elevated arterial P_{CO_2}, before and after low-dose injections of acetazolamide up to an average cumulative low dose of 4.64 ± 0.22 mg·kg^{-1}. The second group of rabbits was also studied first during spontaneous breathing. Thereafter, the animals were vagotomized, paralysed and mechanically ventilated. The apneic threshold Pa_{CO_2} was approached by hyperventilation. The CO_2-response of the phrenic nerve activity was tested at different steady-state levels of Pa_{CO_2} above the individual threshold value (Fig. 1C). The role of the sodium/proton exchanger type 3 (NHE3) was studied during infusion of the brain-permeant NHE3 inhibitor S8218 (Sanofi-Aventis), at plasma levels maintained in the range of 0.3 µg/ml (10^{-6} M). Under these conditions, measurements at apneic threshold and at the different CO_2 levels above were repeated and compared to controls.

Evaluation. Pulmonary ventilation ($\dot{V}$) was obtained under steady state conditions, calculated as $60 \cdot V_T/(T_I+T_E)$. The arbitrary units of the integrated tidal phrenic nerve amplitude (IPNA) were normalized to V_T under initial eucapnic conditions in each individual, thus phrenic minute activity (Fig. 1C) being the analogue of $\dot{V}$ (Fig. 1A). Dead space ventilation was estimated from the anatomical plus equipmental dead space, along with the non-perfused portion of the alveolar space, given by the arterio-endtidal P_{CO_2} difference in relation to Pa_{CO_2} [Kiwull-Schöne and Kiwull, 1997]. The metabolic CO_2 production $(\dot{V}_{CO_2})_{STPD}$ is then given by the relationship between alveolar ventilation $(\dot{V}_A)_{BTPS}$ and Pa_{CO_2}, representing the metabolic hyperbola.

Statistics. Non-linear regression analysis was performed of either ventilation or phrenic minute activity as functions of arterial P_{CO_2} under steady state conditions. By this method, individual Pa_{CO_2} values at apneic threshold ($Pthr_{CO_2}$) were determined, as well as the set point P_{CO_2} (Psp_{CO_2}) at the intersection of the CO_2 response curve and the metabolic hyperbola. At this intercept of both curves, the steady state loop gain (G) of the control system was assessed as ratio of the slope of the respiratory CO_2 response (S) and that of the metabolic hyperbola (S_L), whereby $S_L = 863 \cdot (\dot{V}_{CO_2})_{STPD} /(Psp_{CO_2})^2$ [Honda et al., 1983; Khoo, 2005]. Means ±SEM were calculated for the two groups of rabbits. Differences between controls and treatments were analyzed by paired t-tests and regarded as being significant at least for $P \leq 0.05$. Statistical analysis was in part carried out by SPSS 8.0 for Windows software (SPSS Inc. Chicago, IL, USA).

3. RESULTS

3.1 CA Inhibition

The table shows that application of low-dose acetazolamide significantly reduced the set point P_{CO_2} (Psp_{CO_2}) at the intercept of the CO_2 response curve and the metabolic hyperbola (Fig. 1A). The overall loop gain was lowered by a higher slope of the metabolic hyperbola and a lower slope of the CO_2 response curve, whereas the metabolic CO_2 production remained unaffected. Fig. 1B shows the mean values of G before and after CA inhibition, as well as the theoretical impact of both, Psp_{CO_2} level and CO_2 sensitivity on the loop gain at constant metabolic rate.

Table 1. The role of carbonic anhydrase and Na^+/H^+ exchanger 3 for respiratory control

	Control	Inhibition of Carbonic Anhydrase	Control	Inhibition of Sodium/Proton Exchanger
$Pthr_{CO_2}$ [mmHg]	26.7±1.7	19.5±2.8**	22.9±1.4	19.5±1.3**
Psp_{CO_2} [mmHg]	33.1±1.8	30.3±1.4**	27.3±1.6	23.8±1.4*
$(\dot{V}_{CO_2})$ [ml·min⁻¹ STPD]	17.4±0.9	17.3±1.0	20.1±0.8	20.5±0.8
S_L [ml·min⁻¹ BTPS· mmHg⁻¹]	-14.5±1.6	-16.4±1.4*	-24.9±3.2	-30.9±4.4*
S [ml·min⁻¹ BTPS· mmHg⁻¹]	152±29	111±31*	271±55	272±61
Loop gain (G=S/S_L)	10.4±1.6	6.3±1.6**	11.4±2.6	8.5±1.0

The table shows means ±SEM (N=7). Abbreviations: Arterial P_{CO_2} at apneic threshold ($Pthr_{CO_2}$) and at the set point (Psp_{CO_2}), metabolic CO_2 production ($\dot{V}_{CO_2}$), slope of the metabolic hyperbola (SL), slope of the ventilatory CO_2 response curve (S).

3.2 NHE3 Inhibition

As can be seen from Table 1 and Fig. 1C, NHE3 inhibition likewise enhanced the slope (S_L) of the metabolic hyperbola at the intercept with the CO_2 response curve by shifting Psp_{CO_2} towards lower values, while it had no significant effect on the metabolic rate. In contrast to acetazolamide, there was no effect on the slope S of the CO_2 response curve, so that the small decrease in overall loop gain G did not reach the level of significance. In analogy to Fig. 1B, the role of both, Psp_{CO_2} and CO_2 sensitivity on the loop gain is shown in Fig.1D, together with mean values of G before and after NHE3 inhibition.

4. DISCUSSION

By theoretical reasons, the steady state loop gain of the respiratory control system can be reduced by three factors, namely by an elevated metabolic rate, a

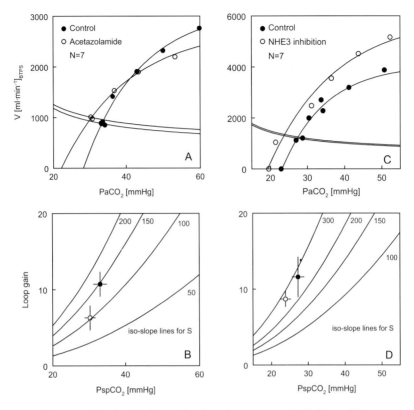

Figure 1. Factors affecting the loop gain of respiratory control. (A) Mean CO_2 response curves of pulmonary ventilation and metabolic hyperbolas before and after low-dose application of acetazolamide. (B) Mean loop gain ±SEM calculated from individual slopes at the intersection of these curves. (C) Mean CO_2 response curves of phrenic minute activity (normalized to spontaneous breathing) and metabolic hyperbolas of mechanically ventilated rabbits before and after application of the NHE3 inhibitor. (D) Mean loop gain ±SEM of these animals calculated in analogy to diagram B. The CO_2 response curves are obtained during hyperoxia and based on data from Kiwull-Schöne et al., 2001a,b. Diagrams B and D illustrate how the loop gain G depends on both, set point P_{CO_2} (P_{spCO_2}) and slope of the CO_2 response (S), the latter indicated by iso-slope lines for different theoretical values of S [ml·min^{-1}$_{BTPS}$·mmHg^{-1}] at constant metabolic rate.

lowered set point P_{CO_2} and a diminished ventilatory CO_2 sensitivity. A low loop gain implies protective properties against the incidence of periodic breathing and apnea, e.g. during sleep [Longobardo et al., 1982; Honda et al., 1983; Khoo, 2000; Dempsey et al., 2004].

Carbonic anhydrase inhibition by low-dose acetazolamide significantly reduced the loop gain first, by shifting the set point P_{CO_2} towards a steeper part (S_L) of the metabolic hyperbola and second, by attenuating the slope (S) of the CO_2 response curve (controller function), while metabolic rate remained unchanged.

As discussed elsewhere in detail [Kiwull-Schöne et al., 2001a], the leftward shift of the CO_2 response curve may have been caused by peripheral chemoreflex activation due to a slight renal metabolic acidosis or by inhibition of a membrane-bound carbonic anhydrase in brain capillaries, leading to a rise in tissue P_{CO_2}. As far as the slope of the CO_2 response curve is concerned, the same study showed clearly that the CA inhibitor acetazolamide considerably reduced the effectiveness of the CO_2-induced phrenic neural drive to mediate an appropriate rise in tidal volume. Thus, interpretation of the ventilatory CO_2 response, besides central chemosensitivity, has also to consider an attenuating component of acetazolamide on respiratory muscle function.

Selective inhibition of the Na^+/H^+ exchanger 3 likewise shifted the set point P_{CO_2} towards a steeper part (S_L) of the metabolic hyperbola without any change in metabolic rate, but did not attenuate the slope of the neural respiratory CO_2 response. Thus, the slight decrease in loop gain that could be discerned did not reach the level of significance. Although experiments with vagotomized rabbits on mechanical ventilation are useful to exclude secondary effects on central CO_2 sensitivity [Kiwull-Schöne et al., 2001b], interpretation of results is then merely restricted to the central neural chemoreflex output. Up to now, it is unknown whether NHE3 inhibitors may also influence the transmission of neural respiratory drive into lung functions, as has been shown for acetazolamide. Likewise, nothing is known about possible additive effects of NHE3 inhibitors on fluid balance of the bronchial mucous membrane and/or pulmonary vagal sensory nerve endings. Nevertheless, recent observations further underline the role of NHE3 for the adjustment of set point Pa_{CO_2} also in awake rabbits. The level of NHE3 mRNA expression in the medulla oblongata was inversely correlated with alveolar ventilation at constant metabolic rate, thus being involved in setting the baseline P_{CO_2} [Wiemann et al., 2005].

Generally, factors that influence the loop gain of the respiratory control system should be further integrated in physiological and pharmacological experimentation for better understanding the development and treatment of irregular breathing disorders.

REFERENCES

Dempsey JA, Smith CA, Przybylowski T, Chenuel B, Xie A, Nakayama H, Skatrud JB. The ventilatory responsiveness to CO_2 below eupnoea as a determinant of ventilatory stability in sleep. *J Physiol* 2004, 560:1-11.

Honda Y, Hayashi F, Yoshida A, Ohyabu Y, Nishibayashi Y, Kimura H. Overall "gain" of the respiratory control system in normoxic humans awake and asleep. *J Appl Physiol* 1983, 55:1530-1535.

Khoo MCK. Determinants of ventilatory instability and variability. *Respir Physiol* 2000, 122:167-182.

Kiwull-Schöne H, Kiwull P. Effectiveness of the peripheral chemoreflex control system in the adjustment of arterial O_2 pressure and O_2-Hb saturation. *Adv Exp Med Biol* 1997, 428: 433-438.

Kiwull-Schöne HF, Teppema LJ, Kiwull PJ. Low-dose acetazolamide does affect respiratory muscle function in spontaneously breathing anesthetized rabbits. *Am J Resp Crit Care Med* 2001*a*, 163: 478-483.

Kiwull-Schöne H, Wiemann M, Frede S, Bingmann D, Wirth KJ, Heinelt U, Lang H-J, Kiwull P. A novel inhibitor of the Na^+/H^+ exchanger type 3 activates the central respiratory CO_2 response and lowers the apneic threshold. *Am J Resp Crit Care Med* 2001b, 164: 1303-1311.

Longobardo GS, Gothe B, Goldman MD, Cherniack NS. Sleep apnea considered as a control system instability. *Respir Physiol* 1982, 50: 311-333.

Swenson ER, Leatham KL, Roach RC, Schoene RB, Mills WJ, Hackett PH. Renal carbonic anhydrase inhibition reduces high altitude sleep periodic breathing. *Respir Physiol* 1991, 86: 333-343.

Tepel M, Sanner BM, Van der Giet M, Zidek W. Increased sodium-proton antiporter activity in patients with obstructive sleep apnea. *J Sleep Res* 2000, 9:285-291.

Tojima H, Kunitomo F, Kimura H, Tatsumi K, Kuriyama T, Honda Y. Effects of acetazolamide in patients with the sleep apnea syndrome. *Thorax* 1988, 43: 113-119.

Verbraecken J, Willemen M, De Cock W, Coen E, Van de Heyning P, De Backer W. Central sleep apnea after interrupting longterm acetazolamide therapy. *Respir Physiol* 1998, 112: 59-70.

Wiemann M, Frede S, Bingmann D, Kiwull P, Kiwull-Schöne H. Sodium/proton exchanger 3 in the medulla oblongata and set point of breathing control. *Am J Resp Crit Care Med* 2005 (in press)

Adrenaline Increases Carotid Body CO_2 Sensitivity: An *in vivo* Study

PETER D. MASKELL, CHRIS J. RUSIUS, KEVIN J. WHITEHEAD* AND PREM KUMAR

*Department of Physiology, *Department of Pharmacology, The Medical School, University of Birmingham, Edgbaston, Birmingham, B15 2TT. UK.*

1. INTRODUCTION

Alveolar ventilation rises proportionally with metabolic rate during exercise and thus arterial Pco_2 remains constant or may even fall slightly. The mechanism underlying this isocapnic hyperpnea, by which ventilation is coupled so precisely to metabolism, however, remains unclear. We have shown recently (Bin-Jaliah et al., 2004), that an increased metabolic rate, induced by insulin infusion, could produce an isocapnic hyperpnoea in an anaesthetized rat and subsequently, we showed that this hyperpnoea was correlated with an increase in the CO_2 sensitivity, or gain, of the carotid body such that ventilation could be increased without hypercapnia. Low glucose can stimulate catecholamine release from carotid body tissue (Pardal & Lopez Barneo, 2002) but we demonstrated that the effect we observed *in vivo* could not be due to an insulin-induced fall in blood glucose concentration (Bin-Jaliah et al., 2005). We speculated that some other blood borne factor may be involved, and we in this present study, we evaluated the role of circulating adrenaline in the augmentation of chemoreceptor gain. Adrenaline has long been mooted as a possible feed forward factor involved in exercise hyperpnoea (Linton et al., 1992) and is know to be released in both hypoglycaemic states (Vollmer et al., 1997) and during exercise (Christensen et al., 1983).

2. MATERIALS & METHODS

2.1 Animal and Surgical Preparation

Adult male Wistar rats (326.57 ± 5.6 g, $n = 16$) were anaesthetized with halothane (3 - 4% in oxygen), anaesthesia was then maintained with 650 mg.kg^{-1} of 25% w/v ethyl carbamate (Urethane) via an indwelling cannula located in the left femoral vein. Circulating plasma adrenaline levels were measured using a dialysis probe (CMA/12, Microdialysis, Chelmsford, MA) inserted into the left femoral artery. The probe was continuously perfused with saline throughout the experimental procedure. Dialysate samples were collected into ice-cold tubes containing 10 ml of 0.25 M acetic acid at a rate of 1 ml.min^{-1} for 10 min. Blood pressure was recorded via an indwelling cannula located in the left femoral artery. The trachea was exposed and cannulated. Body temperature was monitored using a rectal thermometer and maintained at 36.0 ± 0.5°C using a small homeothermic blanket system. The carotid sinus nerves (CSNs) were either sectioned bilaterally (CSNX) or left intact (sham) as described previously (Bin-Jaliah et al., 2005). Denervation was confirmed by absence of hypoxia-induced hyperventilation prior to experiments. At the end of the experiment, animals were killed with a pentobarbitone overdose and cervical dislocation. Adequacy of anaesthesia was confirmed throughout the procedure by continuous measurement of arterial blood pressure and lack of cardiovascular response to a strong paw pinch. Supplementary anaesthesia, urethane (50 mg kg^{-1}; I.V.) was given as required.

Data are expressed as means ± S.E.M. and significance (P<0.05) tested with a paired t test.

2.2 Re-Breathing Techniques

Animals were allowed 30 - 60 min to recover from surgery, when baseline cardiovascular and respiratory variables were stable, experimental protocols consisted of a determination CO_2 sensitivity using a modified Read re-breathing technique (Read, 1967) in hyperoxia (Pa,$_{O2}$ >300 mmHg) as previously described (Bin-Jaliah et al., 2005). Three re-breathes were performed to give a baseline CO_2 sensitivity with a 15 minute recovery period between each re-breathe to allow the animal to recover. Adrenaline was infused for 10 minutes (10 μg.kg^{-1}.min^{-1}) with or without concomitant propranolol infusion (0.3 mg.kg^{-1}.min^{-1}, a non specific β-adrenoceptor antagonist) or with CSNX. Ventilatory CO_2 chemosensitivity was determined at the end of each adrenaline infusion. Each infusion and ventilatory CO_2 chemosensitivity was repeated twice. In all groups of animals, partial pressures of O_2 and CO_2 as well as pH and glucose concentration of the arterial blood were sampled before and after each re-breathe.

2.3 HPLC Adrenaline Analysis

The dialysate plasma samples were assayed using High Performance Liquid Chromatography (HPLC) with electrochemical detection. Adrenaline was separated on a 3mm, 100 x 2.0 mm, C18 Reverse Phase Column. The mobile phase consisted of Ammonium acetate (50mM), Glacial acetic acid (1.25%), sodium dodecyl sulphate (4mM), EDTA (0.27mM) and Methanol (25%). Adrenaline was oxidised during exposure to a glassy carbon electrode set to a potential of 0.6V versus Ag/AgCl. Data was acquired with the use of ChromPerfect software and calculations were based on peak areas.

3. RESULTS

3.1 Adrenaline Infusion Increases Baseline Ventilation and Plasma Adrenaline Levels

Venous infusion of adrenaline (10 $\mu g.kg^{-1}.min^{-1}$) caused a significant and considerable increase in plasma adrenaline concentration from 45.8 ± 11.6 fmol to 1213.5 ± 258.1 fmol ($P < 0.05$, n = 4). This increase in plasma adrenaline was associated with an increase in basal V_E from 461.5 ± 18.2 ml.min^{-1}.kg^{-1} to 673.01 ± 18.6 ml.min^{-1}.kg^{-1} ($P < 0.05$, n = 3) in spontaneously breathing rats (see fig. 1).

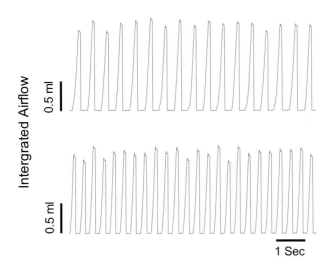

Figure 1. Representative traces of integrated inspiratory tracheal airflow (inspiration upwards) from a sham animal taken just before adrenaline infusion (upper trace) and 10 minutes after the infusion of adrenaline began (lower trace).

3.2 Effect of Adrenaline Infusion Upon Arterial Blood Parameters

The increase in plasma adrenaline was associated with an increase in blood glucose from the basal control levels 8.7 ± 0.34 mmol l^{-1} to 11.37 ± 0.75 mmol l^{-1} ($P < 0.05$, n = 4). Blood pH (from 7.48 ± 0.01 to 7.45 ± 0.01, n = 4), PaCO$_2$ (from 32.6 ± 1.7 mmHg to 33.0 ± 2.5 mmHg, n = 4), PaO$_2$ (from 84.1 ± 2.4 to 76.6 ± 5.4 mmHg, n = 4), HCO$_3^-$ (from 24.7 ± 1.0 mmol l^{-1} to 22.8 ± 1.1 mmol l^{-1}, n = 4), lactate (from 1.31 ± 0.08 mmol l^{-1} to 1.5 ± 0.55 mmol l^{-1}, n = 4) and K$^+$ (from 3.88 ± 0.15 mmol l^{-1} to 3.90 ± 0.20 mmol l^{-1}, n = 4) were not significantly altered from basal levels.

3.3 Ventilatory CO_2 Chemosensitivity Increased During Adrenaline Infusion in Sham but not in CSNX or Propranolol Infused Animals

The relation between V_E and the percentage of inspired CO_2 during rebreathing was used to determine ventilatory CO_2 chemosensitivity. Pre-adrenaline infused levels of ventilatory CO_2 chemosensitivity were not significantly different between sham and CSNX animals (data not shown). On application of adrenaline the ventilatory CO_2 chemosensitivity was significantly increased by ~40%, an effect abolished by CSNX or by concomitant application of propranolol (0.3mg kg^{-1}min^{-1}).

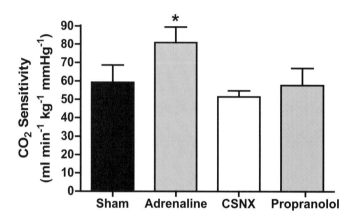

Figure 2. Means ± SEM of CO_2 chemosensitivity in Sham (pre-adrenaline infusion, n = 7), adrenaline infused (10 µg kg^{-1}min^{-1} for 10 minutes, n = 5), CSN section (CSNX) (n = 3) and adrenaline infusion with co-application of propranolol (0.3mg kg^{-1}min^{-1}) (n = 3). A significant increase in CO_2 sensitivity was observed on infusion of adrenaline. This significant increase was abolished upon either CSNX nerve section or the concomitant application of propranolol with adrenaline. * P<0.05 compared to sham (one way ANOVA).

4. DISCUSSION

Exercise hyperpnoea acts to maintain $PaCO_2$ constant during mild to moderate, dynamic exercise (Dempsey, 1995) and a combination of neural and humoral control appears to be involved, with a high degree of redundancy existing between the various systems (Paterson, 1994). The initial studies of the effect of catecholamine infusion upon ventilation in man, were performed over 80 years ago (Tompkins et al., 1919) although the effect in rat has not previously been reported. We now show that the infusion of adrenaline, at 10µg.kg^{-1}.min^{-1}, caused a 26-fold increase in rat arterial adrenaline levels as determined by the HPLC analysis of arterial dialysates. This is similar to the 22-fold increase in adrenaline levels seen in the insulin mediated hypoglycemia study of Vollmer (1997), where blood glucose levels fell to approx 2 mmol.l^{-1} in conscious rats. We demonstrated that this increase in [adrenaline] was associated with a ~50% increase in basal ventilation with no significant change in $PaCO_2$ suggesting that the observed elevated ventilation was an appropriate hyperpnoeic response presumably due to an elevated metabolism. These findings are thus confirmatory of previous reports of adrenaline or isoprenaline infusions where, in cats, it has been shown to increase carotid chemoreceptor discharge and ventilation, an effect abolished by carotid sinus nerve section (Joels and White, 1968) and propranolol (Folgering et al., 1982).

One of the possible elements controlling exercise hyperpnoea is an increase in chemoreceptor gain (Weil et al 1972; Hickam et al., 1951). During dynamic exercise, an increase in sympatho-adrenergic activity occurs in which arterial plasma concentrations of adrenaline increase linearly with exercise duration and exponentially with intensity (Kjaer, 1992). Our present study demonstrates an adrenaline mediated, carotid body-dependent increase in CO_2 chemosensitivity, via a β-receptor mediated mechanism, that might account for our previous findings in the insulin-mediated hypoglycaemic rat model (Bin-Jaliah et al. 2005). In contrast, Folgering et al. (1982) showed, in cats and rabbits, that isoprenaline infusion could induce an upward, but parallel shift in the position of the VE-CO_2 relation, i.e. no increase in gain. Whether this is a species effect is not known.

Linton et al (1992) showed that the elevated ventilation observed during adrenaline infusion might be accounted for either by a concurrent elevation in arterial K^+ levels and/or via an increase in the amplitude or slope of CO_2 arterial blood gas oscillations derived from the increased metabolism. In our previous study (Bin-Jaliah et al., 2004) we excluded any effect of elevated K^+ as this was decreased by the insulin and so could not account for the observed hyperpnea. It has not yet proved possible to measure chemoreceptor discharge oscillations *in vivo* in the rat, but this should be studied, as it cannot be discounted in our studies.

In conclusion, our data demonstrates that ventilation can be increased isocapnically in the rat by infusion of adrenaline through a stimulatory effect at the carotid body and we suggest that this may play a role in the hyperpnea of exercise.

ACKNOWLEDGEMENTS

We wish to thank the British Heart Foundation, Birmingham University and the Royal Society for their support.

REFERENCES

Bin-Jaliah I., Maskell P.D., Kumar P. Carbon dioxide sensitivity during hypoglycaemia-induced, elevated metabolism in the anaethetized rat. J Physiol 2005; 563:883-93.

Bin-Jaliah I., Maskell P.D., Kumar P. Indirect sensing of insulin-induced hypoglycaemia by the caroid body in the rat. J Physiol 2004; 556: 255-66.

Christensen N.J., Galbo H. Sympathetic nervous activity during exercise. Ann Rev Physiol 1983; 45:139-53.

Dempsey J.A. Exercise hyperpnea. Chairman's introduction. Adv Exp Med Biol 1995; 393: 133-6.

Folgering H., Ponte J., Sadig T. Adrenergic mechanisms and chemoreception in the carotid body of the rat and cat. J Physiol 1982; 325:1-21.

Hickham J.B., Pryor W.W., Page E.B., Atwell R.J. Respiratory regulation during exercise in unconditioned subjects. J Clin Invest 1951; 30:503-16.

Joels N. and White H. (1968). The contribution of the arterial chemoreceptors to the stimulation of respiration by adrenaline and noradrenaline in the cat. *J Physiol.* **197**:1-23.

Kjaer M. Regulation of hormonal and metabolic responses during exercise in humans. Ex Sport Sci Rev 1992; 20:161-184.

Linton R.A.F., Band D.M. Wolff C.B. Carotid chemoreceptor discharge during epinephrine infusion in anesthetized cats. J Appl Physiol 1992; 73:2420-24.

Pardal R., Lopez-Barneo J. Low glucose-sensing cells in the carotid body. Nat Neurosci 2002 ;5:197-8.

Paterson D.J. Respiratory control during exercise. Can J Appl Physiol. 1994; 19:289-304.

Read D.J.C. A clinical method for assessing the ventilatory response to carbon dioxide. Australas Ann Med 1967; 16:20-32.

Tompkins E.H., Sturgis C.C., Wearn J.T. Studies on epinephrine. Arch Intern Med 1919; 24:269-71.

Vollmer R.R., Balcita J.J., Sved A.F., Edwards D.J. Adrenal epinephrine and norepinephrine release to hypoglycaemia measured by microdialysis in conscious rats. Am J Physiol Regul Integr Comp Physiol 1997; 273:1758-63.

Weil J.V., Byrne-Quinn E., Sodal I.E., Kline J.S., McCullough R.E., Filley G.F. Augmentation of chemosensitivity during mild exercise in normal man. J Appl Physiol 1972; 33: 813-9.

Peripheral Chemoreceptor Activity on Exercise-Induced Hyperpnea in Human

SHINOBU OSANAI, TORU TAKAHASHI, SHOKO NAKAO, MASAAKI TAKAHASHI, HITOSHI NAKANO AND KENJIRO KIKUCHI

First Department of Medicine, Asahikawa Medical College 2-1-1-1 Midorigaoka Higashi, Asahikawa 078-8510, Japan

1. INTRODUCTION

It has been reported that hypoxic ventilatory response can be enhanced by an increase in work rate of exercise (Weil et al., 1972; Grover et al., 2002). However, it is still vague about how the muscular exercise produces the enhanced ventilatory responsiveness to hypoxia. Previous studies suggested that afferent nerve activity from carotid body was stimulated by some humoral factors related to muscular exercise, i.e. circulating catecholamine, lactate (Wasserman et al., 1986) and K^+ (Band et al., 1985).

This study was designed to estimate the contribution of peripheral chemoreceptor activity (PCA) related to O_2 sensing on exercise-induced hyperpnea in human. In addition, the relationship between PCA and the humoral factors related to muscular exercise was investigated.

2. METHODS

The PCA was studied in ten healthy volunteers (38 +/- 9 yr, 75 +/- 16 kg, mean +/- SD) during rest and exercise. The study was approved by the institute committee of human studies, and informed consent was obtained from each subject. They performed cycle ergometer exercise increasing by 25 W every 10 min in the stepwise manner. Arterial blood was sampled to analyze blood gases and to measure serum $[K^+]$, [lactate] and [noradrenaline] at rest, and in early phase (3 min) and in late phase (10 min) of each step of exercise.

The PCA related to O_2 sensing was estimated by measurement of decline of minute ventilation (ΔV'I) during 5 to 15 sec after transient two breaths of pure oxygen (Fig. 1). In this test, the high inspired O_2 fractions suppress peripheral chemosensory contribution to ventilatory drive before change the condition of central chemoreceptor (Dejours, 1964). The measurements of ΔV'I were practiced at rest, and in early phase and in late phase of each step of exercise. The contribution of PCA in entire ventilatory gain (%ΔV'I) was calculated as follows:

%ΔV'I = ΔV'I / V'I before two breaths of pure oxygen x 100

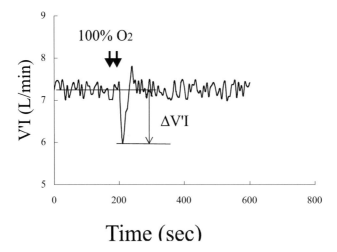

Figure 1. Time course of inspiratory minute ventilatory responses (V˙I) to the abrupt inhalation of 100% O2 test under normoxic conditions for a typical subject.

Data were analyzed using ANOVA with repeated measures, and Student-Newman-Keuls' test for multiple comparisons when appropriate.

3. RESULTS

During the exercise test, arterial blood gases were not changed significantly (Table 1). Therefore, it was unlikely that the alternation of arterial blood gases during exercise affected PCA in the present study.

Table 1. Arterial blood gases during rest and exercise.

	Rest	25 W	50 W	75 W
pH	7.44 ± 0.01	7.42 ± 0.01	7.41 ± 0.01	7.41 ± 0.01
PaCO2 (Torr)	35.8 ± 0.9	37.9 ± 1.2	38.7 ± 0.7	38.7 ± 1.2
PaO2 (Torr)	96.8 ± 3.1	101.1 ± 4.0	98.1 ± 2.7	94.1 ± 2.7

Mean +/- SEM.

Increases of serum [K^+], [lactate] and [noradrenaline] were correlated to the increase of exercise load (Fig. 2a). Also, the increases of humoral factors were tightly correlated to the increase of V'I (Fig. 2b, c, d).

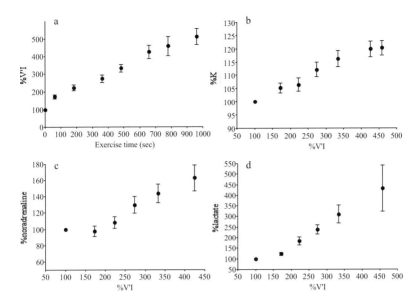

Figure 2. Relationships between increase of minute ventilation and humoral factors during the exercise test. %K, percent of serum [K$^+$] at rest; %lactate, percent of serum [lactate] at rest; %noradrenaline, percent of serum [noradrenaline] at rest; %VI, percent of inspiratory minute ventilation at rest. Mean +/- SEM.

The %ΔV'I in early phase was decreased with the increase of exercise (Fig. 3a). However, the %ΔV'I in late phase of exercise was not changed with the increase of exercise (Fig. 3b). Accordingly, there was difference between the contribution of PCA in the early phase of exercise and that of the late phase of exercise. On the other hand, neither serum [K$^+$], [lactate], nor [noradrenaline] was correlated to %ΔV'I.

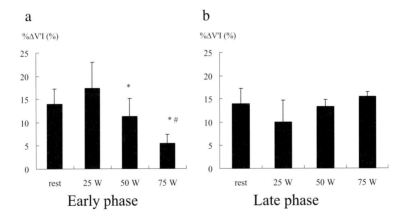

Figure 3. Results of peripheral chemoreceptor activity in early phase (3 min) of each exercise step and that in late phase (10 min) of each exercise step. Mean +/- SEM. *, P < 0.05 vs. 25 W. #, P <0.05 vs. 50 W.

4. DISCUSSION

In the present study, the contribution of PCA in the early phase of exercise is different from that in the late phase of exercise. Since the PCA in the early phase of exercise was reduced with the rise of muscle activity, it is likely that the role of the PCA is insignificant in this phase. In contrast, the PCA in the late phase of exercise was nearly constant regardless of exercise intensity. Accordingly, the PCA in the late phase of exercise was proportionally augmented with the exercise intensity.

In the previous report, it has been demonstrated that the PCA seems to have little effect on initial and early phases of exercise hyperpnea in humans (Cunningham, 1974; Dejours, 1975). However, late phase of exercise hyperpnea seems to be modulated by the carotid bodies in humans because stimulation of carotid chemosensitivity can accelerate the kinetics of hyperpnea (Whipp and Ward, 1991). Also, the contribution of PCA appears to amount to approximately 20% of the steady state of exercise hyperpnea response (Whipp, 1994). Our data was corresponding to those previous findings.

On the other hand, there was no significant relationship between the PCA during exercise and the humoral factors which were rised by the intensity of exercise load. Therefore, our data suggested that there is little effect of those factors on the PCA during exercise.

So far, various hypotheses about exercise-induced hyperpnea have been given (Wasserman, 1986). However, there remains much controversy in those hypotheses. Recently, it has been showed the the hypermetabolism stimulates the PCA and augments ventilation (Bin-Jaliah, 2005). Because the ventilation is tightly link to metablism (Casaburi 1978), this mechanism might be one possible mechanism of exercise-induced hyperpnea. Hereafter, the relationship between PCA and metabolism during exercise should be studied in detail.

5. CONCLUSION

In conclusion, there is differences between the PCA in the early phase of exercise and that in the late phase of exercise. Also, It is unlikely that the serum [K^+], [lactate] and [noradrenaline] affect the PCA related to O_2 sensing.

REFERENCES

Band D.M., Linton R.A.F. The effect of potassium on carotid-body chemoreceptor discharge in the anesthetized cat. J Physiol 1986; 381:39–47.
Bin-Jaliah I., Maskell P.D., Kumar P. Carbon dioxide sensitivity during hypoglycaemia- induced, elevated metabolism in the anaesthetized rat. J Physiol 2005; 563:883-93.
Casaburi R., Whipp B.J., Wasserman K., Koyal S.N. Ventilatory and gas exchange response to cycling with sinusoidal varying pedal rate. J Appl Physiol 1978; 44:97-103.
Cunningham D.J.C. "Integrative aspects of the regulation of breathing: a personal view." In MTP international review of science: physiology, series 1, vol 2. (A.C. Guyton and J.G.Widdicombe, eds.), University Park Press, Baltimore, pp.303–69, 1974.

Dejours P. "Control of respiration in muscular exercise." In Handbook of physiology: respiration, vol. I. (W.O. Fenn and H. Rahn, eds.), American Physiological Society, Washington DC, pp. 631–48, 1964.

Grover R.F., Cruz J.C., Grover E.B., Reeves J.T. Exercise-dependent ventilatory sensitivity to hypoxia in Andean natives. Respir Physiol Neurobiol 2002; 133:35–41.

Wasserman K., Whipp B.J., Casaburi R. "Respiratory control during exercise." In Handbook of Physiology: The Respiratory System. Control of Breathing, sect. 3, vol. II, pt. 2. (A.P. Fishman eds.), American Physiological Society, Bethesda, pp. 595–620, 1986.

Wasserman K., Whipp B.J., Koyal S.N., Cleary MG. Effect of carotid body resection on ventilatory and acid-base control during exercise. J Appl Physiol 1975; 39:354–8.

Weil J.V., Byrne-Quinn E., Sodal I.E., Kline J.S, McCullough R.E., Filley G.F. Augmentation of chemosensitivity during mild exercise in normal man. J Appl Physiol 1972; 33:813–819.

Ward S.A. Assessment of peripheral chemoreflex contributions to ventilation during exercise. Med Sci Sports Exerc 1994; 26:303–10.

Whipp B.J, Ward S.A. "The coupling of ventilation to pulmonary gas exchange during exercise." In Pulmonary physiology and pathophysiology of exercise (Whipp BJ, Wasserman K, eds.), Dekker, New York, pp.271–307, 1991.

Whipp B.J. Peripheral chemoreceptor control of the exercise hyperpnea in humans. Med Sci Sports Exerc 1994; 26:337–47.

Effects of Low-Dose Methazolamide on the Control of Breathing in Cats

J.H.L. BIJL, B. MOUSAVI GOURABI, A. DAHAN, L.J. TEPPEMA

Department of Anesthesiology Leiden University Medical Center Leiden, The Netherlands

1. INTRODUCTION

Inhibitors of carbonic anhydrase (CA) have complex effects on respiration. Many cells and tissues that are involved in the control of breathing contain various isoforms of CA, *e.g.*, red cells, carotid bodies, lung and brain capillary endothelial cells, muscle and neurons closely associated with central chemoreceptors (1-9). In human and cats, low intravenous doses of acetazolamide have both stimulatory and inhibitory effects on the control of breathing. (10, 11). One of the inhibitory effects applies to the peripheral chemoreceptors because acetazolamide has been shown to reduce the hypoxic response and also the O_2-CO_2 interaction that is known to reside the carotid bodies (10,12,13). The mechanism by which this occurs is unclear: however, due to its physical-chemical properties acetazolamide does not easily cross biological membranes (1, 2) so that at low dose this inhibiting effect is unlikely due to inhibition of an intracellular isoform of CA in the carotid bodies. Methazolamide, another CA inhibitor with an about equal affinity for sulfonamide-sensitive CA isoforms is much more lipophilic and rapidly permeates into cells (1, 2). Therefore, this agent would be a suitable tool to study the effect to intracellular CA inhibition on carotid body-mediated responses. Another difference between acetazolamide and methazolamide refers to their effects on large-conductance Ca^{2+}- dependent potassium (BK) channels: while acetazolamide specifically opens these channels, methazolamide is without any stimulating effect on them (14). Because BK channels may play a crucial role in the hypoxic response of type-I carotid body cells (15), it is therefore interesting to compare the effects of both agents on the carotid body responses to both hypoxia and hypercapnia.

Dynamic end-tidal CO_2 forcing (DEF) is a suitable means to study the separate effects of pharmacological agents on the CO_2 sensitivity of the peripheral and central chemoreflex loops (16). In this study we have applied this technique to study the effects of low-dose methazolamide on the control of breathing in the cat.

2. METHODS AND DATA ANALYSIS

Animals and Measurements. Experiments were performed in nine female adult cats (weight 2.5-4.1 kg). The Ethical Committee for Animal Experiments of the University of Leiden approved the use of animals. The animals were sedated with 10 mg kg^{-1} ketamine hydrochloride (i.m.). Anaesthesia was induced with 2% sevoflurane in 30 % O_2 in N_2. The right femoral artery and vein were cannulated, 20 mg kg^{-1} alpha-chloralose and 100 mg kg^{-1} urethan were slowly administered intravenously, and the volatile anaesthetic was gradually withdrawn. About 1 h later, an infusion of an alpha-chloralose-urethan solution was started at a rate 1.0-1.5 mg kg^{-1} h^{-1} alpha-chloralose and 5.0-7.5 mg kg^{-1} h^{-1} urethan.

The trachea was cannulated at midcervical level and connected to a respiratory circuit. Tidal volume was measured electronically by integrating airway gas flow obtained from a pneumotachograph (number 0 flow transducer, Fleisch) connected to a differential pressure transducer (Statham PM 197). The respiratory fractions of O_2 and CO_2 were continuously measured with a Datex gas monitor (Multicap), which was calibrated with gas mixtures of known composition. The inspiratory gas concentrations were made with computer-steered mass flow controllers (type AFC 260, Advanced Semiconductor Materials). Arterial blood pressure was measured using a Statham pressure transducer ((P23ac). Arterial blood samples were taken for blood gas analysis (Radiometer ABL 700).

Experimental design. Using the technique of end-tidal CO_2 forcing, we performed step changes in end-tidal P_{CO_2} before and after intravenous infusion of 3 mg.kg^{-1} methazolamide (purchased from Sigma), dissolved in 0.1 N NaOH and 0.1 N HCl (pH was adjusted to 7-3-7.4). End-tidal PO_2 was kept constant throughout the experiments at a nomoxic level of 14 kPa. Both before and after methazolamide administration, 2-4 dynamic end-tidal forcing (DEF) runs were performed and the dynamic ventilatory responses analysed (see below). The $P_{ET}CO_2$ pattern during a DEF run was as follows. After a 10- to 15 min period of steady state ventilation at constant end-tidal PCO_2 (about 0.5 kPa above the apnoeic threshold), the $P_{ET}CO_2$ was increased by 1-1.5 kPa in a step-wise fashion and kept constant for 7 min. thereafter; the $P_{ET}CO_2$ was returned to its previous value and maintained for another 7 min.

Dynamic end-tidal forcing. The steady-state relation of inspiratory ventilation V_I to $P_{ET}CO_2$ at constant $P_{ET}O_2$ can be described by:

$$V_I = (S_P + S_c)(P_{ET}CO_2 - B)$$

Where S_P = the carbon dioxide sensitivity of the peripheral chemoreflex loop, S_c = the carbon dioxide sensitivity of the central chemoreflex loop, and B = the apnoeic threshold or extrapolated $P_{ET}CO_2$ at zero V_I. The sum of S_P and S_c is the overall carbon dioxide sensitivity.

For the analysis of the dynamic response of ventilation to a step-wise change in $P_{ET}CO_2$ we used a two-compartment model (16):

$$V_P(t) + \tau_p \, d/dt \, V_P(t) = S_P (P_{ET}CO_2 [t - T_p] - B)$$

$$V_c(t) + \tau_c \, d/dt \, V_c(t) = S_c (P_{ET}CO_2 [t - T_c] - B)$$

Where τ_p and τ_c = the time constants of the peripheral and central chemoreflex loops, respectively, $V_P(t)$ and $V_C(t)$ = the outputs of the peripheral and central chemoreflex loops, respectively, $P_{ET}CO_2 [t - T_P]$ = the stimulus to the peripheral chemoreflex loop delayed by the peripheral transport delay time (T_P), and $P_{ET}CO_2 [t - T_c]$ = the stimulus to the central chemoreflex loop delayed by the central transport delay time (T_c).

To allow the time constant of the ventilatory on transient to be different from that of the off transient τ_c is written as:

$$\tau_c = x \cdot \tau_{on} + (1-x) \tau_{off}$$

τ_{on} = the time constant of the ventilatory on transient, τ_{off} = the time constant of the off transient, and $x = 1$ when $P_{ET}CO_2$ is high, while $x = 0$ when $P_{ET}CO_2$ is low.

In most experiments a small drift in ventilation was present. We therefore included a drift term ($C \cdot t$) in our model. The total ventilatory response $V_I(t)$ is made up of the contributions of the central and peripheral chemoreflex loops and $C \cdot t$

$$\dot{V}_I(t) = \dot{V}_P(t) + \dot{V}_C(t) + C \cdot t$$

The parameters of the model were estimated by fitting the model to the breath-by-breath data with a least-squares method. To obtain optimal time delays, a grid search was applied, and all combinations of T_P and T_c, with increments of 1 s and with T_p smaller or equal than T_c, were tried until a minimum in the residual sum of squares was obtained. The minimum time delay was chosen, arbitrarily, to be 1 s, the τ_p was constrained to be at least 0.3 s.

Statistical analysis. To compare the means of the values obtained from the analysis of the DEF runs in the control situation with those obtained after methazolamide infusion, analysis of variance was performed on individual data. The level of significance was set at $P = 0.05$. Results are given as means of the means per cat ± S.D.

3. RESULTS

The dose of 3 mg.kg^{-1} methazolamide did not cause an appreciable arterial-to-end-tidal PCO_2 gradient indicating the absence of effective erythrocytic CA inhibition ($P_{(a-ET)}CO_2$ differences were −0.14 ± 0.27 kPa in control and 0.18 ± 0.27 kPa after methazolamide, $P = 0.11$). Altogether 58 DEF runs (32 before and 25 after methazolamide) were analysed and the results are summarized in table 1.

Table 1. Effects of methazolamide on respiratory parameters. Sp and Sc are the CO_2 sensitivities of the peripheral and central chemoreflex loops, respectively (in l.min^{-1}.kPa^{-1}). B is the intercept on the $P_{ET}CO_2$ axis of the CO_2 ventilatory response curve. Base Excess in mM. Values are means of the means per cat ± S.D. (n =9)

	B	Sc	Sp	Base Excess
Control	3.60 ± 0.72	0.68 ± 0.27	0.08 ± 0.04	-6.65 ± 1.75
Methazolamide	1.77 ± 1.41	0.44 ± 0.22	0.06 ± 0.03	-7.84 ± 1.90
P	0.00006	0.013	0.13	0.009

Methazolamide reduced the apnoeic threshold B and the CO_2 sensitivity Sc of the central chemoreflex loop. The CO_2 sensitivity of the peripheral chemoreflex loop was not significantly reduced. The individual data are shown in figure 1. Time constants, delays and drift were not significantly influenced by methazolamide (data not shown).

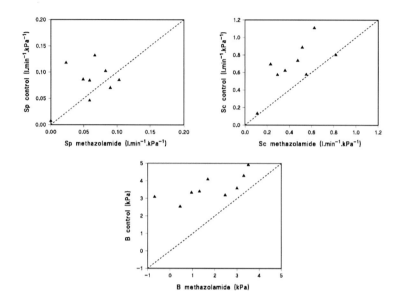

Figure 1. Scatterdiagrams of the respiratory effects of methazolamide. Sp and Sc are the CO_2 sensitivities of the peripheral and central chemoreflex loops, respectively. B is the intercept on the $P_{ET}CO_2$ axis of the CO_2 ventilatory response curve.

4. DISCUSSION

Our data show a clear decrease in apnoeic threshold and central CO_2 sensitivity by low-dose methazolamide. However, we could not demonstrate a significant decrease in Sp in our animals. The agent caused a small but significant decrease in Base Excess.

Previously, we have attributed the decrease in apnoeic threshold by low-dose acetazolamide to a possible effect on the relationship between cerebral blood flow and brain tissue PCO_2 ($PtCO_2$; 11). Similar to our previous experiments with low-dose acetazolamide, a significant arterial-to-end-tidal PCO_2 gradient was also absent after methazolamide, albeit at a somewhat lower dose. Methazolamide, however, is a much more permeable inhibitor with an about equal affinity for carbonic anhydrase II (1,2), so that compared to acetazolamide (4 mg.kg^{-1}) after 3 mg.kg^{-1} methazolamide the fractional inhibition of intracellular CA can be expected to be at least equal if not larger. Avoiding complete inhibition is important to prevent large tissue acidosis, which then could have explained the large decrease in apnoeic threshold. Our data show that the decrease in mean B was considerably larger than after low-dose acetazolamide, and this may be caused by additional inhibition of intracellular CA in brain capillary cells.

A change in the CBF- $PtCO_2$ relationship may also result in a change in CO_2 sensitivity of the central chemoreflex loop. Compared to acetazolamide, we now find a smaller decrease in Sc. After low-dose methazolamide, however, inhibition of CNS carbonic anhydrase could have altered central chemoreceptors CO_2 sensitivity for at least two reasons. First, carbonic anhydrase has been shown to be present in rostroventrolateral medullary structures associated with central chemoreceptors (8,9) and second, in a previous study we showed that a high dose methazolamide caused *an increase* in CO_2 sensitivity independently of extracellular and erythrocytic CA inhibition (17). Inhibition of brain carbonic anhydrase is followed by decease in intracellular buffer capacity against CO_2 (18). Intracellular pH changes play a crucial role in central CO_2 chemoreception (references see 19). In conclusion, the lower decrease in Sc by methazolamide compared to that by acetazolamide may be caused by a combined effect of the former on the CBF-$PtCO_2$ relationship (tending to decrease the CO_2 sensitivity of the central chemoreflex loop) and on the buffer capacity of central chemoreceptors (tending to increase their sensitivity to changes in PCO_2).

An interesting finding was that, in contrast to acetazolamide (11) methazolamide did not significantly reduce the CO_2 sensitivity of the peripheral chemoreflex loop. The carotid bodies contain several CA isoforms (4-6; 20). Because 3 mg.kg^{-1} methazolamide will be sufficient for complete inhibition of sulfonamide-sensitive carbonic anhydrases within the carotid bodies (but not in erythrocytes due to their very large CA content – 1,2), its failure to reduce Sp may seem surprising. Causing extracellular rather than extracellular *and* intracellular CA inhibition, low-dose acetazolamide induces a clear reduction in Sp, while the steady state hypoxic response is reduced by 50% (11,12). Our findings are reminiscent of data obtained from *in vitro* carotid body preparations in which complete CA appeared to reduce the fast initial rather than the steady state CO_2 response (21). One possible explanation of the different effects of methazolamide and acetazolamide on the peripheral chemoreflex loop may be related to a specific effect of acetazolamide on Ca^{2+}-dependent large-conductance potassium (BK) channels that is not shared by methazolamide (14). While acetazolamide has a specific, powerful stimulating effect on these channels (*i.e.* BK channels from skeletal muscles of K^+-deficient rat), methazolamide entirely lacks such an opening effect (14). As recently shown by Williams et al (15) BK channels may play a crucial role in the response of type-I carotid body cells to hypoxia. Preliminary data from our lab indicate that in contrast to acetazolamide, low-dose methazolamide does *not* reduce the steady state hypoxic response in the cat indicating that BK channels may indeed be involved in the inhibiting effect of acetazolamide and that CA inhibition in the carotid bodies not necessarily reduces their steady state response to changes in PO_2 and PCO_2.

REFERENCES

1. Maren T.H., 1967. Carbonic anhydrase: Chemistry, Physiology and inhibition. Physiol. Rev. 47: 595-781.
2. Maren T.H., 1977. Use of inhibitors in physiological studies of carbonic anhydrase. Am. J. Physiol. 232:F291-297.

3. Effros R.M., Chang R.S., Silverman P.,1978. Acceleration of plasma bicarbonate conversion to carbon dioxide by pulmonary carbonic anhydrase. Science 199: 1292-1298
4. Ridderstråle Y., Hanson M.A., 1984.Histochemical localization of carbonic anhydrase in the cat carotid body. Proc. N.Y. Acad. Sci. 429: 398-400.
5. Nurse C.A., 1990. Carbonic anhydrase and neuronal enzymes in cultured glomus cells of the carotid body of the rat. Cell Tissue Res. 261:65-71.
6. Rigual C., Iñiguez C., Carreres J., Gonzales C., 1985. Carbonic anhydrase in the carotid body and the carotid sinus nerve. Histochem. 82:577-580
7. Geers C., Gros G., 2000. Carbon dioxide transport and carbonic anhydrase in blood and muscle. Physiol. Rev. 80:681-715.
8. Ridderstråle Y., Hanson M.A., 1985. Histochemical study of the distribution of carbonic anhydrase in the cat brain. Acta Physiol. Scand. 124:557-564.
9. Torrance R.W., 1993. Carbonic anhydrase near central chemoreceptors. Adv. Exp. Med. Biol. 337: 235-239.
10. Swenson E.R., Hughes J.M.B., 1993. Effects of acute and chronic acetazolamide on resting ventilation and ventilatory responses in man. J. Appl. Physiol. 74:230-237.
11. Wagenaar M., Teppema L.J., Berkenbosch A., Olievier C.N., Folgering H., 1996. The effect of low-dose acetazolamide on the ventilatory CO_2 response curve in the anaesthetized cat. J. Physiol. (Lond.) 495:227-237.
12. Teppema,L.J., Dahan A., 2004. Low-dose acetazolamide reduces the hypoxic ventilatory response in the anaesthetized cat, Respir. Physiol. Neurobiol. 140:43-51.
13. Teppema L.J., Dahan A., Olievier C.N., 2001. Low-dose acetazolamide reduces CO_2-O_2 stimulus interaction within peripheral chemoreceptors in the anaesthetized cat. J. Physiol. (Lond.) 537: 221-229.
14. Tricarico D., Barbieri M., Mele, A., Carbonara G., Camerino D.C., 2004. Carbonic anhydrase inhibitors are specific openers of skeletal muscle BK channel of K^+ deficient rats. FASEB J. 18:760-761.
15. Williams S.E., Wootton P, Mason MS, Bould J, Iles D.E., Riccardi D., Peers C., Kemp P.J., 2004. Hemoxygenase-2 is an oxygen sensor for a calcium-sensitive potassium channel. Science. 17: 306(5704):2093-2097..
16. DeGoede J., Berkenbosch A., Ward D.S., Bellville J.W., Olievier C.N., 1985. Comparison of chemoreflex gains obtained with two different methods in cats. J. Appl. Physiol. 59:170-179.
17. Teppema L., Berkenbosch A., DeGoede J., Olievier C., 1995. Carbonic anhydrase and control of breathing: different effects of benzolamide and methazolamide in the anaesthetized cat. J. Physiol. (Lond.) 488:767-777.
18. Kjällquist A., Messeter K., Siesjö B.K., 1970. The *in vivo* buffer capacity of the rat brain tissue under carbonic anhydrase inhibition. Acta Physiol. Scand. 78: 94-102.
19. Teppema L.J., Dahan A., 2005. "Central chemoreceptors". In: *Pharmacology and Pathophysiology of the Control of Breathing* , D.S. Ward, A. Dahan, L.J. Teppema eds. Series: Lung Biology in Health and Disease. Marcel Dekker Inc. New York 2005.
20. Yamamoto Y., Fujimura M., Nishita T., Nishijima K., Atoji Y., Suzuki Y., 2003. Immunohistochemical localization of carbonic anhydrase isoenzymes in the rat carotid body. J. Anat. 202: 573-577.
21. Iturriaga R., Mokashi A., Lahiri S., 1933. Dynamics of carotid body responses in vitro in the presence of CO_2-HCO_3: role of carbonic anhydrase. J. Appl. Physiol. 75:1587-1594.

Stimulus Interaction between Hypoxia and Hypercapnia in the Human Peripheral Chemoreceptors

TORU TAKAHASHI, SHINOBU OSANAI, SHOKO NAKAO, MASAAKI TAKAHASHI, HITOSHI NAKANO, YOSHINOBU OHSAKI, KENJIRO KIKUCHI

First Department of Medicine, Asahikawa Medical College 2-1-1-1 Midorigaoka Higashi, Asahikawa 078-8510, Japan

1. INTRODUCTION

The combined effects of hypoxia and hypercapnia on ventilation were demonstrated synergistic in mammalian. (Neilson 1952). Although stimulus interaction between hypoxia and hypercapnia in peripheral chemoreceptors has been clearly demonstrated in experimental animals, there have been limited studies targeted in human. It is difficult to distinguish peripheral chemoreception from central chemoreception in steady-state ventilatory response to hypercapnia, since inspired hypercapnic gas stimulates both peripheral and central chemoreceptor. On the contrary, "two breaths method" has been used to estimate the peripheral chemoreceptor activity in human. In this method, transient alteration of ventilation with two breaths of hypoxic gas or hypercapnic gas may show the peripheral chemoreceptor activity. In the present study, we evaluate the ventilatory response to combined effects of hypoxia and hypercapnia in the peripheral chemosensitivity in healthy human by using two breaths method.

2. METHODS

The subjects were all healthy adult male-volunteers (25-39 yr.). They were explained of the study design and gave informed consent to participate in the study.

Subjects seated quietly with nose clip and breathed though a mouthpiece attached with one-way valve. End-tidal partial pressure of carbon dioxide ($P_{ET}CO_2$) and oxygen ($P_{ET}O_2$), respiratory rate (RR), inspiratory tidal volume (TV), inspiratory minute volume (V'_I), and arterial oxygen saturation (S_aO_2) were continuously monitored.

Two breaths of hypercapnic gas (12% CO_2) were performed in three levels of S_aO_2 (approximately 80%, 95% and 100%), which were controlled by regulation of F_IO_2. V'_I, RR and TV were measured during 5 to 15 sec after two breaths of hypercapnic gas.

2.1 Data Analysis

All results were presented as means ± S.E.M. Statistical significance was assessed using the Student's two-tailed paired *t*-test for comparison between hypoxic challenge and hypoxic hypercapnic challenge. Analysis of variance was used in order to perform a multiple comparisons among each challenge. If a significant value was indicated, Scheffe's F test was done. Statistical significance was assumed at P < 0.05.

3. RESULTS

Increments of V_I responded to CO_2 ($\Delta V'_I$) at the three levels of S_aO_2 were 1.0 l/min (at hyperoxia), 1.8 l/min (at normoxia) and 2.3 l/min (at hypoxia), respectively (Figure 1). The $\Delta V'_I$ were significantly increased by the decrease of S_aO_2. Slope of peripheral hypercapnic response ($\Delta V'_I/\Delta P_{ET}CO_2$) were also increased at each degree of S_aO_2 (Figure 2). The physiological value of peripheral chemosensitivity of this study was 0.21 l/min/mmHg. It was demonstrated that the increments of V_I response to both hypoxia and hypercapnia depended on TV, but not RR (Figure 3). Stimulus effects of hypoxia and hypercapnia on peripheral chemoreception increased V'_I elicited by increasing of TV.

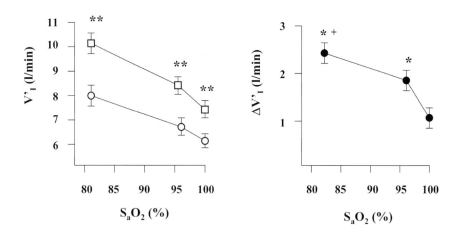

Figure 1. Hypoxic challenges under both normocapnia (open circles) and hypercapnia (open squares) by two-breath method. $\Delta V'_I$ is differences between normocapnic hypoxia and hypercapnic hypoxia at each S_aO_2 (closed circles). *: P<0.05 vs. hyperoxia. **: P<0.01 vs. normocapnia. $^+$: P<0.05 vs. normoxia.

Interaction between Hypoxia and Hypercapnia 265

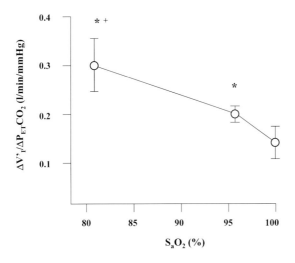

Figure 2. Interactions between hypoxia and hypercapnia presented by slope of peripheral hypercapnic response ($\Delta V'_I/\Delta P_{ET}CO_2$). *: $P<0.05$ vs. hyperoxia. $^+$: $P<0.05$ vs. normoxia.

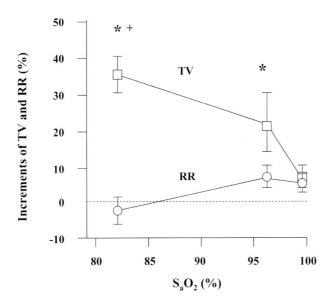

Figure 3. Analysis of increments of tidal volume (TV) and respiratory rate (RR) response to both hypoxia and hypercapnia. RR: open circles and TV: open squares. *: $P<0.05$ vs. hyperoxia. $^+$: $P<0.05$ vs. normoxia.

4. DISCUSSION

We studied the interaction between hypoxia and hypercapnia on the peripheral chemosensitivity in the human by using two breaths method. The present study showed that two breaths method can provide a useful assessment to estimate the peripheral chemosensitivity.

The modulations of peripheral chemosensitivity are implicated in various diseases, including sleep-disordered breathing, congestive heart failure, and certain forms of hypertension (Osanai 1999, Prabhakar 2004). But, it is still controversial whether the chemoreception is augmented or not. Hereafter, the interaction between hypoxia and hypercapnia on peripheral chemosensitivity in those patients should be studied for solution of this issue.

REFERENCES

Neilson M and Smith H. Studies on the regulation of respiration in acute hypoxia. Acta Physiol. Scand 24: 293-313, 1952.

Miller JP, Cunningham DJC, Llyod BB and Young JM. The transient respiratory effects in man of sudden changes in alveolar CO_2 in hypoxia and in high oxygen. Respir Physiol 20: 17-31, 1974.

Osanai S, Akiba Y, Fujiuchi S, Nakano H, Matsumoto H, Ohsaki Y and Kikuchi K. Depression of peripheral chemosensitivity by a dopaminergic mechanism in patients with obstructive sleep apnoea syndrome. Eur Respir J 13: 418-423, 1999.

Prabhaker NR and Peng YJ. Peripheral chemoreceptiors in health and disease. J Appl Pysiol 96: 359-366, 2004.

Gene Expression and Signaling Pathways by Extracellular Acidification

NORIAKI SHIMOKAWA, MARINA LONDOÑO AND NORIYUKI KOIBUCHI

Department of Integrative Physiology, Gunma University Graduate School of Medicine, Maebashi, Japan

1. INTRODUCTION

The respiratory response to extracellular acidosis by hypercapnia is mediated by central chemoreceptor neurons in the medulla oblongata [1]. There are actually two defined groups of respiratory neurons. The dorsal group of neurons is located in and near the nucleus of the tractus solitarius and their activity is regulated by changes in the arterial partial pressure of CO_2 (Pco_2), O_2 (Po_2) or H^+. The ventral group is a long column of neurons that extends through the nucleus ambiguous and retroambiguous in the ventrolateral medulla. In addition to reacting to peripheral stimuli, the ventral neurons detect changes in the H^+ and/or CO_2 concentrations in the cerebrospinal fluid (CSF) and brain interstitial fluid [2]. The capacity to detect these changes is called central chemosensitivity.

According to the finding that in cats the discharge frequency of ventral medullary surface (VMS) neurons is increased with lowered pH of CSF, the H^+ concentration seems to be a stimulant of central chemosensitivity [3-6]. Carbon dioxide readily penetrates membranes, whereas H^+ and HCO_3^- penetrate slowly. The CO_2 that enters the brain and CSF is promptly hydrated. The H_2CO_3 dissociates, so that the local H^+ concentration rises. Therefore, the effects of CO_2 on respiration are mainly due to CO_2 movement into the CSF, where it increases the H^+ concentration and stimulates receptors/sensors for H^+. Thus, the direct stimulant of the central chemosensitive neurons may be H^+ rather than CO_2 [7, 8].

It is still not clear how H^+ excites the H^+-sensitive (chemosensitive) neurons in the VMS. There is some evidence for H^+-sensing ionic channels in sensory neurons: H^+ activates Na^+ conductance in small neurons of the rat trigeminal ganglion [9]; H^+ activates Ca^{2+} channel in rat sensory neurons [10]; a stepwise reduction in extracellular pH induced an increase in Na^+ current in small dorsal root ganglion cells of the frog [11]; and H^+ and capsaicin share a common mechanism of neuronal activation in rat dorsal root ganglion cells [12]. The H^+-sensitive neurons in the VMS may also have H^+-sensing ionic channels/sensors or similar mechanisms for reacting to extracellular H^+ changes. Few studies have investigated the identification of chemosensitive molecules responsible for respiratory regulation in the VMS. Not long ago, Waldmann et al. succeeded in cloning the H^+-gated cation channel (ASIC, for acid-sensing ionic channel) that belongs to the amiloride-sensitive Na^+ channel/degenerin family of ion channels [13]. ASIC is expressed in dorsal root ganglia and is also distributed widely throughout the brain. The H^+-gated cation cannel is activated transiently by rapid extracellular acidification and induces cation (Na^+, Ca^{2+}, K^+) influx. More recently, it has been shown that ovarian cancer G-protein-coupled receptor 1 (OGR1), previously described as a receptor for sphingosylphosphorylcholine, acts as an H^+-sensing receptor stimulating inositol phosphate (IP) formation [14]. The receptor is stabilized in an inactive state at pH 7.8, and fully activated at pH 6.8. Pertussis toxin did not inhibit IP formation measured at pH 7.0, indicating that OGR1 acts through Gq. Ovarian cancer G-protein-coupled receptor 4 (OGR4) also responds to pH changes, the receptor promotes cAMP formation through Gs. ASIC, OGR1 and 4 are candidates for chemosensitive molecules responsible for respiratory regulation in the VMS. However, it has been no evidence that these channel/receptors are involved in the central chemosensitivity for respiratory regulation.

Finally, several advances have been made in understanding central chemosensitivity at the cellular and functional levels. Here, we describe the detection of genes that were stimulated by acidosis after hypercapnia (increased arterial CO_2) and the intracellular signal transduction by extracellular acidosis.

2. ANALYSIS OF ACIDOSIS-INDUCED GENES

To elucidate the cellular response at molecular level to the acidosis of neuronal cells, we screened genes that were stimulated by acidosis after hypercapnic stimulation. We applied the differential display technique to adult rat brains and compared gene expression under normocapnic and hypercapnic conditions. Differential display is an established technique that provides the microanalysis of transcriptional changes occurring in a given cell or tissue [15]. Rats inhaled either air (normocapnic stimulation, 0.04% CO_2) or air containing 7% CO_2 (hypercapnic stimulation) for 5 min to stimulate the medullary chemosensitive neurons [16]. The RNA derived from the VMS after either air or 7% CO_2 inhalation was used for polymerase chain reaction (PCR) amplification. Over 11,500 PCR products were generated, and 14 (0.12%) of the observed bands exhibited gene profiles of high expression as a result of hypercapnic stimulation. We found that the sequences of eight clones were novel and that six other clones had already been reported to be genes readily induced by hypercapnia (Table 1). Several representative clones are also shown below.

2.1 Fos/Jun Family

At the end of the 1980s, proto-oncogene c-fos was found to serve as a marker for activated neurons. Several investigators have shown that the c-fos expression is rapidly and transiently activated after hypercapnic stimulation. Sato et al. [17] found that the c-Fos protein was expressed in the neurons of the VMS and NTS in rats after the animals inhaled 10-15% CO_2. Since then, increasing evidence has shown that the c-Fos protein is expressed in neurons of the VMS and putative chemosensitive sites in response to hypercapnic stimulation [18-20]. Miura et al. have reported a topological map of acidosis-sensitive neurons after hypercapnic stimulation in the VMS and the morphological and immunochemical properties of these neurons by the c-Fos immunohistochemistry [16]. We also detected genes for Fos/Jun by differential display technique for the identification of high-expressed genes at hypercapnic stimulation in the VMS and therefore, the technique indicates reliable for screening of acidosis-induced genes.

2.2 Maf Protein Family

We found that hypercapnic stimulation induced the gene expression of *maf*G [21], a member of the Maf protein family of basic leucine zipper (bZIP) transcription factors. It has been reported that MafG forms heterodimers with c-Fos at each leucine zipper structure and that the Maf complex recognizes two palindromic DNA sequences, TGCTGACTCAGCA and TGCTGACGTCAGCA [22]. The middle parts of the two consensus binding sequences for Maf are identical with two binding sequences for Fos/Jun (TGACTCA and TGACGTCA). MafG may be involved in the signal transduction of H^+-sensitivity and respiration with Fos/Jun protein, either competing for binding sites or interacting directly with Fos/Jun. MafG-2 is a novel splice variant of MafG and differs from MafG by an insertion of 27 amino acids [23]. Sequence analysis of the protein has shown that the basic domain for DNA binding and the leucine zipper structure are conserved in MafG-2. Because expression of *maf*G-2 mRNA increases when extracellular pH is decreased gradually from 7.40 to 7.20, both MafG and MafG-2 may be involved in signal transduction of extracellular pH change.

Recently, we analyzed the properties of gene expression, protein interaction and DNA binding activity of Maf and FosB during extracellular acidification [24]. When cells were incubated in low pH medium, the expression of small Maf proteins (MafG, MafK and MafF) and FosB were clearly increased in an extracellular pH-dependent manner showing a peak after 1-2 h of stimulation. Immunofluorescence and protein binding studies indicated that MafG was partially co-localized with FosB in the nucleus and MafG

Table 1. Acidosis-induced genes.

Gene	Characteristics
Fos/Jun	Transcriptional regulator of genes having AP-1 site
MafG	Basic leucine zipper transcription factor
MafG-2	Novel splice variant of MafG
Rhombex-29	New member of PLP/DM20-M6 family
Past-A	pH-dependent sugar transporter

can form heterodimers with FosB at extracellular pH 7.40. Moreover, we found that MafG-FosB complexes are able to bind to AP-1 consensus sequence, TGACTCA. To investigate whether extracellular acidification influences to dimerization and DNA binding activity of MafG and FosB, extracellular pH of cultured cells was decreased from 7.40 to 6.80. The decrease in extracellular pH gave rise to an enhanced dimerization of MafG with FosB leading to augmentation of the DNA binding activity of the heterodimer to AP-1 consensus sequence. Taken together, these results suggest that MafG-FosB complexes are involved in transcriptional regulation in response to extracellular acidification.

2.3 Past-A

Lately, we cloned a new sugar transporter and named as Past-A (proton-associated sugar transporter-A), which is induced in the brain after hypercapnic stimulation [25]. Past-A mRNA is expressed highly in the VMS, moderately in the cerebral cortex and cerebellum, extremely poorly in the heart and kidneys. In response to hypercapnic stimulation, the number of Past-A immunoreactive neurons in the VMS was about four times greater than that of after air inhalation.

The cDNA of Past-A contains an open reading frame encoding a sequence of 751 amino acids, and the relative molecular weight of the residues was calculated as approximately 82 kDa. Analysis of the predicted amino acid sequence suggested the presence of 12 putative membrane-spanning helices with a long cytoplasmic loop between transmembrane (TM) helices TM6 and TM7, and with cytoplasmically oriented NH_2 and COOH termini (Fig. 1). Primary structure analysis indicated that the Past-A protein belongs to a sugar transporter family in the major facilitator superfamily (MFS) [26]. In addition, Past-A has the same membrane topology as the glucose transporters conserving the important motifs for glucose transport activity such as the RXGRR motif (where X denotes any amino acid), the PESP, and the QLS motif. Supporting this, transient transfection of Past-A in COS-7 cells leads to the expression of a membrane-associated 82 kDa protein that possesses a glucose transport activity. The acidification of extracellular medium facilitated glucose uptake, whereas the addition of carbonyl cyanide m-chlorophenylhydrazone, a protonophore, inhibited glucose import.

Expression of Past-A is significantly increased in neurons of the VMS in response to decreased extracellular pH following hypercapnia. Our results also indicate that Past-A has a functional role in controlling glucose uptake along the pH gradient, suggesting that hypercapnia may stimulate the uptake of glucose, the primary energy source, into acidosis-stressed neurons of VMS. Signal transduction of Past-A-related sugar homeostasis in neuronal cells after hypercapnia is under investigation. AMP-activated protein kinase (AMPK) is a metabolic stress sensing protein kinase that plays a key role in regulation of energy homeostasis [27], more specifically, in glucose uptake. The kinase is activated by an elevated AMP:ATP ratio due to cellular and environmental stress, such as hypoxia, ischemia and heat shock [28]. It is intriguing to speculate that the promotion of the gene expression and translocation to membrane of Past-A is regulated through the AMPK signaling pathway. Future functional analysis of Past-A may provide new insights into the biochemical regulation of glucose-sensing mechanisms in the brain.

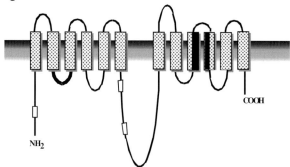

Figure 1. Proposed structural model of the Past-A protein. Shaded areas show the amino acid segment forming a putative transmembrane region. Three proline-rich regions are indicated by open rectangles. The leucine repeats are marked with black. The sucrose-H^+ transport motif is indicated with bold line.

3. SIGNAL TRASDUCTION OF ACIDOSIS-INDUCED c-JUN EXPRESSION

As mentioned above, an increase of the H^+ concentration in the cerebrospinal fluid and brain interstitial fluid by hypercapnic stimulation induces expression of several bZIP transcription factors, such as c-Jun, Fos and small Maf proteins, in specific nuclei in the central nervous system [16-19, 21]. Among such proteins, c-Jun immunoreactive neurons are distributed in the respiration-related motor nuclei of the medulla oblongata and spinal cord, and in the central chemoreceptive area of the ventral medullary surface of the medulla after hypercapnic stimulation. In *in vitro* study, we found significant increase in *c-jun* mRNA with decrease in extracellular pH from 7.40 to 7.20 [29]. The acidosis-induced *c-jun* mRNA expression was inhibited with the Ca^{2+}/calmodulin inhibitor trifluoperazine, indicating that the expression of *c-jun* mRNA by an increase in extracellular H^+ is mediated partly by this system. Calmodulin, a ubiquitous intracellular Ca^{2+} receptor, binds to short peptide sequences of many target proteins upon binding Ca^{2+}. Such interaction is thought to induce a conformational change on the target, resulting in its activation, e.g. phosphorylation. Targets include the Ca^{2+}/calmodulin-dependent protein kinases (CaM kinases) and Calcineurin [30]. This study may clarify the mechanisms of our findings that the hyperventilatory response to the CO_2 inhalation is abolished when the intercellular Ca^{2+} chelator BAPTA-AM (1, 2-bis [2-amino-4-fluorophenoxy] ethane-*N, N, N', N'*-tetraacetic acid, tetraacetoxymethyl ester) is applied to the VMS [31]. On the other hand, Kuo et al. [32] reported that protein kinase C_α (PKC_α) is a possible mediator in H^+-induced c-Fos expression, and that PKC_α may be activated by Ca^{2+}. Activated PKC_α could lead to an increase in the phosphorylation of Raf-1 kinase, which in turn activates mitogen-activated protein (MAP) kinases so enhancing the expression of *c-fos* mRNA. Taken together, it is highly probable that an increase in the concentration of extracellular H^+ induces c-Jun/Fos through Ca^{2+}/calmodulin and MAP kinase pathways.

In a recent study, we found another intracellular signaling pathway for extracellular H^+-induced c-Jun expression in several cell lines (HEK293, COS-7 and PC12)[33]. When cells were incubated in low pH medium, the phosphorylation of JNK and expression of c-Jun were clearly observed in cells in an extracellular pH- and time-dependent manner. We also demonstrated that after being phosphorylated in the cytoplasm due to an increase in extracellular H^+ concentration, the JNK is accumulated in the nucleus. Many tumor cells have relatively acidic extracellular pH and are killed by intracellular acid-induced injury. The acid-induced cell death depends on bax, a proapoptotic binding partner of bcl-2, and on JNK signaling pathways [34]. Recently, Yamamoto et al. reported that acidification of the cytoplasm using cycloprodigiosin hydrochloride (cPrG·HCl), a novel H^+/Cl^- symport drug, gives rise to apoptosis in cancer cells through up-regulation of Fas ligand, JNK and caspase [35]. Taken together, an increase in concentration of extracellular H^+ leads to phosphorylation of JNK, and then the phosphorylated JNK translocates to the nucleus to augment the transcriptional activity of c-Jun.

On the other hand, electrophysiological studies showed that a rapid shift in extracellular pH from 7.4 to 6.9 caused an inward current, probably due to an increase in Na^+ and K^+ permeability across the membrane [36], and that an increase in the concentration of extracellular H^+ induced H^+-gated Na^+ current in the hypothalamic neurons [37]. H^+-induced Na^+ current produces depolarization of the membrane and then provokes influx of extracellular Ca^{2+} via voltage-gated Ca^{2+} channels [38, 39]. From these observations, we examined whether the Ca^{2+} influx is involved in extracellular H^+-induced JNK phosphorylation and c-Jun expression using nimodipine, pharmacological agent known as a blocker of voltage-gated Ca^{2+} channels [33]. Nimodipine prevented partly phosphorylation of JNK and expression of c-Jun. This result indicates that extracellular H^+-induced JNK phosphorylation and c-Jun expression are mediated partially by the increase of intracellular Ca^{2+} concentration. Taken together, our results suggest a novel pathway to acidosis-induced c-Jun expression: an increase of extracellular H^+ provokes Ca^{2+} influx by depolarization through Na^+ influx, and the increase of intracellular Ca^{2+} concentration induces c-Jun expression through phosphorylation of JNK. Critical molecules and pathways involved in signal transduction for extracellular acidification are shown in Fig. 2.

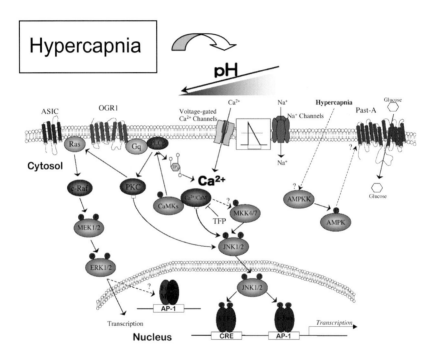

Figure 2. Signaling pathways in extracellular acidification. Extracellular acidification induces JNK phosphorylation and c-Jun expression via partly extracellular Ca^{2+} influx through voltage-gated Ca^{2+} channels. Elevation of intracellular Ca^{2+} concentration after IgE- and antigen-dependent stimulation in rat basophilic leukemia mast cells increases JNK activity, possibly through the calmodulin pathway [40]. Activation of the CaMK pathway increased JNK kinase activity through regulating the PKC pathway [41]. Increase of extracellular H^+ concentration activates dimerization and DNA binding activities between FosB and MafG and also induces nuclear transcription factors such as Fos and Maf through PKC-MAP kinases and Ca^{2+}/calmodulin system. AMPK plays a key role in regulation of glucose homeostasis. AMPK phosphorylation at Thr172 by the upstream kinase AMPKK is required for its activation [42].

4. CONCLUSIONS AND FUTURE PERSPECTIVES

How neuronal cells sense and respond to changes in pH in their environment is one of fundamental questions in neurophysiology. Recent findings indicate the existence of different mechanisms elicited by different types of neurons involved in respiratory regulation. It is also becoming clear that several intracellular signal transduction pathways are implicated in the regulation of gene transcription that provides the appropriate set of early responsive genes to acidosis by hypercapnia. These novel discoveries may provide new insights into the design for new drugs and understanding of neurological disorders.

REFERENCES

[1] Loeschcke, H.H. *J. Physiol.,* 1982, *332*, 1.
[2] Ganong, W.F. Regulation of respiration; In *Review of Medical Physiology*; Ganong, W.F., ed. Appleton and Lange: Connecticut, 1999; pp 640-649.
[3] Fukuda, Y.; Honda, Y. *Pfluegers Arch.,* 1976, *364*, 243.
[4] Schlaefke, M.E.; Pokorski, M.; See, W.R.; Prill, R.K.; Loeschcke, H.H. *Physiopathol. Respir.,* 1975, *11*, 277.
[5] Fukuda, Y.; Honda, Y. *Nature,* 1975, *256*, 317.
[6] Schlaefke, M.E.; See, W.R.; Loeschcke, H.H. *Respir. Physiol.,* 1970, *10*, 198.
[7] Cragg, P.; Patterson, L.; Purves, M.J. *J. Physiol.,* 1977, *272*, 137.
[8] Mitchell, R.A.; Loeschcke, H.H.; Massion, W.H.; Severinghaus, J.W. *J. Appl. Physiol.,* 1963, *18*, 523.
[9] Krishtal, O.A; Pidoplichko, V.I. *Neurosci. Lett.,* 1981, *24*, 243.
[10] Kovalchuk, Y.; Krishtal, O.A.; Nowycky, M.C. *Neurosci. Lett.,* 1990, *115*, 237.
[11] Akaike, N.; Krishtal, O.A.; Maruyama,T. *J. Neurophysiol.,* 1990, *63*, 805.
[12] Bevan, S.; Geppetti, P. *Trends Neurosci.,* 1994, *17*, 509.
[13] Waldmann, R.; Champigny, G.; Bassilana, F.; Heurteaux, C.; Lazdunski, M. *Nature,* 1997, *386*, 173.
[14] Ludwig, M.G.; Vanek, M.; Guerini, D.; Gasser, J.A.; Jones, C.E.; Junker, U.; Hofstetter, H.; Wolf, R.M.; Seuwen, K. *Nature,* 2003, *425*, 93.
[15] Liang, P.; Pardee, A.B. *Science,* 1992, *257*, 967.
[16] Miura, M.; Okada, J.; Kanazawa, M. *Brain Res.,* 1998, *780*, 34.
[17] Sato, M.; Severinghaus, J.W.; Basbaum A.I. *J. Appl. Physiol.,* 1992, *73*, 96.
[18] Haxhiu, M.A.; Yung, K.; Erokwu, B.; Cherniack, N.S. *Respir. Physiol.,* 1996, *105*, 35.
[19] Teppema, L.J.; Veening, J.G.; Kranenburg, A.; Dahan, A.; Berkenbosch, A.; Olievier, C. *J. Comp. Neurol.,* 1997, *17*, 169.
[20] Belegu, R.; Hadziefendic, S.; Dreshaj, I.A.; Haxhiu, M.A.; Martin R.J. *Respir. Physiol.,* 1999, *117*, 13.
[21] Shimokawa, N.; Okada, J.; Miura, M. *Mol. Cell. Biochem.,* 2000, *203*, 135.
[22] Kataoka, K.; Noda, M.; Nishizawa, N. *Mol. Cell. Biol.,* 1994, *14*, 700.
[23] Shimokawa, N.; Kumaki, I.; Takayama, K. *Cell. Signal.,* 2001, *13*, 835.
[24] Shimokawa N, Kumaki I, Qiu CH, Ohmiya Y, Takayama K, Koibuchi N. *J. Cell. Physiol.,* 2005, in press.
[25] Shimokawa, N.; Okada, J.; Hugland, K.; Dikic, I.; Koibuchi, N.; Miura, M. *J. Neurosci.,* 2002, *22*, 9160.
[26] Pao, S.S.; Paulsen, I.T.; Saier, M.H. *Mol. Biol. Rev.,* 1998, *62*, 1.
[27] Hardie, D.G.; Carling, D.; Carlson, M. *Annu. Rev. Biochem.,* 1998, *67*, 821.
[28] Corton, J.M.; Gillespie, J.G.; Hardie, D.G. *Curr. Biol.,* 1994, *4*, 315.
[29] Shimokawa, N.; Sugama, S.; Miura, M. *Cell. Signal.,* 1998, *10*, 499.
[30] Anthony, R.M. *Mol. Endocrinol.,* 2000, *14*, 4.

[31] Kanazawa, M.; Sugama, S.; Okada, J.; Miura, M. *J. Auton. Nerv. Syst.*, 1998, *72*, 24.
[32] Kuo, N.T.; Agani, F.H.; Haxhiu, M.A.; Chang, C.H. *Respir. Physiol.*, 1998, *111*, 127.
[33] Shimokawa, N.; Qiu, C.H.; Seki, T.; Dikic, I.; Koibuchi, N. *Cell. Signal.*, 2004, *16*, 723.
[34] Zanke, B.W.; Lee, C.; Arab, S.; Tannock, I.F. *Cancer Res.*, 1998, *58*, 2801.
[35] Yamamoto, D.; Uemura, Y.; Tanaka, K.; Nakai, K.; Yamamoto, C.; Takamoto, H.; Kamata, K.; Hirata, H.; Hioki, K. *Int. J. Cancer*, 2000, *88*, 121.
[36] Krishtal, O.A.; Pidoplichko, V.I. *Neuroscience*, 1980, *5*, 2325.
[37] Inoue, H; Kirschner, D.A. *J. Neurosci. Res.*, 1991, *28*, 1.
[38] Ueno, S.; Nakaye, T.; Akaike, N. *J. Physiol.*, 1992, *447*, 309.
[39] Akaike, N.; Ueno, S. *Prog. Neurobiol.*, 1994, *43*, 73.
[40] Funaba, M.; Ikeda, T.; Ogawa, K.; Abe, M. *Cell. Signal.*, 2003, *15*, 605.
[41] Werlen, G.; Jacinto, E.; Xia, Y.; Karin, M. *EMBO J.*, 1998, *17*, 3101.
[42] Hawley, S.A.; Davison, M.; Woods, A.; Davies, S.P.; Beri, R.K.; Carling, D.; Hardie, D.G. *J. Biol. Chem.*, 1996, *271*, 27879.

Hypoxic Modulation of the Cholinergic System in the Cat Carotid Glomus Cell

JEFFREY A. MENDOZA, IRENE CHANG AND MACHIKO SHIRAHATA

Departments of Environmental Health Sciences, Johns Hopkins Bloomberg School of Public Health, Baltimore, USA

1. INTRODUCTION

The carotid body is a primary sensory organ for arterial hypoxia. Chemosensory glomus cells in the carotid body release neurotransmitters, including ACh, in response to hypoxia. The release of neurotransmitters from the glomus cell, a putative chemoreceptor cell, appears to be triggered by an influx of calcium and subsequent increase in intracellular calcium ($[Ca^{2+}]i$). Several reports indicate that L-type and some other types of voltage-gated calcium channels are responsible for neurotranmitter release from glomus cells (Gonzalez et al., 1994). These channels are activated by depolarization of the cell membrane. However, the speed and the degree of depolarization in glomus cells may not be sufficient to activate voltage-gated Ca^{2+} channels at mild hypoxia (Chou et al., 1998), where afferent neural activity from the carotid body starts increasing. This discrepancy led us to search for other mechanisms which elevate $[Ca^{2+}]i$ followed by neurotransmitter release.

Pharmacolgical studies have indicated that Ca^{2+} influx into the glomus cell can occur via nicotinic ACh receptors (nAChRs) (Dasso et al,. 1997; Shirahata et al., 1997), and therefore, nAChRs could play an important role in Ca^{2+} homeostasis of glomus cells. The activation of nAChRs in glomus cells increases the release of neurotransmitters such as catecholamines (Dinger et al., 1985; Gomez-Nino et al., 1990; Obeso et al., 1997). Several subunits of neuronal nicotinic nAChRs are present in the carotid body (Hirasawa et al., 2003). We have also shown that the nAChR subunits expressed in N1E115 neuroblastoma cells are similar to those in glomus cells. Further, the activity of nAChRs in N1E115 neuroblastoma cells is enhanced by mild hypoxia using patch clamp and microfluorometric techniques (Shirahata et al., 2003). A question arises whether the activity of nAChRs in glomus cells is also enhanced by mild hypoxia. If this happens, these receptors could contribute to further release of neurotransmitter from glomus cells at the beginning of hypoxia.

2. METHODS

2.1 Cell Culture

Carotid body cells were cultured as described previously (Shirahata et al., 1994). Cats were euthanized with ketamine (30-50 mg/kg, ip) and pentobarbital (50-100 mg/kg, iv), and decapitated to avoid bleeding in the area of the carotid body. The carotid bodies were harvested, cleaned, and dissociated with collagenase and gentle trituration. The cells are seeded in wells made of a round glass coverslip (bottom) and a plastic cylinder (side). Cells are cultured in a defined medium in a CO_2 incubator (5% CO_2/air) at 37 °C for up to two weeks. The medium is changed twice a week. The basic nutrient solution is a 1:1 mixture of Dulbecco's modified Eagle's medium and Ham's F12 medium supplemented with bovine serum albumin, bovine transferrin, bovine insulin, sodium selenite, 7s-nerve growth factor and pyruvate.

2.2 Fluorescent Imaging

Cells cultured on a glass coverslip were incubated with Fura-2/AM for 90 min. The cells were moved to the recording chamber on an inverted microscope and washed for 5-10 min by superfusing Krebs solution at a flow rate of 1.5ml/minute. The composition of Krebs solution was (in mM): NaCl 118, KCl 4.7, $MgSO_4 \cdot 7H_2O$ 1.2, KH_2PO_4 1.2, $NaHCO_3$ 25, EDTA 0.0016, and glucose 11.1, pH 7.4 equilibrated with 5% CO2/air at 37 °C. Images were collected through a 510 nm interference filter with a cooled CCD camera during alternate excitations at 340 and 380 nm. A PC based computer and ImageMaster software (Photon Technology International) were used for the acquisition of the data. Several cells in one frame were selected. After subtracting background (taken from the area without cells) data for each cell were analyzed by averaging the fluorescent ratio values in all pixels within the selected field. In most analysis relative changes from the control values were used for describing $[Ca^{2+}]i$ changes.

2.3 Protocol for Experimental Set 1

The experiments were designed to test if desensitization of nAChRs is modified by mild hypoxia. The cells were continually superfused with Krebs solution equilibrated with 5% CO_2/ air (normoxic Krebs). Cholinergic agonists (ACh or Nicotine: 100 µM) were added in Krebs and applied for one minute. Cells were exposed 3 times to cholinergic agonists with 10 minute resting periods between exposures. Five minutes prior to the second exposure of drug, Krebs solution either remained the same or was changed to one equilibrated with 5%CO_2/ 10% O_2/ 85% N_2 (hypoxic Krebs) 4 minutes after the exposure to drug, if the oxygen level was changed, it was returned to original levels.

In a subset of experiments, ACh was repeatedly administered as described above with the superfusion of normoxic Krebs. However, 5 minutes prior, during, and 4 minutes after the second exposure to ACh, Dithiothreitol (DTT: 1 mM) was added to the perfusion.

2.4 Protocol for Experimental Set 2

The second set of experiments was conducted similar to the previous set described using cholinergic agonists (ACh, nicotine: 100 µM; muscarine: 300-600 µM). However, the resting interval was increased to 20 minutes between exposures. DTT was not used for this set.

3. RESULTS

3.1 Experimental Set 1

We observed that ACh increases $[Ca^{2+}]i$ of glomus cells, but repeated applications of ACh in normoxic Krebs with 10 minute intervals resulted in graded decreases in the $[Ca^{2+}]i$ response (desensitization). Perfusion with 10% O_2 ($PO_2 \sim 70$mmHg) reduced desensitization of ACh-induced $[Ca^{2+}]i$ response (Figs. 1 & 2). Dithiothreitol (DTT) perfusion during the second exposure to ACh also reduced the desensitization of the second response (data not shown).

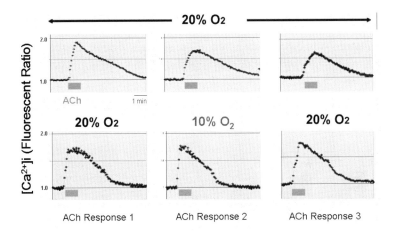

Figure 1. The top panel shows increases in [Ca2+]i from a single cell exposed to ACh under normoxic conditions, and the lower panel shows the responses of another single cell to repeated exposures to ACh when conditions are changed from normoxic to hypoxic for the second exposure.

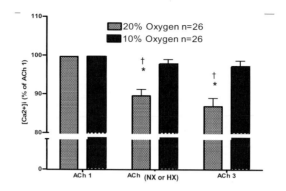

Figure 2. [Ca^{2+}]i responses of glomus cells to repeated exposures to ACh with 10 minute resting periods between exposures. * significantly different from the first exposure. +, siginificantly different between different O$_2$ levels.

Nicotine increased [Ca^{2+}]i of glomus cells. In contrast to ACh response, desensitization of nicotine-induced [Ca^{2+}]i responses was smaller and did not occur until the third exposure during normoxic Krebs superfusion. However, desensitization was clearly observed at the second and the third exposures when the cells were superfused with hypoxic Krebs during the second exposure to nicotine (Fig. 3).

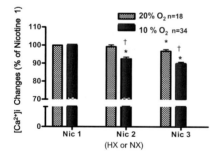

Figure 3. Glomus cell responses to repeated exposures to nicotine with 10 minute intervals. * significantly different from the first exposure. +, siginificantly different between different O$_2$ levels.

3.2 Experimental Set 2

When we increased resting intervals between Ach exposures, we observed that the ACh-induced [Ca^{2+}]i response remained stable in both the normoxic and hypoxia Krebs superfusions (Fig. 4A).

The nicotine-induced [Ca^{2+}]i response was stable with repeated application of nicotine with 20 minute resting intervals and normoxic Krebs superfusion. However, when given hypoxic Krebs during the second nicotine exposure, a small but significant decrease in the [Ca^{2+}]i response was observed (Fig. 4B).

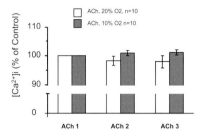
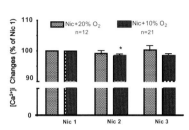

Figure 4. $[Ca^{2+}]i$ responses of glomus cells to ACh and nicotine with 20 minute resting periods between exposures. *, siginificantly different from control.

Muscarine was administered in the same manner as ACh and nicotine. Muscarine-induced $[Ca^{2+}]i$ responses were variable among the three consecutive exposures, and the responses did not show siginificant differences. There was no significant differences in $[Ca^{2+}]i$ responses between cells perfused with normoxic and hypoxic Krebs solutions (data not shown).

4. DISCUSSION

This study showed that desensitization of AChR-induced $[Ca^{2+}]i$ response in glomus cells was attenuated with repeated applications with short intervals. However, when mild hypoxia was introduced during the second application of ACh, the response was maintained. Neuronal nAChRs exist in several different states (Quick and Lester, 2002). The current results suggest that a balance between the active state and the desensitized state of cholinergic receptors shift toward the active state with mild hypoxia. Based on our previous data used neuroblastoma cells (Shirahat et al., 2003), we speculated that nAChRs are responsible for this phenomenon. However, further experiments using nicotine suggest that desensitization of nAChRs with repeated and short interval applications of nicotine is enhanced, not reduced, by mild hypoxia. Further, while ACh-induced $[Ca^{2+}]i$ increase was maintained with repeated and long interval applications at normoxia and hypoxia, nicotine-induced $[Ca^{2+}]i$ increase was again attenuated during hypoxic condition. Thus, the current data did not support our original hypothesis stating that onset of hypoxia (mild hypoxia) increases the activity of nAChRs in glomus cells. Because desensitization of ACh-induced $[Ca^{2+}]i$ response of glomus cells is attenuated by mild hypoxia, muscarinic receptors appear play a role in this phenomenon. Further studies are required to elucidate how muscarinic receptors are involved in reduced desensitization during mild hypoxia and whether different levels of hypoxia modify cholinergic functions.

ACKNOWLEDGEMENTS

This work was supported by NHLBI HL61596.

REFERENCES

Chou C.-L., Schofield B., Sham J.S.K., Shirahata M., 1998, Electrophysiological and immunological demonstration of cell-type specific responses to hypoxia in the adult cat carotid body, 789:229-238.
Dasso L.L., Buckler K.J., Vaughan-Jones R.D., 1997, Muscarinic and nicotinic receptors raise intracellular Ca^{2+} levels in rat carotid body type I cells, *J. Physiol.* 498: 327-338.
Dinger B., Gonzalez C., Yoshizaki K., Fidone S., 1985, Localization and function of cat carotid body nicotinic receptors, *Brain Res.* 339: 295-304.
Gonzalez C., Almaraz L., Obeso A., Rigual R., 1994. Carotid body chemoreceptors: from natural stimuli to sensory discharges. Physiol. Rev. 74: 829-889.
Gomez-Nino A., Dinger B., Gonzalez C., Fidone S.J., 1990, Differential stimulus coupling to dopamine and norepinephrine stores in rabbit carotid body type I cells, *Brain Res.* 525: 160-164.
Hirasawa S., Mendoza J.A., Okumura M., Kobayashi C., Chandrasegaran S., Fitzgerald R.S., Schofield B., Shirahata, M., 2003, Cholinergic receptors in the cat chemosensory unit. *Adv. Exp. Med. Biol.* 536:313-319.
Obeso A., Gomez-Nino M.A., Almaraz L., Dinger B., Fidone S., Gonzalez C., 1997, Evidence for two types of nicotinic receptors in the cat carotid body chemoreceptor cells, *Brain Res.* 754: 298-302.
Quick M.W., Lester R.A.J., 2002, Desensitization of Neuronal Nicotinic Receptors, *J. Neurbiol.* 53(4):457-78.
Shirahata M., Fitzgerald R.S., Sham J.S.K., 1997, Acetylcholine increases intracellular calcium of arterial chemoreceptor cells of adult cats, *J. Neurophysiol.* 78: 2388-2395.
Shirahata M., Higashi T., Mendoza J.A., Hirasawa S.. 2003, Hypoxic augmentation of neuronal nicotinic ACh receptors and carotid body function. *Adv. Exp. Med. Biol.* 536:269-75.
Shirahata M., Schofield B., Chin B.Y., Guilarte T.R., 1994, Culture of arterial chemoreceptor cells from adult cats in defined media. *Brain Res.* 658:60-66

Are There "CO_2 Sensors" in the Lung?

L.Y. LEE[1], R.L. LIN[1], C.Y. HO[2], Q. GU[1] AND J.L. HONG[1]

[1]Department of Physiology, University of Kentucky, Lexington, Kentucky 40536, U.S.A. and
[2]Department of Otolaryngology, Veterans General Hospital and National Yang-Ming University, Taipei, Taiwan, R.O.C.

1. INTRODUCTION

Previous investigators have suggested the existence of "CO_2 sensors" in the lung and an important role of these receptors in detecting the increase in venous CO_2 flux and in regulating ventilatory response to meet the metabolic demand during exercise (38). However, no direct and definitive evidence has been established in identifying the CO_2 receptor in the lung.

When the conduction of myelinated fibers in both vagus nerves was selectively blocked by differential cooling (29) or by anodal hyperpolarization (32), the increase in respiratory rate during the hypercapnic challenge persisted, suggesting the involvement of bronchopulmonary C-fibers in the hyperpneic response to CO_2 (29, 32). Recent studies have shown that pulmonary C-fibers are consistently activated when pH in the arterial blood (pulmonary venous blood) is lowered to ~7.1 by a bolus intravenous injection of acid solution (e.g., lactic acid, formic acid) (19, 24). Thus, an increase in alveolar CO_2 during hypercapnia challenge may lead to a decrease in the pulmonary interstitial pH, which can then activate the pulmonary C-fibers. This study was, therefore, carried out to determine whether pulmonary C-fibers are activated by an increase in the CO_2 concentration of alveolar gas.

Recent studies have clearly demonstrated that airway mucosal inflammation induces a pronounced increase in the sensitivity of pulmonary C-fiber afferents to various stimuli including hydrogen ion (25, 26). Since both airway inflammation and hypercapnia are common symptoms encountered in patients who suffer from either acute or chronic obstructive pulmonary diseases, the second aim of this study was to determine if the CO_2 sensitivity of these afferents is altered during airway inflammation and, if so, whether the action is mediated through the production of hydrogen ions.

2. EXPERIMENTAL PROCEDURES AND PROTOCOLS

Male Sprague-Dawley rats (body weight, 320–470 g) were anesthetized with intraperitoneal injection of α-chloralose (100 mg/kg) and urethane (500 mg/kg). The right femoral artery and vein, and the left jugular vein were cannulated for recording arterial blood pressure (ABP), infusion and injection of various chemical agents, respectively. The trachea was cannulated, and the lungs were artificially ventilated with a respirator. Tidal volume (V_T) and respiratory frequency were set at 8–10 ml/kg and 50 breaths/min, respectively, to mimic those of unilaterally vagotomized rats. The end-tidal CO_2 concentration, monitored by a CO_2 gas analyzer (Novametrix 1260), was maintained within normal physiological range (4.5–5.1%). A mid-line thoracotomy was performed, and the expiratory outlet of the respirator was placed under water to maintain a 3-cmH_2O pressure and a near-normal functional residual capacity. Body temperature was maintained at ~36 °C throughout the experiment by a heating pad placed under the animal.

Single-unit pulmonary C-fiber activity was recorded from sectioned right vagus nerve as previously described (14, 19, 20). Pulmonary C-fibers were identified by: 1) an immediate (delay < 1 s) response to bolus injection of capsaicin (0.5–1.0 μg/kg) into the right atrium (e.g., Fig. 1A); 2) a mild response to hyperinflation of the lung (3–4 × V_T); and 3) the physical location of the receptor field in the lung structures. To induce an abrupt increase in alveolar CO_2 concentration and to avoid any lingering systemic effects of hypercapnia (HPC), a transient HPC was induced by connecting a balloon containing a CO_2-enriched gas mixture (25–30% CO_2, 21% O_2, balance N_2) to the inlet of the respirator for 5–8 consecutive breaths (average ~6 breaths).

Four series of experiments were carried out with the following aims: *Study 1*, to determine whether bronchopulmonary C-fibers were stimulated by transient HPC under control conditions; *Study 2*, to investigate if the C-fiber response to HPC was altered by airway mucosal inflammation; *Study 3*, to determine if the effect of HPC on C-fiber afferents is mediated through the action of hydrogen ions; *Study 4*, to investigate the relative involvements of the transient receptor potential vanilloid type 1 (TRPV1) channel (5) and the amiloride-sensitive acid-sensing ion (ASIC) channels (36, 37) in the C-fiber response to CO_2.

3. RESULTS

A total of 97 vagal bronchopulmonary C-fibers were studied in 87 rats. The locations of these C-fiber endings are: upper lobe, 13; middle lobe, 33; lower lobe, 31; and accessory lobe, 3. Two other receptors were found in the left lung. The locations of the remaining 15 fibers were not identified.

Study 1: During the transient HPC challenge, the end-tidal (alveolar) CO_2 concentration increased rapidly and progressively to near or above 13% at the end of the HPC challenge (e.g., Fig. 1B). The arterial blood pH (pH_a) decreased from 7.43 ± 0.05 at base-line to 7.10 ± 0.05 at the end of the HPC challenge ($P < 0.01$; $n = 7$) (Fig. 1C). However, transient HPC did not generate a significant stimulatory effect on these C-fibers (ΔFA = 0.21 ± 0.11 imp/s; $P > 0.05$; $n = 77$; Fig. 1D). The HPC challenge evoked only a mild stimulation (ΔFA > 0.5 imp/s) in 10 of the 77 (12.9%) bronchopulmonary C-fibers tested; an example is shown in Fig. 1B. In these 10 fibers, the discharges began to emerge when the end-tidal CO_2 concentration exceeded 10%, and ceased shortly after the termination of HPC. A mild hypotension frequently developed toward the end of the HPC challenge (e.g., Fig. 1B), but returned to control after 20–40 s.

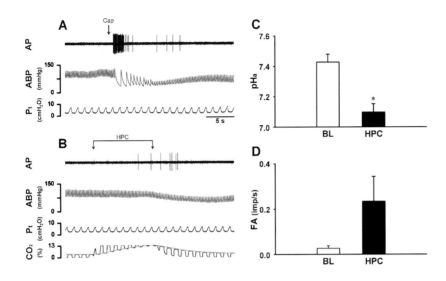

Figure 1. Response of pulmonary C-fibers to transient alveolar hypercapnia (HPC). *A*: Experimental records illustrating the response to a right-atrial bolus injection of capsaicin (1 µg/kg) of a pulmonary C-fiber arising from the ending in the right upper lobe of an anesthetized, open-chest rat (370 g). *B*: response to HPC challenge, which was induced by administering a CO_2-enriched gas mixture (30% CO_2, 21% O_2, balance N_2) via the respirator for 8 breaths (between the two arrows). AP, action potentials; P_t, tracheal pressure; CO_2, CO_2 concentration measured in the tracheal cannula; because the display limit of the CO_2 monitor was only 13%, the inspired CO_2 concentration (30%) during the HPC challenge was not expressed to the full scale. A gray line was added to the CO_2 trace to depict the continuous change in end-tidal (alveolar) CO_2 concentration during the HPC challenge. *C*: Group data showing the change in arterial blood pH (pH_a) measured from blood samples drawn during the base-line (BL) and at the end of the HPC challenge (HPC) in 7 rats. *D*: Group response indicated no statistically significant increase in the C-fiber activity (FA) generated by the HPC (25–30% CO_2) challenge ($P > 0.1$; n = 77). BL, base-line FA averaged over 20 s; HPC, the peak FA during the HPC challenge (2-s average). Data are means ± SE. *Significantly different from the base-line data.

Study 2: The sensitivity of bronchopulmonary C-fibers to transient HPC was clearly elevated by airway mucosal inflammation (e.g., Fig. 2), which is distinctly different from the weak and inconsistent HPC responses observed in *Study 1*. For example, after the intratracheal instillation of poly-L-lysine (PLL; 0.25 mg/ml, 0.1 ml), a synthetic cationic protein known to induce mucosal inflammation (7, 34), transient HPC activated 7 of the 8 (87.5%) pulmonary C-fibers tested; the peak response of these afferents to the same HPC challenge was markedly enhanced ($\Delta FA = 0.06 \pm 0.06$ imp/s at control and 6.59 ± 1.78 imp/s after PLL; $P < 0.01$; n = 8; Fig. 3). This enhanced response to HPC gradually returned to control after 60–120 min (e.g., Fig. 2*D*). Similarly, after acute exposure to O_3 (2.5 ppm; 30–45 min), an oxidant that is known to cause airway inflammation and hyperresponsiveness in a number of species including humans (10, 16, 17, 40), the same HPC challenge evoked a consistent and pronounced stimulatory effect on pulmonary C-fibers ($P < 0.05$, n = 6; Fig. 4*A*). Furthermore, when prostaglandin E_2 (PGE_2; 1–2 μg/kg/min) and adenosine (Ado; 40–120 μg/kg/min) were administered, the C-fiber responses to HPC were also clearly augmented (Fig. 4B and 4C); both PGE_2 and Ado are common inflammatory mediators in the airways, and have been shown to enhance the sensitivity of these afferents (12, 14, 20, 22, 23).

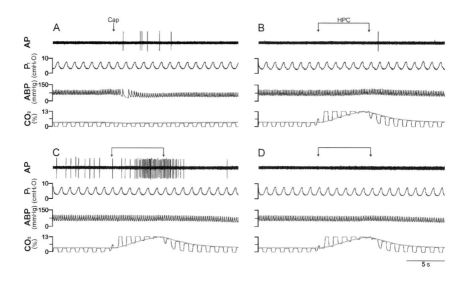

Figure 2. Experimental records illustrating the effect of intratracheal instillation of poly-L-lysine (PLL) on the response to HPC in a pulmonary C-fiber. *A*: response to an intravenous bolus injection of capsaicin (1 μg/kg). *B*, *C* and *D*: response to a HPC challenge (between the two arrows) before, 20 and 100 min after the administration of PLL (0.25 mg/ml; 0.1 ml), respectively. Receptor location: right lower lobe. Rat weight: 430 g. See legend of Fig. 1 for further explanation. (Modified from reference 27).

Study 3: The C-fiber responses to the same HPC challenges were compared in the same fibers between before and during a constant infusion of sodium bicarbonate ($NaHCO_3$; 1.82 mmol/kg/min) in order to prevent the acidosis caused by the HPC challenge. Infusion of $NaHCO_3$, completely abolished the enhanced C-fiber response to HPC induced by PLL ($\Delta FA = 4.16 \pm 2.5$ imp/s and 0.56 ± 0.26 imp/s, respectively, before and during infusion of $NaHCO_3$ ($P < 0.05$, $n = 6$; Fig. 3*A*). Similar attenuating effect of $NaHCO_3$ was also found when the C-fiber sensitivity to HPC was augmented by adenosine infusion ($n = 8$) (27).

Before the instillation of PLL, the HPC challenge caused the base-line pH_a to decrease from 7.42 ± 0.01 to 7.16 ± 0.03 ($n = 6$); after PLL, HPC decreased pH_a from 7.40 ± 0.02 to 7.13 ± 0.03. During infusion of $NaHCO_3$, the base-line pH_a was elevated to 7.54 ± 0.03 after PLL, and was only reduced to 7.31 ± 0.03 at the peak of HPC challenge (Fig. 3*B*).

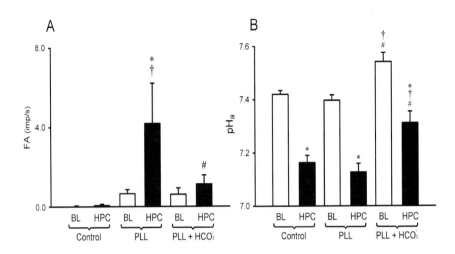

Figure 3. Effect of PLL on response to HPC challenge. *A*: Effect of intratracheal instillation of PLL (0.25 mg/ml; 0.1 ml) on pulmonary C-fiber responses to HPC challenge (n = 8; 30% CO_2, 21% O_2, balance N_2; 5–8 breaths) before and during infusion of $NaHCO_3$ (1.82 mmol/kg/min; 35 s). FA, fiber activity; BL, base-line FA averaged over 20 s; HPC, the peak FA during the HPC challenge (2-s average). *B*: Corresponding changes in pH_a (n = 6) in response to the same HPC challenge before and during infusion of $NaHCO_3$. BL and HPC are pH_a measured during the base-line and at the end of HPC challenge, respectively. *Significantly different from the corresponding base-line. †Significantly different from the corresponding control response. #Significantly different between the corresponding data of before and during $NaHCO_3$ infusion. Data are means ± SE. (Modified from reference 27)

Study 4: During the infusion of capsazepine (0.3 mg/kg/min, 5 min, i.v.), a selective antagonist of TRPV1, the increase in sensitivity to HPC challenge induced by PLL was reduced to ~43% in the same C-fiber afferents (ΔFA = 2.50 ± 0.45 imp/s and 1.33 ± 0.30 imp/s, respectively, before and after capsazepine; $P < 0.05$; $n = 12$; Fig. 5*A*). The attenuating effect of capsazepine subsided 20–40 min after termination of the infusion. On the other hand, infusion of amiloride (2–10 mg/kg/min, 5 min, i.v.), a blocking agent of the ASIC channels (36), did not attenuate the PLL-induced increase in sensitivity of pulmonary C-fibers to HPC (ΔFA = 2.33 ± 0.46 imp/s and 1.83 ± 0.35 imp/s, respectively, before and after amiloride; $P > 0.05$; $n = 9$; Fig. 5*B*).

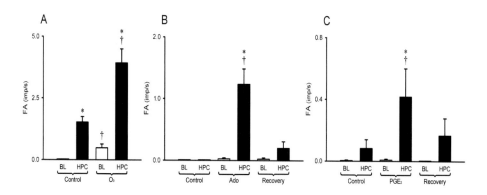

Figure 4. Effect of airway inflammation on pulmonary C-fiber responses to HPC challenge. Effects of ozone exposure (O_3; 2.5 ppm for 30–45 min; panel *A*), intravenous infusion of adenosine (Ado; 40–120 μg/kg/min for 2 min; panel *B*) and prostaglandin E_2 (PGE_2; 1–2 μg/kg/min for 3 min; panel *C*) on pulmonary C-fiber responses to the HPC (25–30% CO_2, 21% O_2, balance N_2; 5–8 breaths) challenge. FA, fiber activity; BL, base-line FA averaged over 20 s; HPC, the peak FA during the HPC challenge (2-s average). Numbers of C-fibers studied were 6, 15 and 12 for O_3, Ado and PGE_2, respectively. Recovery data were obtained 20–30 and 20–30 min after the treatments of Ado and PGE_2, respectively; the recovery data were not obtained in the study of O_3. *Significantly different from the corresponding base-line. †Significantly different from the corresponding control response. (Modified from reference 27).

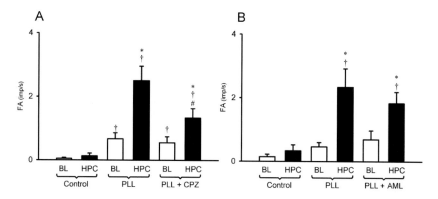

Figure 5. Effects of capsazepine (CPZ) and amiloride (AML) on the enhanced sensitivity of pulmonary C-fibers to HPC generated by PLL. *A*: Effect of intratracheal instillation of PLL (0.25 mg/ml; 0.1 ml) on pulmonary C-fiber responses to HPC challenge (30% CO_2, 21% O_2, balance N_2; 7 breaths) before and during infusion of CPZ (0.3 mg/kg/min; 5 min). *B*: Effect of PLL on pulmonary C-fiber responses to the same HPC challenge before and during infusion of AML (2-10 mg/kg/min; 5 min). FA, fiber activity; BL, base-line FA averaged over 20 s; HPC, the peak FA during the HPC challenge (2-s average). *Significantly different from the corresponding base-line. †Significantly different from the corresponding control response. #Significantly different between the corresponding data of before and during infusion of CPZ. Data are means ± SE of 12 and 9 fibers in panels *A* and *B*, respectively.

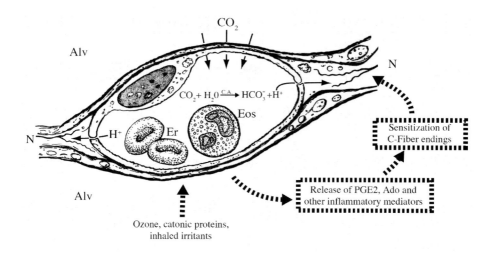

Figure 6. Schematic illustration of the mechanisms possibly involved in augmenting the action of CO_2 on pulmonary C-fiber endings during airway inflammation. H^+, hydrogen ion. HCO_3^-, bicarbonate ion. C.A., carbonic anhydrase. N, C-fiber ending. Alv, alveoli. Eos, eosinophil. Er, erythrocyte. PGE_2, prostaglandin E_2. Ado, adenosine.

4. DISCUSSION

Results of this study suggest that vagal bronchopulmonary C-fibers do not act as the CO_2 sensor in the lung under normal physiological conditions. This conclusion is based upon our observation that only ~13% of these afferents exhibit sensitivity to CO_2; even in those CO_2-sensitive C-fibers, their responses to an intense alveolar hypercapnia (alveolar CO_2 concentration > 13%) were relatively weak. In sharp contrast, the CO_2 sensitivity of the same afferent endings is drastically and consistently elevated during airway inflammation, making them a potential candidate in detecting a high CO_2 concentration in the lung under these pathophysiological conditions. The stimulatory effect of CO_2 observed in this study was probably mediated through the action of H^+ ions on the terminal membrane of these C-fiber afferents because the enhanced C-fiber sensitivity to CO_2 was significantly attenuated by infusion of HCO_3^- that prevented the HPC-induced acidosis in the pulmonary venous blood. Presumably, after CO_2 enters the pulmonary capillary from alveoli, its hydration releases H^+ in the blood; this reaction is catalyzed by the enzymatic action of carbonic anhydrase (Fig. 6). In addition, during the HCO_3^- infusion, the same HPC challenge produced a higher arterial CO_2 partial pressure because of the additional CO_2 generated by the reaction between H^+ and HCO_3^- ions in the blood. Thus, our results further indicated that CO_2 itself was not responsible for the stimulatory effect of HPC on these afferents.

Ventilation increases during exercise in proportional to the intensity level of exercise or the rate of CO_2 production. Interestingly, there is, in general, no detectable increase in the level of arterial CO_2 partial pressure during exercise, which led to the speculation that "CO_2 sensors" are present in the systemic venous or pulmonary circulatory systems and can therefore detect the increase in CO_2 in the venous blood (39). Based upon the observations made in the experiments after the myelinated fibers in the vagus nerves were selectively blocked, previous investigators have suggested an important role of bronchopulmonary C-fibers in the increase in respiratory rate during hypercapnia (29, 32). Therefore, a plausible explanation could be that the increase in the level of venous blood PCO_2 during exercise decreases the pulmonary interstitial pH, which in turn activates the pulmonary C-fibers. In support of this hypothesis, Delpierre et al. reported a paradoxical stimulatory effect of CO_2 on pulmonary C-fibers in cats (9). In comparison, our study showed a much smaller percentage of pulmonary C-fibers activated by CO_2 in rats under normal conditions.

The finding that transient HPC does not stimulate pulmonary C-fiber endings during control (without inflammation) is somewhat surprising because a decrease in arterial pH generated by right atrial injection of various acid solutions (e.g., lactic acid, formic acid, etc.) to approximately the same level (pH_a at ~7.1) consistently induced a pronounced stimulatory effect on these afferents (19, 24). We speculate that several factors may have contributed to this discrepancy: 1) the pH in the systemic arterial blood (pulmonary venous blood) may not accurately represent the local pH in the tissue or interstitial fluid surrounding these sensory terminals of which the afferent activity was recorded. Because these C-fiber endings are located either in the walls of alveoli and intrapulmonary airways (Fig. 6), a direct and precise measurement of the local interstitial fluid pH was not feasible. Alternatively, we measured the changes of pH in mixed pulmonary "venous" blood (i.e., systemic arterial blood) as an estimation of the overall change of interstitial fluid pH in the whole lung. Our assumption is based upon the fact that H^+ exchanges freely between the blood and the interstitial fluid through the pores (the inter-endothelial junctions) existing in the capillary wall as the blood flows through the pulmonary capillary (8). We further assume that the diffusion of H^+ between capillary blood and interstitial fluid reaches an "equilibrium" before the blood leaves the capillary; whether the equilibrium is reached during transient HPC remains to be determined. 2) The production of H^+ from CO_2 may also take place in pulmonary interstitial fluid, but the rate of reduction in pH there may be substantially slower because the rate of CO_2 hydration is critically dependent on the activity of carbonic anhydrase that is present at a higher level in erythrocytes than in other pulmonary tissues (13). 3) It is also possible that the sensory terminals of these afferents are located upstream to the pulmonary capillaries (e.g., at or proximal to the pulmonary arterioles) or remote form the capillaries (e.g., innervating the wall of small airways) (1, 3). If this is the case, they are expected to be relatively insensitive to the change in blood pH induced by the increase in CO_2 content in the alveolar gas during the HPC challenge in this study.

In sharp contrast, pulmonary C-fibers were consistently activated by the same level of alveolar hypercapnia after acute exposure of the lungs to either PLL or O_3. Both of these agents have been shown to induce airway mucosal inflammation and injury, accompanied by airway hyperresponsiveness (7, 10, 16, 17, 34). The possible involvement of vagal pulmonary C-fiber afferents in the airway hyperresponsiveness induced by PLL or O_3 has been suggested (7, 11, 15, 39). Our study furthered showed that administration of individual inflammatory mediators such as PGE_2 and adenosine also significantly enhanced the CO_2

sensitivity in pulmonary C-fibers. PGE_2 is a potent autacoid derived from arachidonic acid metabolism through the enzymatic action of cyclooxygenase and PGE synthase. The airway epithelium, which is the main target of initial assault by the inhaled irritants, is also the primary cell type that releases this autacoid (18). Adenosine is a purine nucleoside product of ATP metabolism and is produced by virtually all metabolically active cells, particularly when the energy demand cannot be matched by oxygen supply such as during hypoxia or tissue inflammation. Published evidence has strongly suggested the involvement of PGE_2 and adenosine as inflammatory mediators in the pathogenesis of airway hyperreactivity (18, 28, 30).

The mechanisms underlying the action of H^+ ions on the pulmonary C-fiber endings cannot be determined in this study, but several possible transduction mechanisms should be considered. The fact that pretreatment with capsazepine markedly attenuated the PLL-induced CO_2 sensitivity in these afferents suggests the involvement of TRPV1. In agreement with our contention, a recent study using an isolated airway-nerve preparation demonstrated that an activation of TRPV1 receptor plays a part in the stimulation of airway C-fiber afferents by sustained acidification of the airway tissue (21). Indeed, it is known that hydrogen ion can modulate the channel properties of the TRPV1 and enhance its sensitivity to capsaicin (5). Alternatively, it is also possible that certain chemical mediators (e.g., lipoxygenase metabolites, anandamide, etc.) are released from the surrounding tissue upon the action of H^+, which in turn can either activate the TRPV1 receptor or lower its activation threshold. Indeed, the observation that the CO_2 sensitivity of pulmonary C-fibers is enhanced during airway inflammation in this study is consistent with a recent report demonstrating that the excitability of TRPV1 can be elevated by PGE_2 via an activation of the cAMP/protein kinase A transduction pathway (22, 26). On the other hand, the expression of acid-sensitive ion channels (ASICs) have been reported in rat dorsal root ganglion nociceptors (36, 37), the counterpart of bronchopulmonary C-fiber afferents in other organ systems. The ASICs are H^+-gated cation channels and members of the amiloride-sensitive Na^+ channel/degenerin family (36); the ASICs have an activation threshold of pH around or slightly below 7.0 and are rapidly inactivated following the activation by H^+. A non-inactivating subunit of the ASICs that is less sensitive to H^+ (pH threshold <6.0) and expressed selectively in nociceptor neurons has also been identified (37). Interestingly, tissue inflammation has been shown to up-regulate the expression of ASICs in nociceptors (35). Nevertheless, the possible role of amiloride-sensitive ASICs in the stimulatory effect of CO_2 can be ruled out in this study because pretreatment with amiloride did not significantly attenuate the CO_2 sensitivity of these afferents (Fig. 5B).

An alternative explanation is that the observed effect of H^+ was on the voltage-sensitive ion channels which became active when membrane depolarization occurred as a result of activation of ligand-gated channels (e.g., TRPV1). Indeed, changes in intracellular H^+ have been shown to modify several types of voltage-gated ion channels, including Ca^{2+} channels (9), inward rectifier K^+ (Kir) channels (33), delayed rectifier K^+ channels (38) and Na^+ channels (4). Although more than one channel species are probably involved in the regulation of membrane excitability during HPC, K^+ channels, especially the Kir channels, may play a potentially significant role (31). Recent studies have shown that several Kir channels are inhibited by high CO_2 (31, 33, 41). The Kir channels play an important role in the regulation of resting membrane potential; inhibition of these K^+ channels leads to depolarization and increase of membrane excitability. The change in intracellular pH during HPC could not be determined in our study, but presumably was in parallel with that in the extracellular fluid.

Bronchopulmonary C-fibers represent > 75% of vagal afferents innervating the respiratory tract (2). Responses evoked by activating these afferents are mediated by both central reflex pathways and by local or axon reflexes involving the release of tachykinins from sensory endings (25, 26). The overall responses to C-fiber stimulation include bronchoconstriction, hypersecretion of mucus, edema of airway mucosa and cough (6, 25, 26). Considering the high probability of simultaneous occurrence of airway inflammation and alveolar hypercapnia in various types of obstructive lung diseases, a potential involvement of this stimulatory effect on C-fiber afferents in the pathogenesis of airway dysfunctions should be further explored.

In summary, under normal conditions only ~13% of the vagal bronchopulmonary C-fiber afferents exhibit mild sensitivity to transient alveolar hypercapnia. However, both the percentage of C-fibers activated and the response of the same afferents to CO_2 to the same HPC challenge markedly increased when mucosal inflammation was induced in the airways. The enhanced stimulatory effect of CO_2 on these afferents is mediated through the action of H^+, and an activation of TRPV1 is partially responsible.

ACKNOWLEDGEMENTS

The authors thank Robert F. Morton and Dr. Ting Ruan for their technical assistance. This study was supported by grants from the National Institutes of Health (HL58686, HL67379).

REFERENCES

1. Adriaensen D, Timmermans JP, Brouns I, Berthoud HR, Neuhuber WL, and Scheuermann DW. Pulmonary intraepithelial vagal nodose afferent nerve terminals are confined to neuroepithelial bodies: an anterograde tracing and confocal microscopy study in adult rats. *Celi Tissue Res* 293: 395-405, 1998.
2. Agostoni E, Chinnock JE, De Daly MB, and Murray JG. Functional and histological studies of the vagus nerve and its branches to the heart, lungs and abdominal viscera in the cat. *J Physiol* 135: 182-205, 1957.
3. Baluk P, Nadel JA, and McDonald, DM. Substance P-immunoreactive sensory axons in the rat respiratory tract: a quantitative study of their distribution and role in neurogenic inflammation. *J Comp Neurol* 319: 586-598, 1992.
4. Brodwick MS and Eaton DC. Sodium channel inactivation in squid axon is removed by high internal pH or tyrosine-specific reagents. *Science* 200: 1494-1496, 1978.
5. Caterina MJ, Schumacher MA, Tominaga M, Rosen TA, Levine JD, and Julius D. The capsaicin receptor: a heat-activated ion channel in the pain pathway. *Nature* 389: 816-824, 1997.
6. Coleridge JC and Coleridge HM. Afferent vagal C fibre innervation of the lungs and airways and its functional significance. *Rev Physiol Biochem Pharmacol* 99: 1-110, 1984.
7. Coyle AJ, Perretti F, Manzini S, and Irvin CG. Cationic protein-induced sensory nerve activation: role of substance P in airway hyperresponsiveness and plasma protein extravasation. *J Clin Invest* 94: 2301-2306, 1994.
8. Crone C and Levitt DG. Capillary permeability to small solutes. In: *Handbook of Physiology*. Bethesda, MD: Am. Physiol. Soc, 1984, sect. 2, vol. IV, chapt. 10, p. 411-466.

9. Delpierre S, Grimaud C, Jammes Y, and Mei N. Changes in activity of vagal broncho-pulmonary C fibres by chemical and physical stimuli in the cat. *J Physiol* 316: 61-74, 1981.
10. Gordon T, Venugopalan CS, Amdur MO, and Drazen JM. Ozone-induced airway hyperreactivity in the guinea pig. *J Appl Physiol* 57: 1034-1038, 1984.
11. Gu Q and Lee LY. Hypersensitivity of pulmonary C fibre afferents induced by cationic proteins in the rat. *J Physiol* 537: 887-897, 2001.
12. Gu Q, Ruan T, Hong JL, Burki N, and Lee LY. Hypersensitivity of pulmonary C-fibers induced by adenosine in anesthetized rats. *J Appl Physiol* 95: 1315-1324, 2003.
13. Heming TA, Stabenau EK, Vanoye CG, Moghadasi H, and Bidani A. Roles of intra- and extracellular carbonic anhydrase in alveolar-capillary CO_2 equilibration. *J Appl Physiol* 77: 697-705, 1994.
14. Ho CY, Gu Q, Hong JL, and Lee LY. Prostaglandin E_2 enhances chemical and mechanical sensitivities of pulmonary C-fibers. *Am J Respir Crit Care Med* 162: 528-533, 2000.
15. Ho CY and Lee LY. Ozone enhances excitabilities of pulmonary C-fibers to chemical and mechanical stimuli in anesthetized rats. *J Appl Physiol* 85: 1509-1515, 1998.
16. Holtzman MJ, Cunningham JH, Sheller JR, Irsigler GB, Nadel JA, and Boushey HA. Effect on ozone on bronchial reactivity in atopic and nonatopic subjects. *Am Rev Respir Dis* 120: 1059-1067, 1979.
17. Holtzman MJ, Fabbri LM, O'Byrne PM, Gold BD, Aizawa H, Walters EH, Alpert SE, and Nadel JA. Importance of airway inflammation for hyperresponsiveness induced by ozone. *Am Rev Respir Dis* 127: 686-690, 1983.
18. Holtzman MJ. Sources of inflammatory mediators in the lung: the role of epithelial and leukocyte pathways for arachidonic acid oxygenation. In: *Lung Biology in Health and Disease Series. Mediators of Pulmonary Inflation*, edited by Bray MA and Anderson WH. New York: Dekker, 1991, vol. 54, chapt. 6, p. 279-325.
19. Hong JL, Kwong K, and Lee LY. Stimulation of pulmonary C fibres by lactic acid in rats: contributions of H^+ and lactate ions. *J Physiol* 500: 319-329, 1997.
20. Hong JL, Ho CY, Kwong K, and Lee LY. Activation of pulmonary C fibres by adenosine in anaesthetized rats: role of adenosine A_1 receptors. *J Physiol (Lond)* 508: 109-118, 1998.
21. Kollarik M and Undem BJ. Mechanisms of acid-induced activation of airway afferent nerve fibres in guinea-pig. *J Physiol* 543: 591-600, 2002.
22. Kwong K and Lee LY. PGE_2 sensitizes cultured pulmonary vagal sensory neurons to chemical and electrical stimuli. *J Appl Physiol* 93: 1419-1428, 2002.
23. Lee LY, and Morton RF. Pulmonary Chemoreflexes are potentiated by Prostaglandin E_2 in anesthetized rats. *J Appl Physiol* 79: 1679-1686, 1995.
24. Lee LY, Morton RF, and Lundberg JM. Pulmonary chemoreflexes elicited by intravenous injection of lactic acid in anesthetized rats. *J Appl Physiol* 81:2349-2357, 1996.
25. Lee LY and Pisarri TE. Afferent properties and reflex functions of bronchopulmonary C-fibers. *Respir Physiol* 125: 47-65, 2001.
26. Lee LY and Undem BJ. Bronchopulmonary vagal sensory nerves. Chapter 11 in: *Advances in Vagal Afferent Neurobiology*. Ed. by Undem BJ and Weinreich D. *Frontiers in Neuroscience Series*, CRC Press, 2005.
27. Lin RL, Gu Q, Lin YS, and Lee LY. Stimulatory effect of CO_2 on vagal bronchopulmonary C-fiber afferents during airway inflammation. *J Appl Physiol* (In press, 2005)
28. Nyce JW and Metzger WJ. DNA antisense therapy for asthma in an animal model. *Nature* 385: 721-725, 1997.
29. Phillipson EA, Fishman NH, Hickey RF, and Nadel JA. Effect of differential vagal blockade on ventilatory response to CO_2 in awake dogs. *J Appl Physiol* 34: 759-763, 1973.
30. Polosa R, Rorke S, and Holgate ST. Evolving concepts on the value of adenosine hyperresponsiveness in asthma and chronic obstructive pulmonary disease. *Thorax* 57: 649-654, 2002.
31. Qu Z, Zhu G, Yang Z, Cui N, Li Y, Chanchevalap S, Sulaiman S, Haynie H, and Jiang C. Identification of a critical motif responsible for gating of Kir2.3 channel by intracellular protons. *J Biol Chem* 274: 13783-13789, 1999.

32. Russell NJW, Raybould HE, and Trenchard D. Role of vagal C-fiber afferents in respiratory response to hypercapnia. *J Appl Physio* 56: 1550-1558, 1984.
33. Tucker SJ, Gribble FM, Zhao C, Trapp S, and Ashcroft FM. Truncation of Kir6.2 produces ATP-sensitive K^+ channels in the absence of the sulphonylurea receptor. *Nature* 387: 179-83, 1997.
34. Uchida DA, Ackerman SJ, Coyle AJ, Larsen GL, Weller PF, Freed J, and Irvin CG. The effect of human eosinophil granule major basic protein on airway responsiveness in the rat in vivo. A comparison with polycations. *Am Rev Respir Dis* 147: 982-988, 1993.
35. Voilley N, de Weille J, Mamet J, and Lazdunski M. Nonsteroid anti-inflammatory drugs inhibit both the activity and the inflammation-induced expression of acid-sensing ion channels in nociceptors. *J Neurosci* 21: 8026-8033, 2001.
36. Waldmann R, Champigny G, Bassilana F, Heurteaux C, and Lazdunski M. A proton-gated cation channel involved in acid-sensing. *Nature* 386: 173-177, 1997.
37. Waldmann R, Bassilana F, de Weille J, Champigny G, Heurteaux C, and Lazdunski M. Molecular cloning of a non-inactivating proton-gated Na^+ channel specific for sensory neurons. *J Biol Chem* 272: 20975-20978, 1997.
38. Wanke E, Carbone E, and Testa PL. K^+ conductance modified by a titratable group accessible to protons from intracellular side of the squid axon membrane. *Biophys J* 26: 319-324, 1979.
39. Wasserman K, Whipp BJ, Casaburi R, and Beaver WL. Carbon dioxide flow and exercise hyperpnea. Cause and effect. *Am Rev Resp Dis* 115: 225-237, 1977.
40. Wu ZX, Morton RF, and Lee LY. Role of tachykinins in ozone-induced airway hyperresponsiveness to cigarette smoke in guinea pigs. *J Appl Physiol* 83: 958-965, 1997.
41. Xu H, Cui N, Yang Z, Qu Z, and Jiang C. Modulation of Kir4.1 and Kir5.1 by hypercapnia and intracellular acidosis. *J Physiol (Lond)* 524: 725-735, 2000.

Nitric Oxide in Brain Glucose Retention after Carotid Body Receptors Stimulation with Cyanide in Rats

[1,2]S. A. MONTERO, [1]J. L. CADENAS, [1]M. LEMUS, [1]E. ROCES DE ÁLVAREZ-BUYLLA., AND [1]R. ÁLVAREZ-BUYLLA†

[1]Centro Universitario de Investigaciones Biomédicas and [2]Facultad de Medicina, Universidad de Colima, Colima, México.

1. INTRODUCTION

In contrast to most other tissues, which exhibit considerable flexibility with respect to the nature of the substrates for their energy metabolism, the normal brain is restricted almost exclusively to glucose due to its distinguishing characteristics *in vivo*. Actual glucose utilization is 31 µmol/100 g tissue/min, in the normal, conscious human brain, indicating that glucose consumption is in excess for total oxygen consumption (Sokoloff, 1991). Although present in low concentration in brain (3.3 mmol/kg in rat), glycogen is a unique energy reserve for initiation of its metabolism. However, if glycogen concentration in the brain were the sole supply, normal energetic requirements would be maintained for less than 5 min (Sokoloff, 1991). While the brain contains insulin receptors, and insulin-responsive glucose transporters, the role of insulin in the regulation of brain glucose metabolism is controversial (Obici et al., 2002). The carotid body receptors (CBR) are sensitive to glucose (Alvarez-Buylla and Alvarez-Buylla, 1988, 1994, Pardal and López Barneo, 2002) and play an important role in the insulin-induced counterregulatory response to mild hypoglycemia (Koyama et al., 2000). Local stimulation of CBR by cyanide (NaCN), or local low glucose levels in the isolated carotid sinus (CS), have been shown to promptly increase the activity in the carotid sinus nerve, that in turn trigger an enhancement in glucose retention by the brain (BGR) (Alvarez-Buylla et al., 1994). In contrast, this effect is not observed in animals with denervated carotid bodies (Alvarez-Buylla and Alvarez-Buylla, 1988). The central mechanism that mediates the previously mentioned glycemic responses is unknown, but other studies from our laboratory suggest the participation of arginine-vasopressin (AVP), the endogenous ligand for the V1a vasopressin receptor, as the effector mediator in this response (Montero et al., 2003). AVP is widely synthesized in the brain, including the paraventricular, supraoptic and suprachiasmatic nuclei of the hypothalamus, and has been related to nitric oxide (NO) function in brain (Kadekaro et al., 1998). There are evidences that NO, an intercellular signaling molecule and neuromodulator, synthesized from arginine via Ca^{2+} activation of nitric oxide synthase (NOS), widely distributed in the central nervous system (CNS), plays a role in the glucose homeostasis in rats (Tong et al., 1997; Higaki et al., 2001). NO induces transient decreases in cellular ATP in nervous tissue, modulating metabolic pathways such as glycolysis (Almeida et al., 2005). NO is produced, as well, in the carotid body as an inhibitory neuromodulator in hypoxic chemoreception (Prabhakar et al., 1994; Wang et al., 1995; Trzebski et al., 1995). The perfusion of the carotid body with NO donors (*in situ* and *in vitro*) such as nitroglycerine (NG) or sodium nitroprusside (SNP), reduces the chemosensory discharges in the carotid sinus nerve to hypoxia (Wang et al., 1995; Iturriaga et al., 2000), to the contrary, NO donors increase basal firing in CBR in normoxia (Iturriaga et al., 2000). On the other hand, a NOS inhibitor such as N-nitro-L-arginine methyl ester (L-NAME) enhances the hypoxic response in the carotid body *in situ* (Iturriaga et al., 1998). Taking the aforementioned data into consideration, it can be assumed that NO in the CNS may modulate information arising from CBR to induce BGR and control the energetic metabolism. We now explore whether or not a NO donor or a NOS inhibitor applied intracisternally, modify hyperglycemic response and BGR after CBR stimulation with NaCN.

2. METHODS

2.1 Animal and Surgical Procedures

Experiments were performed on adult male Wistar rats (250-300 g weight) fasted for 12 h with free access to water, and maintained in 12 h light-dark cycle. Anesthesia was induced by intraperitoneal administration of sodium pentobarbital (3 mg/100 g) which was supplemented by a continuous intraperitoneal infusion of 0.063 mg/min in 0.9 % NaCl. Under these conditions no pain responses were observed, but the eye palpebral reflex was present. Respiration and body temperature were artificially maintained. All procedures performed on animals were in accordance with the National Research Council Guide for the Care and Use of Laboratory Animals (NCR Pub., 1996). To obtain blood samples, catheters were inserted into the femoral artery and jugular sinus (via the right external jugular vein) (Alvarez-Buylla et al., 2003). The correct placement of the catheters was verified at the end of each experiment during autopsy. The injection time was considered as t=0 (indicated as an arrow in the figures). For glucose determinations, 5 samples of 0.15 mL of arterial blood and 5 samples of 0.15 mL of venous blood were simultaneously collected at each time in the same rat: two basal samples (t=-4 min and t=-2 min), and 3 experimental samples (t=4 min, t=8 min and t=16 min). Brain glucose retention (BGR) was determined between glucose concentration in the arterial blood and glucose concentration in the venous blood returning from the brain.

2.2 CBR Stimulation and Intracerebro Injections

The circulation in the carotid sinus (CS) was isolated from general circulation during injections of NaCN (5 µg/100 g in 0.1 ml saline) as a bolus through a 27 gauge needle connected to a catheter (Clay Adams PE-10) placed in the common carotid artery below the isolated carotid sinus, to avoid baroreceptors stimulation. Briefly, both the left external carotid artery (beyond the lingual branch) and the internal carotid near the jugular foramen were temporarily occluded (15-20 sec) to prevent NaCN solution from entering the brain (Alvarez-Buylla and Alvarez-Buylla, 1988).

Intracerebro injections were done into the cisterna magna (CM). Once the rat was placed on a plastic platform, the head and neck were positioned over the edge at an angle of approximately 90°. with respect to the body axis in order to open the atlanto-occipital space (Hudson et al., 1994). The membrane over the CM was exposed, and a micromanipulator, holding a 23-gauge butterfly needle with attached tubing, was moved to the position adjacent to the platform and was connected to a syringe pump for infusions (Baby Bee, BAS, Lafayette, IN); then, the hand control of the manipulator was used to slowly penetrate the membrane and enter the cistern. The correct position was verified by the flow of cerebrospinal fluid (CSF), which entered the tubing at the end of each experiment.

2.3 Drugs

The drugs used in this study were: sodium cyanide (NaCN) (see CBR stimulation); nitroglycerine (NG, Scherer GMBH Eberbach Baden), 3 µg/5 µL, as a NO donor; N-nitro-L-arginine methyl ester (L-NAME, Sigma St. Louis, MO), 250 µg/5 µL, as a NOS inhibitor. Compounds were dissolved in artificial cerebrospinal fluid (aCSF, 145 mM NaCl, 2.7 mM KCl, 1.2 mM $CaCl_2$ and 1.0 mM $MgCl_2$, ascorbate 2.0 mM and NaH_2PO_4 2.0 mM at pH 7.3-7.4) (Mitchell and Owens, 1996) immediately before application. In control experiments the same volume of aCSF or saline alone were injected.

2.4 Analytical Methods

Glucose concentration in blood samples was measured by the glucose-oxidase method (Beckman Autoanalyzer, Fullerton CA) in mg/dL. The data are expressed as means±SEM, the statistical comparisons were performed using Student's paired t-test between basal (mean of t=-2 min and t=-4 min) and postinjection values in each protocol, and analysis of variance (ANOVA) for repeated measurements using the Scheffé test for multiple comparisons between different protocols. $P<0.05$ was considered as statistically significant.

2.5 Experimental Protocol

Animals were subjected to one of the following procedures: (a)aCSF infusion for 30 sec into CM, simultaneously with saline injection (0.1 mL) into

CS (n=5) (control 1); (b)aCSF infusion into CM, simultaneously with CBR stimulation (n=5) (control 2); (c)NG infusion into CM, simultaneously with saline injection into the CS (n=5); (d)NG infusion into CM, simultaneously with CBR stimulation (n=5); (e)L-NAME infusion into CM, simultaneously with saline injection into CS (n=5); (f)L-NAME infusion into CM, simultaneously with CBR stimulation (n=5).

3. RESULTS

After a direct infusion of aCSF (5 µL/30 sec) into CM simultaneously with saline injection into the CS (control 1), no significant changes were observed neither in blood glucose concentration nor in BGR values (n=5) (Fig. 1A). When a direct infusion of aCSF (5 µL/30 sec) into CM was simultaneously made with CBR stimulation in normal rats (n=5) (control 2), arterial and venous blood glucose concentrations increased significantly (P<0.05) (Fig. 1B). In this case, the increase in glucose concentration in arterial blood was higher than in venous blood, resulting in a subsequent increase in BGR levels at t=4 min, t=8 min and t=16 min after the stimulus (Fig. 1C). When comparing the results obtained in BGR values in control (1) with those obtained in control (2), significant differences were observed (P<0.05).

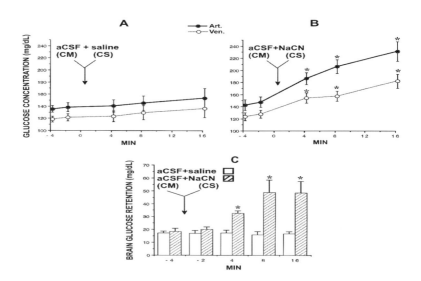

Figure 1. Changes in plasma glucose concentrations (**A** and **B**), and brain glucose retention (**C**). In **A**: after saline injections into the vascularly isolated carotid sinus (CS) simultaneously with artificial cerebrospinal fluid (aCSF) infusion into the cisterna magna (CM) (n=5); in **B**: after carotid body receptor stimulation with NaCN, simultaneously with aCSF infusion into the CM (n=5). Art, arterial glucose concentration; Ven, venous glucose concentration; NaCN, sodium cyanide. The values are means ± SEM. *Significantly different (P<0.05, *t*-test and ANOVA for repeated measurements).

3.1 CBR Stimulation Simultaneously with NG Infusion into CM

In order to investigate whether NO is involved in the previously described results, we simultaneously infused a NO donor into the CM (instead of aCSF), with saline injection in the CS or CBR stimulation. When a NO donor such as NG (3 μg/5 μL aCSF) was simultaneously infused in the CM with saline injection in the CS in normal rats (n=5), arterial glucose concentrations increased significantly ($P<0.05$) at t=4 min and t=8 min (Fig. 2A) ($P<0.05$). No significant changes were observed in venous glucose concentrations, but BGR levels increased significantly at the same times indicated above ($P<0.05$) (Fig. 2C). To the contrary, simultaneous NG infusion and CBR stimulation, did not alter either blood glucose concentrations or BGR levels (Fig. 2B and 2C).

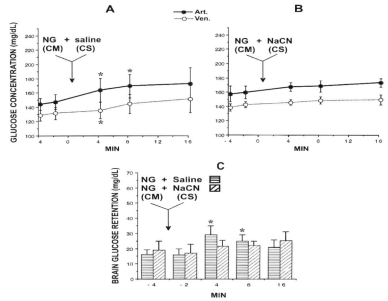

Figure 2. Changes in plasma glucose concentrations (**A** and **B**), and brain glucose retention (**C**). In **A**: after saline injections into the vascularly isolated carotid sinus (CS), simultaneously with nitroglycerine (NG) infusion into the cisterna magna (CM) (n=5); in **B**: after carotid body receptor stimulation with NaCN, simultaneously with NG infusion into the CM (n=5). Art, arterial glucose concentration; Ven, venous glucose concentration; NaCN, sodium cyanide. The values are means ± SEM. *Significantly different ($P<0.05$, *t*-test and ANOVA for repeated measurements).

3.2 CBR Stimulation Simultaneously with L- NAME Infusion into CM

When a NOS inhibitor such as L-NAME (250 μg/5 μL aCSF), was simultaneously infused into the CM with saline injections in CS (without CBR stimulation) in normal rats (n=5), no significant changes were observed in the parameters studied (Fig. 3A and 3C). However simultaneous L-NAME infusion into the CM, and CBR stimulation (n=5), increased arterial blood glucose concentration at t=4 min, t =8 min and t=16 min (P<0.05); and BGR levels at t=4 min and t =8 min (P<0.05) (Fig. 3B and 3C).

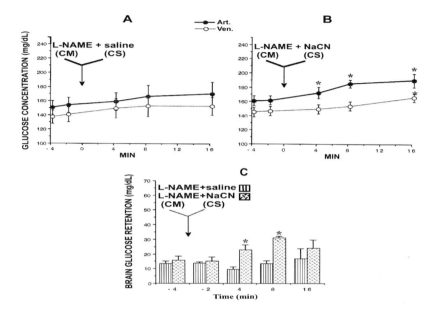

Figure 3. Changes in plasma glucose concentrations (**A** and **B**), and brain glucose retention (**C**). In **A**: after saline injections into the vascularly isolated carotid sinus (CS), simultaneously with N-nitro-L-arginine methyl ester (L-NAME) into the cisterna magna (CM) (n=5); in **B**: after carotid body receptors stimulation with NaCN, simultaneously with L-NAME infusion into the CM (n=5). Art, arterial glucose concentration; Ven, venous glucose concentration.; NaCN, sodium cyanide. The values are means ± SEM. *Significantly different (P<0.05, *t*-test and ANOVA for repeated measurements).

4. DISCUSSION

The results presented in the first series of experiments show that NO released by a NO donor, like NG, infused intracisternally, increases significantly arterial blood glucose and brain glucose retention, when the anoxic stimulus produced by CBR stimulation was not present. After the CBR stimulation, the same NG infusion into the CM did not alter either of the parameters studied. Taking into account Prabhakar's (1994) concept of the carotid body as a mini-brain, and knowing that the carotid body is sensitive not only to oxygen tension, but also to blood glucose concentration (Álvarez-Buylla and Roces de Álvarez-Buylla, 1994; Pardal and López-Barneo, 2002), the present results obtained from the whole animal, could be compared with previous observations obtained in carotid body preparations *in vitro* (Chugh et al., 1994; Iturriaga et al., 2000). In effect, in these preparations, NO donors such as SNAP and NOC-9, act as inhibitors of the increased chemosensory discharge during hypoxia, while in normoxia, they produce the opposite effect. (Iturriaga et al., 2000). In addition, large doses of

NG abolish the chemosensory excitation induced by hypoxia (Iturriaga et al. 2000) and also NO concentrations in the carotid bodies superfused with L-arginine, induce a dose-dependent inhibition of CS nerve discharges during hypoxia via actions on receptor elements as well as in their associated vessels (Buerk and Lahiri, 2000).

The results presented in the second series of experiments show that the infusion of L-NAME into the CM significantly increased arterial blood glucose concentration as well as BGR levels during the hypoxic stimulus. Similar results are obtained in perfused cat carotid bodies with progressive decreases in NO levels after L-NAME (Buerk and Lahiri, 2000). Although the mechanism involved in this glucose retention response is not known, inhibition of central production of NO affects secretion of AVP and cerebral metabolic responses (Kadekaro et al., 1998). Centrally infused AVP also increases blood glucose levels and BGR levels (Montero et al., 2003). NO within the brain aparently increases sympathetic nerve activity through caudal ventrolateral medulla and paraventricular nucleus (Kantzides and Badoer, 2005), stimulating adrenal secretion of epinephrine (Montero et al., 2003) to subsequently elevate glucose levels in peripheral blood (Uemura et al., 1997). In the hypothalamus, NO also activates cAMP–protein kinase to further secrete AVP (Yamaguchi and Hama, 2003), increasing hepatic glycogenolysis (McCrimmon et al., 2004). However, our results cannot rule out, at least in part, the possibility that the changes observed in glucose variables are vascularly mediated (Wang et al., 1998).

In summary, the present study demonstrates that centrally infused NO, during normoxia, participates in glucose homeostasis increasing arterial glucose levels and BGR. During a hypoxic state, such as that which occurs after CBR stimulation, central NO administration does not enhance the glucose parameters studied.

ACKNOWLEDGEMENTS

We thank Dr. Arturo Álvarez-Buylla, UCSF, USA, for revision of the manuscript. This project was supported by Grants: FRABA 2003, University of Colima, and CONACYT 29569N, México.

REFERENCES

Almeida A., Cidad P., Delgado-Esteban M., Fernandez E., Garcia-Nogales P., Bolaños J.P. Inhibition of mitochondrial respiration by nitric oxide: its role in glucose metabolism and neuroprotection. J Neurosci Res 2005; 79:166-171.

Álvarez-Buylla R., Álvarez-Buylla E. Carotid sinus receptors participate in glucose homeostasis. Respir Physiol 1988; 72:347-360.

Álvarez-Buylla R., Roces de Álvarez-Buylla, E. Changes in blood glucosa concentration in the carotid body-sinus modify brain glucose retention. Brain Res 1994; 654:167-170.

Álvarez-Buylla R., Huberman A., Montero S. Lemus M., Valles V. Roces de Alvarez-Buylla E. Induction of brain glucose uptake by a factor secreted into cerebrospinal fluid. Brain Res 2003; 994:124-133.

Buerk D.G., Lahiri S. Evidence that nitric oxide plays a role in O_2 sensing from tissue NO and PO_2 measurements in cat carotid body. Adv. Exp. Med. Biol 2000; 475:337-347.

Chugh D.K, Katayama M., Mokashi A., Debout D.E., Ray D.K. Lahiri S. Nitric oxide-related inhibition of carotid chemosensory activity in the cat. Respir Physiol 1994; 97:147-152.

Higaki Y., Hirshman M.F., Fujii N., Goodyear L.J. Nitric oxide increases glucose uptake through a mechanism that is distinct from the insulin and contraction pathways in rat skeletal muscle. Diabetes 2001; 50:241-247.

Hudson L.C., Hughes C.S., Bold-Fletcher N.O., Vaden, S.L. (1994). Cerebrospinal fluid collection in rats: modification of a previous technique. *Lab Animal Sci*, 44: 358-361.

Iturriaga R., Alcayaga J., Rey S. Sodium nitroprusside blocks the cat carotid chemosensory inhibition induced by dopamine, but not that by hyperoxia. Brain Res 1998; 799:26-34.

Iturriaga R., Villanueva S., Mosqueira M. Dual effects of nitric oxide on cat carotid body chemoreception. J Appl Physiol 2000; 89:1005-1012.

Kadekaro M., Terrell M. L., Liu H., Gestl S., Bui V., Summy-Long, J.Y. Effects of L-NAME on cerebral metabolic, vasopressin, oxytocin, and blood pressure responses in hemorrhaged rats. American Journal of Physiology 1998; 274:1070-1077.

Koyama Y., Cocker R.H., Stone E.E., Lacy D.B., Jabbour K., Williams P.E. Wasserman D.H. Evidence that carotid bodies play an important role in glucoregulation in vivo. Diabetes 2000; 49:1434-1442.

Kantzides A., Badoer E. nNOS-containing neurons in the hypothalamus and medulla project to the RVLM. Brain Res 2005; 1037:25-34.

McCrimmon R.J., Fan X., Ding Y, Zhu W., Jacob RJ., Sherwin R.S. Potential role for AMP-activated protein kinase in hypoglycemia sensing in the ventromedial hypothalamus. Diabetes 2004; 53:1953-1958.

Mitchell D.H., Owens, B. Replacement therapy: arginine-vasopressin (AVP), growth hormone (GH), cortisol, thyroxine, testosterone and estrogen. J. Neurosci Nurs 1996; 28:140-154.

Montero S.A., Yarkov A., Lemus M., Mendoza H., Valles V., Álvarez-Buylla E. Álvarez-Buylla R. Enhancing effect of vasopressin on the hyperglycemic response to carotid body chemoreceptor stimulation: role of metabolism variables. Adv Exp Med Biol 2003; 536:95-107.

Obici S., Zhang B.B, Karkanias G., Rossetti, L. Hypothalamic insulin signaling is required for inhibition of glucose production. Nat Med 2002; 8:1376-1382.

Pardal, R., López-Barneo, J. Low glucose sensing cells in the carotid body. Nature Neuroscience 2002; 5:197-198.

Prabhakar N.R. Neurotransmitters in the carotid body. Adv Exp Med Biol 1994; 360:57-69.

Sokoloff, Louis. "Measurement of local cerebral glucose utilization and its relation to local functional activity in the brain". In *Fuel Homeostasis and the Nervous System*, M. Vranic and et al. eds. New York: Plenum Press, 1991; pp. 21-42

Tong Y.C., Wang C.J., Cheng J.T. The role of nitric oxide in the control of plasma glucose concentration in spontaneously hypertensive rats. Neurosci Lett 1997; 233:93-96.

Trzebski A., Sato Y., Susuki A., Sato A. Inhibition of nitric oxide synthesis potentiates the responsiveness of carotid chemoreceptors to systemic hypoxia in the rat. Neurosci Lett 1995; 190:29-33.

Uemura K., Tamagawa T., Chen Y., Maeda N., Yoshioka S., Itoh K., Miura H., Iguchi A., Hotta N. NG-methyl-L-arginine, an inhibitor of nitric oxide synthase, affects the central nervous system to produce peripheral hyperglycemia in conscious rats. Neuroendocrinol 1997; 66:136-144.

Wang Z.Z., Stensaas L.J., Dinger B.G., Fidone, S.J. Nitric oxide mediates chemoreceptor inhibition in the cat carotid body. Neurosci 1995; 65:217-29.

Yamaguchi K., Hama H.A. Study on the mechanism by which sodium nitroprusside, a nitric oxide donor, applied to the anteroventral third ventricular region provokes facilitation of vasopressin secretion in conscious rats. Brain Res 2003; 968:35-43.

Pulmonary Nociceptors are Potentially Connected with Neuroepithelial Bodies

J. YU, S.X. LIN, J.W. ZHANG, J.F. WALKER

Pulmonary Medicine, University of Louisville, Louisville, KY 40292 USA

1. INTRODUCTION

Airway sensory receptors regulate cardiopulmonary function by providing constant information about the mechanical and chemical status of the lung to the central nervous system (CNS). There are at least three airway sensor types: slowly adapting receptors (SARs), rapidly adapting receptors (RARs), and C-fiber receptors (CFRs) [1]. We recently identified additional A-delta fiber receptors in intact rabbits that are different from SARs and RARs. Having a high mechanical threshold, they respond to hypertonic saline and are termed high threshold A-delta receptors (HTARs) [2]. SARs and RARs monitor airway mechanical changes, whereas HTARs and CFRs sense chemical alterations and may serve as nociceptors. As with nociceptors in other tissue, the latter are activated during lung inflammatory processes [3-6]. Also, the airway houses neuroendocrine cells aggregated in organoids called neuroepithelial bodies (NEBs). NEBs are richly innervated by nerve fibers from different origins. Similar in structure to the carotid bodies, NEBs are believed to be sensors, with at least some sensory fibers that have cell bodies in the nodose ganglia. Therefore, they may serve CNS reflex functions. Strategically located at airway bifurcations, NEBs may signal the chemical composition of or presence of irritants in the air. This study intends to explore the possibility that NEBs are associated with nociceptors.

2. RESULTS AND DISCUSSION

2.1 Neuroepithelial Bodies Connected with the Vagal Afferents to the Nodose Ganglia but not the Mechanoreceptors in the Rabbit

In the airway, vagal afferent cell bodies are believed to be housed in the nodose ganglia. At least some NEB cell bodies are also found there, based on evidence that sectioning the vagus nerve below the nodose ganglia, but not above it, denervates NEBs in the rat. NEBs can be visualized by injecting an antegrade tracer into the ganglion [7]. It has been suggested that NEBs are connected to one of the known vagal receptors [2;7;8]; however, which one is debated. Recently, we provided evidence that NEB afferents do not run with SARs in the rat [9]. Furthermore, we characterized the airway receptors in the rabbit, including SARs, with histochemical [10;11] and neural tracer [12] techniques. Briefly, a neural tracer, carbocyanine dye DiI (1,1' - dioleyl - 3,3,3',3' tetramethylindo carbocyanine) was used. The rabbit was anesthetized and the nodose ganglion was exposed to inject 10 µl of DiI (25mg/ml in methanol, Molecular Probes, Eugene, Ore, USA) in multiple sites through a fine tipped (30 µm) glass micropipette. The injections were made by pressure pulses (5 ms at approximately 60 psi) delivered from a PicoSpritzer III pressure system (Parker Instrumentation). After the injection, the rabbits were returned to the cage for four weeks, allowing the transportation of DiI to the nerve terminal targets in the lung. Then the animal was anesthetized and sacrificed. Airway tissues were fixed for 24 hours in a phosphate-buffered saline solution, containing 4% paraformaldehyde at a pH of 7.4. Whole mount preparations were made on a gelatin-coated glass slide and examined under a Laser Scanning Confocal Microscope (Zeiss510) equipped with a helium-neon laser (543 nm) and fitted with an appropriate filter for the detection of DiI signals. Under the confocal microscope, NEBs were labeled by DiI (Fig. 1), showing ball-like or oval shapes. The structures are similar to the NEB described in rats with DiI labeling or with histochemical labeling. However, these structures are clearly different from the receptor structures identified in muscular and submucosal layers, and also differ from epithelial receptors, which are leaf-like in appearance [10]. Our results are consistent with the recent report in the rat and support the contention that NEBs are not connected with SARs [9]. Since mechanoreceptors (RARs and SARs) may share the same afferents [13], and NEB afferents do not run with SAR afferents, they will not run with RAR afferents. If NEBs do transmit information through vagal sensory receptor afferents, HTARs and CFRs afferents are potential candidates. In view of the NEB's chemical sensory structure and abundant afferent innervation, its activation may trigger or modify signals transmitted in HTARs as well as CFRs [2].

2.2 Airway Nociceptors Activated by a Known Mediator Released from NEBs

HTARs are regarded as nociceptors because they are stimulated by hypertonic saline and hydrogen peroxide. NEBs contain vesicles storing 5-HT[14;15] which is an inflammatory mediator. In the present study, we evaluated whether 5-HT, a mediator released by NEBs in the rabbit, can stimulate

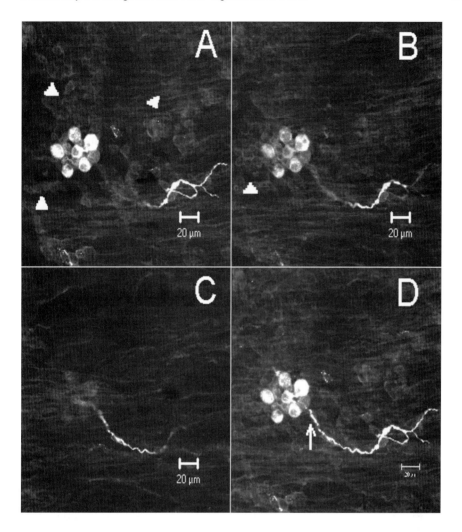

Figure 1. Micrographs from the confocal microscope show antegrade labeling of a neuroepithelial body (NEB) in a medium sized airway (about the 4th or 5th generation). Antegrade neural tracer (DiI) was injected into the nodose ganglia of the rabbit and transported to the lung periphery to label the structure. A, B, and C are representative optical slices from inside (epithelium, A) to outside (lamina propria, C) of the bronchial wall. D is a projection image of stacks of six optical sections that are 2.6 µm thick each. The axon fiber runs in the lamina propria and penetrates into the epithelial layer, where the NEB is located. The axon fiber is 2.4 µm in diameter at the site indicated by an arrow in D, which falls into the myelinated fiber range. Six neuroendocrine cells are clearly lighted up (A, B, and D). In addition, many groups of epithelial cells can be observed (indicated by 3 arrow heads in A and an arrow head in B) and many other epithelial cells are located between the positive cells (A and B). The NEB structure is different from airway mechanosensors, which are knob-like endings and buried in the airway smooth muscle layer.

nociceptors such as HTARs and CFRs (Fig.2). In anesthetized, open-chest and mechanically ventilated rabbits, single unit activity of HTARs was recorded, with identification based on discharge pattern [low background activity and insensitivity to changes in lung mechanics] and confirmed by measuring conduction velocities (1.5 to 20 m/sec for HTARs, and less than 1.3 m/sec for CFRs). At baseline, HTAR activity was 0.24±0.06 imp/s and was stimulated by microinjection of 5-HT (10μM, 20μl) into the receptive field. Activity increased to 2.53±0.86 imp/s during peak response at 25.0±7.6 seconds after injection (n=8, $P<0.05$). HTAR activity was still elevated (0.89±0.26 imp/s) one minute after injection. CFRs responded to 5-HT similarly. This is in agreement with previous reports that CFRs can be activated by 5-HT [4]. At this point, it is still unknown whether NEBs are connected with nociceptors, and, if so, whether they are connected with myelinated (HTARs), unmyelinated (CFRs) vagal afferents, or both. Our results are consistent with the hypothesis that nociceptors may transmit signals that can be triggered or modified by NEBs.

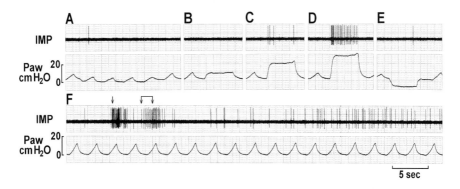

Figure 2. High threshold A-delta receptor (HTAR) activities recorded from the left cervical vagus nerve in an anesthetized, open-chest and mechanically ventilated rabbit. The traces are: IMP, impulses (sensory activity); Paw, airway pressure recorded from the trachea tube opening. A-E are the receptor responses to removal of positive-end-expiratory pressure (PEEP, A), to lung inflation at constant pressures of 10 (B), 20 (C) and 30 (D) cmH$_2$O, and to lung deflation at a constant pressure of –7 cmH$_2$O (E). F shows stimulation of the receptor by injecting 5-HT (10μM, 20μl) into the receptive field, which is located at the left lower lobe. The first arrow denotes poking the receptive field during insertion of the needle. The following two connected arrows denote the injection period. This receptor has a very low background activity (0.1 imp/s), does not respond to lung deflation either by PEEP removal of four ventilator cycles (A) or by negative pressure (E), has very high threshold to lung inflation (about 20 cmH$_2$O in this particular case). The conduction velocity of the sensory afferent is 4 m/s. Thus, it is a typical HTAR.

Increasing evidence shows NEBs are richly innervated by sensory nerves [8]. They are also closely related to molecules of inflammation and nociception. For example, they are associated with inflammatory mediators such as substance P and calcitonin gene-related peptide (CGRP), in addition to 5-HT. They also possess vanilloid type 1 receptors (VR-1) and purinergic receptors (P2X3) that operate under the influence of nociceptor activators such as capsaicin and adenosine. At this point, there is no question that NEBs are associated with nociception. Chemicals released from NEBs may exert paracrine effects on nociceptors, if they are not directly connected with them. The sensory afferents may participate in a local reflex, or mediate reflexes through sympathetic afferents. It is also possible that NEBs may connect directly with nociceptors. Regardless, it is of growing interest to figure out how NEBs perform nociception.

3. CONCLUSIONS

Based on the present results, three conclusions can be made: 1) NEBs do connect with afferents to the nodose ganglia; 2) these afferents do not supply the mechanoreceptors because the NEB structure is different from and does not associate with them; and 3) like CFRs, HTARs can be activated by NEB-released mediator, 5-HT. If NEBs do directly send the signal through the vagal afferent to the nodose ganglia, the nociceptors are their likely targets.

ACKNOWLEDGEMENTS

Work supported by grants from NIH HL-58727 and Rett Syndrome Research Foundation.

REFERENCES

1. Coleridge HM, Coleridge JC. Pulmonary reflexes: neural mechanisms of pulmonary defense. Annu Rev Physiol 1994; 56:69-91.
2. Yu J. An overview of vagal airway receptors. Acta Physiologica Sinica 2002; 54(6):451-459.
3. Belvisi MG. Sensory nerves and airway inflammation: role of A delta and C-fibres 1. Pulm Pharmacol Ther 2003; 16(1):1-7.
4. Coleridge JCG, Coleridge HM. Afferent vagal C fibre innervation of the lungs and airways and its functional significance. Rev Physiol Biochem Pharmacol 1984; 99:1-110.
5. Lee LY, Pisarri TE. Afferent properties and reflex functions of bronchopulmonary C-fibers. Respir Physiol 2001; 125(1-2):47-65.
6. Undem BJ, Carr MJ. Pharmacology of airway afferent nerve activity. Respir Res 2001; 2(4):234-244.
7. Van Lommel A, Lauweryns JM, Berthoud HR. Pulmonary neuroepithelial bodies are innervated by vagal afferent nerves: an investigation with in vivo anterograde DiI tracing and confocal microscopy. Anat Embryol (Berl) 1998; 197(4):325-330.
8. Adriaensen D, Brouns I, Van Genechten J et al. Functional morphology of pulmonary neuroepithelial bodies: Extremely complex airway receptors. Anat Rec 2003; 270(1):25-40.
9. Yu J, Zhang J, Wang Y et al. Neuroepithelial bodies not connected to pulmonary slowly adapting stretch receptors. Respir Physiol Neurobiol 2004; 144(1):1-14.
10. Wang Y, Yu J. Structural survey of airway sensory receptors in the rabbit using confocal microscopy. Acta Physiol sinica 2004; 56(2):119-129.
11. Yu J, Wang YF, Zhang JW. Structure of slowly adapting pulmonary stretch receptors in the lung periphery. J Appl Physiol 2003; 95(1):385-393.

12. Wang Y, Cheng Z, Zhang JW et al. A novel approach to investigate pulmonary receptors. 2002: A453.
13. Yu J. Airway mechanosensors. Respir Physiol Neurobiol 2005.
14. Fu XW, Nurse CA, Wong V et al. Hypoxia-induced secretion of serotonin from intact pulmonary neuroepithelial .bodies .in neonatal .rabbit. J Physiol 2002; 539(Pt 2):503-510.
15. Van Lommel A, Lauweryns JM, De Leyn P et al. Pulmonary neuroepithelial bodies in neonatal and adult dogs: histochemistry, ultrastructure, and effects of unilateral hilar lung denervation. Lung 1995; 173(1):13-23.

Modulators of Cat Carotid Body Chemotransduction

R. S. FITZGERALD[1,2,3], M. SHIRAHATA[1,4], I. CHANG[1], A. BALBIR[1]

Department of Environmental Health Sciences, Bloomberg School of Public Health[1], Departments of Physiology[2], of Medicine[3], of Anesthesiology/Critical Care Medicine[4], School of Medicine, The Johns Hopkins University, Baltimore, MD, 21205 USA

1. INTRODUCTION

The Carotid Body (CB) senses hypoxia, hypercapnia, and acidosis in the arterial blood. The resulting increase in CB neural output (CBNO) to the nucleus tractus solitarius in the medulla promotes reflex responses in the respiratory, circulatory, renal, and endocrine systems. Increases in CBNO are commonly thought to be due to the release of neurotransmitters from glomus cells in the CB. Additional to the action of these released transmitters on the postsynaptic afferent neurons which abut on the glomus cells the transmitters act presynaptically on glomus cell autoreceptors. Among the several transmitters contained in the glomus cells there now exists considerable evidence supporting excitatory roles for both acetylcholine (ACh) and ATP and an inhibitory role for dopamine (DA) and norepinephrine (NE) (Fitzgerald, 2000). The release of ACh (Fitzgerald et al., 1999; Kim et al., 2004) and catecholamines (Wang and Fitzgerald, 2002) appears to be influenced by modulators. The present study investigated the action of adenosine (ADO) on the release of ACh, DA, and NE since it has been reported that ADO influences CBNO (McQueen and Ribeiro, 1981) and CB-mediated increases in ventilation (Monteiro and Ribeiro, 1987). The study further investigated the action of nitric oxide (NO) on the release of ACh since NO has been reported to reduce the hypoxia-induced increase in CBNO (Wang, et al., 1994).

2. MATERIALS AND METHODS

Standard procedures were followed for harvesting the paired CBs from anesthetized cats. Procedures for harvesting and euthanizing the cats were in compliance with the policies of the Animal Care and Use Committee of the Johns Hopkins Medical Institutions which policies are completely consonant with the requirements of the National Institutes of Health. In each phase of the study the CB was cleaned of excessive fat and connective tissue rapidly after removal from the animal. They were then given a 45 minute recovery period in hyperoxic Krebs Ringer bicarbonate solution (KRB) at 37°C. After recovery they were immersed in 85μL of KRB and bubbled with 40% O2/5% CO2 for 10-15 min at the end of which the incubation medium was withdrawn, filter-centrifuged, and stored on ice for later determination of ACh, NE, or DA with high pressure liquid chromatography/electrochemical detection (Fitzgerald, et al., 1999; Wang and Fitzgerald, 2002). This hyperoxic control period was followed immediately by immersing the CBs in KRB, bubbled with 4% O2/ 5% CO2 and incubating again at 37°C for 10-15 minutes. A second hyperoxic recovery period ensued. Then followed the above steps in KRB made 100 μM adenosine. The results (n=6) for ACh were from two CBs.

However, preparation of the CBs for DA and NE determination followed the same steps. The major difference here was that two Eppendorf tubes were used, one for each CB and the KRB contained 40 mM L-dihydroxyphenyl alanine (L-DOPA) which served as a precursor for DA (Wang and Fitzgerald, 2002). The results (n=8) for DA and NE were from one CB without the 100 μM adenosine and from one CB exposed to the 100 μM adenosine.

3. RESULTS

The effect of 100μM adenosine (ADO) on the release of the "more classical" transmitters can be seen in Table 1. ADO clearly enhances the hypoxia-induced release of ACh significantly, while significantly reducing the hypoxia-induced release of DA and NE.

Table 1. Transmitter Recovered from Incubation Medium (Mean ± SEM; picomoles/20μL).

	(N=6) ACh (15 min exposures)			(N=8) DA (10 min exposures)		NE	
	Control	Hypoxia	Post-Hypox Hyperox	Control	Hypoxia	Control	Hypoxia
-ADO	0.5±0.1	0.99±0.27	0.6±0.15	0.23±0.05	2.23±0.26	0.61±0.2	3.05±0.78
+ADO	1.7±0.45	2.00±0.40	1.97±0.40	0.50±0.20	1.45±0.38	0.95±0.3	2.15±0.75
	P=0.014	P=0.101	P=0.005		P=0.006		P=0.011

NO has been shown to significantly reduce the hypoxia-induced increase in CBNO (Wang et al., 1994). CBs were harvested from deeply anesthetized cats, and prepared in the manner described above (N=9).

Table 2. Transmitter Recovered From Incubation Medium (Mean±SEM; Femtomoles of ACh/20μL).

	Control	Hypoxia	Post-Hypoxic Hyperoxia
0mM L-arginine	65.2±7.5	72.9±10.1	67.8±5.8
1mM L-arginine	61.3±6.5	55.8±4.8	51.5±4.1
10mM L-arginine	51.1±6.3	46.6±4.1	47.6±5.3

Notable is the response to hypoxia. In the absence of L-arginine there is a modest increase (112% of control) in the release of ACh. But in the presence of L-arginine the response to hypoxia tends to be less than the control (91% of control). Exposure to the last six gas mixtures in the presence of 1 mM and 10 mM L-arginine followed one another with no interval. Not surprisingly, therefore, the release of ACh during control with 10 mM L-arginine (51.1±6.3) is significantly (P=0.001) less than the amount released during control in the absence of L-arginine (65.2±7.5). So also is the release of ACh during hypoxia in 10 mM L-arginine (46.6±4.1) significantly (P=0.009) less than the hypoxic release in the absence of L-arginine (72.9±10.1). Post-arginine tests with a severe hypoxia (6%O2 for 15 min) showed an increase in ACh release of 85% over control. Post-arginine test #2 was to challenge the in vitro CBs with 100 μM nicotine under hyperoxic conditions for six minutes. This challenge in four of the animals produced 119±7.9 femtomoles of ACh/20 μL of incubation medium; this was significantly (P=0.001) more than even the pre-arginine hypoxic challenge (72.9±10.1). Hence, the decrease in ACh release during exposures to L-arginine was not due to the CBs exhausting their supply of ACh, but rather due to the influence of L-arginine.

Bypassing the involvement of nitric oxide synthase (NOS) and using an NO donor, sodium nitroprusside (SNP), we found that exposure of the in vitro cat CBs (N=5) to 5 μM and to 10 μM SNP reduced the amount of ACh recovered from the incubation medium significantly (P=0.039) below ACh recovery from the SNP-free incubation media. So the above data suggest that a loss of ACh could be at least part of the reason that NO produces a reduction in CBNO.

Carbon dioxide acidosis and metabolic acidosis certainly generate an increase in CBNO. What is responsible for this increase is still the subject of investigation. In some ongoing experiments we have measured the behavior of ACh in response to respiratory acidosis, compensated respiratory acidosis, and metabolic acidosis. By varying bicarbonate concentration the degree of acidosis could be determined. Control and Recovery values for PCO_2 and pH were 35 torr and 7.40, respectively. On the basis of the data so far, shown in Table 3 below, one might conclude that what seems to be required for a significant release of ACh is an extracellular acidosis.

Table 3. Transmitter Recovered From Incubation Medium (Mean±SEM;Femtomoles of ACh/20μL).

Condition	PCO_2 (torr)	pH	Control	Acidosis*	Recovery
*Respiratory	~85	~7.03	11.8±1.9	14.1±2.1#	11.9±2.0
*Compensated Respiratory	~85	~7.39	13.8±3.5	14.5±3.2	14.6±3.0
*Metabolic	~35	~7.07	11.0±3.5	13.6±4.0 #	12.2±1.8

#: Significantly greater than control: P<0.02

4. DISCUSSION

The action of ADO on transmitter release in the CNS has been explored. Jin and Fredholm (1997) found that stimulation of the A2a receptor provoked release of ACh from the rat hippocampus but not the striatum and did not affect catecholamine release. Kirk and Richardson (1994) reported that the A2a receptors on cholinergic nerve terminals enhanced the release of ACh, but A2a receptors on the GABAergic terminals inhibited the release of GABA. Hence, ADO appears to have opposing effects on neurotransmitter release from neural tissue in the CNS similar to what has been observed in the CB.

Monteiro and her associates (Monteiro and Ribeiro, 1987) as well as others have reported that ADO stimulates respiration or CBNO (McQueen and Ribeiro, 1981). Given the above results, namely that ADO increases the hypoxia-induced release of ACh while decreasing the hypoxia-induced release of catecholamines, suggests the possibility that the excitatory effect of ADO may be through its modulation of ACh and catecholamine release. Our recent immunocytochemical studies of A2a adenosine receptors in the cat carotid body show a very strong positive signal for the A2a receptor on glomus cells. There does not seem to be strong staining on nerves, certainly not on nerve structures that abut onto the glomus cells. We have been unable to detect a positive signal for the A1 adenosine receptor.

Wang and his colleagues (Wang, et al., 1994) clearly demonstrated an NO-induced reduction in the hypoxia-induced increase in the cat CBNO. This feature of NO interaction was demonstrated with the use of L-NAME. This NOS inhibitor significantly increased the cat's CBNO during hypoxia; the increase was reduced by the addition of L-arginine. Further, 10 mM L-arginine significantly reduced the hypoxia-induced increase in cat CBNO. The above data support the possibility that at least part of the NO-mediated reduction in CBNO is due to the reduction in the release of ACh. NO may have other mechanisms...quite possibly a reduction in the hypoxia-induced release of ATP. NO has often been reported to increase the level of cGMP. But the mechanisms identifying how an increase in cGMP promotes a reduction in CBNO has yet to be reported.

Further studies are needed to determine the effect of NO on the release of both ATP and catecholamines from the cat carotid body.

Intracellular pH was not measured in the acidosis experiments. Very probably intracellular pH decreased significantly during respiratory acidosis. But during compensated respiratory acidosis when the incubation medium was 50 mM HCO_3^- the intracellular pH may not have been so acidotic. CO_2 increases the permeability of some membranes (e.g. blood-brain barrier). Possibly the 50 mM HCO_3^- could pass readily into the glomus cells during the compensated respiratory acidosis. In metabolic acidosis extracellular HCO_3^- was 10 mM. The incubation media for the three initial control periods had PCO2 values of about 35-36 mmHg and a HCO_3^- concentration of 22mM. So on the basis of the data what is common to the increased release of ACh is the extracellular acidosis of respiratory and metabolic acidosis.

ACKNOWLEDGEMENT

This study was supported by awards from the NIH (NHLBI) RO 1 HL-50712 and RO 1 HL-72293.

REFERENCES

Fitzgerald, R.S., 2000.Oxygen and carotid body chemotransduction: the cholinergic hypothesis – a brief history and new evaluation. Respir. Physiol., 120: 89-104.
Fitzgerald, R.S., Shirahata, M., Wang, H-Y., 1999. Acetylcholine release from cat carotid bodies. Brain Res. 841: 53-61.

Jin, S., Fredholm, B.B., 1997. Adenosine A2a receptor stimulation increases release of acetylcholine from rat hippocampus not striatum and does not affect catecholamine release. Naunyn-Schmiedeberg's Arch. Pharmacol.353:48-56.

Kim, D.K., Prabhakar, N.R., Kumar, G.K., 2004. Acetylcholine release from the carotid body by hypoxia: evidence for the involvement of autoinhibitory receptors. J. Appl. Physiol. 96: 376-383.

Kirk, J.P., Richardson, P.J., 1994. Adenosine A2a receptor-mediated modulation of striatal [^{3}H] GABA and [^{3}H] acetylcholine release. J. Neurochem. 62: 960-966.

McQueen, D.S. and Ribeiro, J.A., 1981. Effect of adenosine on carotid chemoreceptor activity in the cat. Br. J. Pharmac. 74: 129-136.

Monteiro, E.C. and Ribeiro, J.A., 1987. Ventilatory effects of adenosine mediated by carotid chemoreceptors in the rat. Naunyn-Schmiedeberg's Arch. Pharmacol., 335: 143-148.

Wang, H-Y., Fitzgerald, R.S., 2002. Muscarinic modulation of hypoxia-induced release of catecholamines from the cat carotid body. Brain Res. 927: 122-137.

Wang, Z.Z., Stensaas, L.J., Bredt, D.S., Dinger, B.G., Fidone, S.J., 1994. Localization and actions of nitric oxide in the cat carotid body. Neuroscience 60: 275-286.

Identification and Characterization of Hypoxia Sensitive Kvα Subunits in Pulmonary Neuroepithelial Bodies

X.W. FU AND E. CUTZ

Division of Pathology, Department of Pediatric Laboratory Medicine, The Research Institute, The Hospital for Sick Children and University of Toronto, Toronto, Canada M5G 1X8

1. INTRODUCTION

Pulmonary neuroepithelial bodies (NEB) are composed of innervated clusters of amine and peptide producing cells and are thought to function as hypoxia sensitive airway chemoreceptors. We have shown previously that the plasma membrane of rabbit fetal NEB in culture expresses an O2 sensing molecular complex composed of O_2 sensitive K^+ channel coupled to an O_2 sensing protein (NADPH oxidase)(Nature, 1993;365:153). A Shaw-like, outward non-inactivating delayed-rectifier type K^+ channel, was recorded from NEB cells in both culture and lung slices. This K^+ channel was decreased by hypoxia (pO2~20 mmHg), and was sensitive to TEA, 4-AP, and H_2O_2. Another whole cell K^+ current recorded from NEB in culture exhibited electrophysiological characteristics of a slowly inactivating K^+ current similar to the one described in *Xenopus oocyte* expressing Kv3.3a channel and this K^+ current was increased by H_2O_2. Here we report findings on A-type K^+ currents recorded from NEB in neonatal rabbit lung slice preparation. This slowly inactivating K^+ current was inhibited by BDS-I (3 µM), specific blocker of Kv3.4 and rheteropodatoxin (HpTx-2; 0.2 µM), specific blocker of Kv4, and also sensitive to hypoxia. Using in situ hybridization method, mRNA for Kv3.4 and Kv4.3 was localized in NEB cells identified by immunostaining for serotonin. Expression of Kv3.4 and Kv4.3 proteins in NEB cells was confirmed by immunohistochemistry using specific antibodies. Multiple subtypes of voltage-dependent K^+ current are expressed in NEB cells that may function as O_2-sensitive K^+ channels.

2. RESULTS AND DISCUSSION

2.1 Oxygen Sensitive K^+ Currents in NEB Cells in Culture and Lung Slice Preparation

In our initial studies, we used primary cultures of NEB cells isolated from late-gestation (26 days) fetal rabbit lung. Whole-cell patch clamp recordings revealed an outward non-inactivating K^+ current that when exposed to hypoxia solution (PO₂: 25-30mmHg), was reversibly inhibited by ~26%. This K^+ current was blocked by DPI, a specific inhibitor of NADPH oxidase. In current clamp recording NEB cells showed an increase in firing frequency with hypoxia. We concluded that O₂ sensitive K^+ current was coupled to an oxygen binding protein, identified as multicomponent NADPH oxidase localized to the plasma membrane of NEB cells and that NEB cells act as transducers of airway hypoxia stimulus (Youngson et al., 1993). These initial findings were supported by the demonstration of expression of mRNA for both Kv3.3a and NADPH oxidase related proteins gp 91phox and p22phox in NEB cells of rabbit fetal and human neonatal lungs using in situ hybridization method (Wang et al., 1996). In the same study, using whole-cell patch clamp recording we have observed a slowly inactivating K^+ current with inactivating properties similar to those described in *Xenopus oocyte* model expressing Kv3.3a mRNA (Vega-Saenz & Rudy, 1992). Furthermore an increase in K^+ current was observed when NEB cells were exposed to depolarizing pulses in the presence of varying concentrations of H_2O_2, suggesting redox modulation of this current (Wang et al., 1996).

In subsequent studies, we have further characterized voltage-activated K^+ currents in NEB cells and their sensitivity to hypoxia using neonatal rabbit lung slice preparation (Fu et al., 1999). Similar O₂- sensitive K^+ current was also observed in NEB cells of mice lung slice model (Fu et al., 2000). The advantage of this model is that NEB cells are studied intact within their "natural" environment without effects of cell dissociation and culture. Using this preparation we observed O₂ sensitive non-inactivating outward K^+ current that was reversibly blocked (34%) by hypoxia (PO₂ ~ 20 mmHg), TEA and 4- AP. Among the O₂-sensitive K^+ currents, two types of K^+ current have been observed, a Ca^{2+}-independent ($I_{K(v)}$) and Ca^{2+}- dependent ($I_{K(Ca)}$ component. Of the O₂-sensitive K^+ current, $I_{K(Ca)}$ accounts for ~55% whereas $I_{K(v)}$ for ~ 45%, suggesting that both components were equally suppressed by hypoxia. The activation threshold of outward non-inactivating O₂-sensitive K^+ current in NEB cells is around -50 mV and when the K^+ channel opens, it's role is to stabilize the membrane potential. The closure of this K^+ channel by hypoxia causes membrane depolarization, leading to the opening of voltageactivated Ca^{2+} channels, increase in intracellular Ca^{2+}, and neurotransmitter release (Fu et al., 1999).

2.2 Identification and Characterization of O2 Sensitive Kv3.4 and Kv4.3 Subunits in Rabbit NEB Cells

In the present study voltage-dependent K+ currents in NEB cells were recorded in rabbit lung slice preparation. After establishing the whole-cell configuration, outward K^+ currents were elicited by depolarization from -90 mV, 15 mV a step with 1 s, before a 250 ms pretest pulse with holding at -90 mV. The cell was at holding -60 mV. These K^+ currents were apparent during steps from a holding potential of -90 mV to potentials between -30 and +30 mV. After application of 3 μM BDS-I, a specific blocker of Kv3.4 current, the peak current

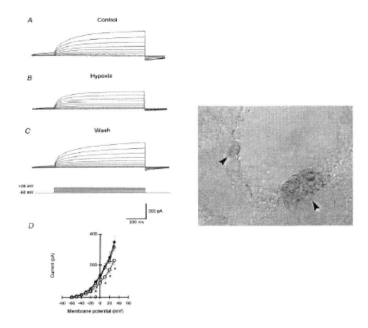

Figure 1. Effect of hypoxia on outward non-inactivate K^+ current in rabbit lung slice.

Left panel Outward delayed-rectifier non-inactivating K^+ current evoked by depolarizing steps from –60 to +30 mV in control normoxia solution (A). Outward current evoked by same voltage steps as in (A) was reversible reduced ~ 40% by hypoxia (PO_2~ 20mmHg) (B and C) D, I-V relationship for the currents in cell demonstrated from A, B, C.
Right panel Neutral red staining of NEB cells (arrowhead) in fresh slice of neonatal rabbit lung.

was reduced by ~ 27% (Fig. 2 A, B). The BDS-I sensitive current (Fig. 2A, D), is based on difference current measurement by subtraction of BDS-I in-sensitive current (Fig. 2A, B) from the control currents (Fig. 2A, A). The BDS-I sensitive current was fast inactivating, the averaged current was 189 ± 14 pA (n=4), and fitted by a two-component exponential functional with time constant of (τ 1) 77 ± 10 ms and (τ 2) 34 ± 3 ms (n=4). After application of 0.2 μM HpTx-2, a specific blocker of Kv4.3 current, the peak was current reduced by ~ 24% (Fig. 2 B, A, B). The HpTx-2 sensitive current (Fig. 2B, D), is based on difference current measurements by subtraction of HpTx-2 in-sensitive current (Fig. 2B, B) from the control currents (Fig. 2B, A). The averaged current of HpTx-2-sensitive was 234 ± 23 pA (n=4) and was fitted by a two-component exponential function with time constants of (τ 1) 82 ± 11 ms and (τ 2) 35 ± 8 ms (n=4).

Exposure to hypoxia (PO_2 ≈20 mmHg) resulted in a rapid and reversible reduction of the peak and sustained Kv currents in NEB cells (Fig. 2 C, *B* and *C*). The reduction of the peak and sustained component of current amplitude was significant at +30 mV test potential. At a test potential of +30 mV, after application of hypoxia solution, peak current decreased by ~ 47% and sustained current decreased by ~ 67%. Fig. 2C *D* shown the hypoxia-sensitive components of Kv current isolated by subtraction of the remaining currents in hypoxia (Fig. 2 *B*) from the control currents recorded in normoxia (Fig. 2C, *A*). This hypoxia-sensitive component is shown as a difference current for each voltage step in Fig. 2C, D. At +30 mV test potential, peak current is 319 ± 14 pA (n=8), and corresponding *I-V* curve is shown in Fig. 2C, E. Effects of hypoxia are greater in sustained current than peak current in NEB cells.

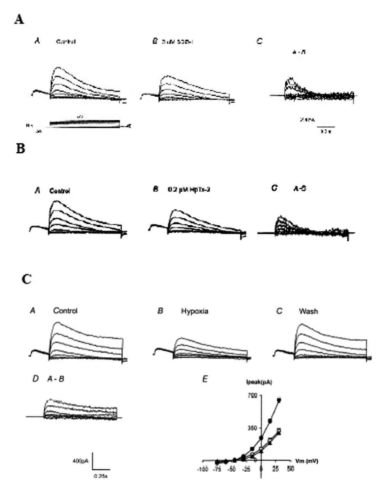

Figure 2. Effect of hypoxia on Kv α current in NEB cells from rabbit lung slice.

A, BDS-I sensitive Kv current A, slowly inactivating Kv α current evoked by depolarizing steps from -90 to +30mv in control Krebs solution. B, slowly inactivating A-type K^+ current was reduced by perfusing 3 μM BDS-I (a specific blocker of Kv3.4 subunit) C, BDS-I sensitive current K^+ was isolated by subtracting currents in B from those in A

B, HpTx-2 sensitive current (A and B) slowly inactivating K^+ was reduced by 0.2 μM HpTx-2. HpTx-2 sensitive current was isolated and shown in C.

C, O_2 sensitive Kv A-type K^+ current evoked by depolarizing steps from -90 to +30mv in control normoxic Krebs solution and K^+ current was reversibly reduced by hypoxia (A, B and C). Hypoxia-sensitive slowly inactivating K^+ current is shown in D. I-V relationship of currents in NEB cells form control normoxic condition (•), after perfusion hypoxia solution (O), and hypoxia-sensitive K^+ current (▲) in (E).

A-type Kv current expressed in NEB cells is one of O_2-sensitive K^+ current component as well, activate threshold around -30 mV, Kv 3.4 and Kv4.3 currents are modulated by hypoxia after cell depolarization during the repolarization. Similar results were reported in the rabbit carotid body where expression of Kv subunits (Kv3.4, Kv4.1 and Kv4.3) which are thought to underlie the fast-inactivating K+ current that is modified by low PO_2 (Sanchez, et al, 2002). Our findings suggest Kvα subunits Kv3.3a, Kv3.4 and Kv4.3 or their heteromultimers underlie the O_2 sensitive K^+ channels in NEB cells of rabbit fetal/neonatal lung.

3. CONCLUSIONS

Based on our previous and current studies, NEB cells express multiple subtypes of voltage-gated K^+ channels.
- O_2-sensitive K^+ channels:
  • Outward non-inactivating delayed-rectifier K^+ current:

(a) Ca^{2+}-independent K^+ current (I k (v))
(b) Ca^{2+} - dependent K^+ current (I k(Ca))

This K^+ current is inhibited by TEA or 4-AP, and modulated by DPI and H2O2.

• Kvα slowly inactivating K^+ current

(a) Kv3.4 current *
(b) Kv4.3 current
    Kvα current is blocked by BDS-I or HpTx-2.

- Kv3.3a current increased by H2O2.
• This slower inactivating K^+ current is modulated by H2O2.
*Kv 3.4 current is also modulated by H2O2 (Vega-Saenz & Rudy 1992).

ACKNOWLEDGEMENTS

This work was supported by grants from Canadian Institutes of Health Research (MOP-12742, MGP 15270).

REFERENCES

Fu, XW., Nurse, CA., Wang, YT., Cutz, E. Selective modulation of membrane currents by hypoxia in intact airway chemoreceptors from neonatal rabbit. 1999, J Phsiol 514: 139-150.

Fu, XW., Nurse, CA., Dinauer, MC., Cutz, E. NADPH oxidase is an O_2 sensor in airway chemoreceptors: Evidence from K+ current modulation in wild-type and oxidase-deficient mice. 2000, PNAS 97, 4374-4379.

Sanchez, D., Lopez-lopez, JR., Pere-Garcia, MT., San-Alfayate, G., Obeso, A.,Ganfornina., MD., Gonzalez., C. Molecular identification of Kvα subunits that contribute to the oxygensensitive K+ current of chemoreceptor cells of the rabbit carotid body. 2002, J Physiol 542:369-382.

Wang, D., Youngson, C., Wong, V., Yeger, H., Dinauer, M de M EV., Rudy, B., Cutz., E. NADPH-oxidase and a hydrogen peroxide-sensitive K+ channel may function as an oxygen sensor complex in airway chemoreceptors and small cell lung carcinoma cell lines. 1996, Pro Natl Acad Sci USA 93: 13182-13187.

Vega-Saenz de Miera E, Moreus H& Rudy B. Modulation of K+ channels by hydrogen peroxide. 1992, Biochem. Biophy. Res. Commun. 186,1681-1687.

Younsgon, C., Nurse, CA., Yeger, H., Cutz, E. Oxygen sensing in airway chemoreceptors. 1993, Nature 365: 153-155.

Voltage-Dependent K Channels in Mouse Glomus Cells are Modulated by Acetylcholine

TOSHIKI OTSUBO, SHIGEKI YAMAGUCHI, MACHIKO SHIRAHATA

Departments of Environmental Health Sciences, Johns Hopkins Bloomberg School of Public Health, Baltimore, USA

1. INTRODUCTION

Several neurotransmitters are present in the carotid body. These neurotransmitters are responsible to evoke action potentials in afferent nerve endings. They also modify the function of glomus cells, by binding to autoreceptors on glomus cells. We have previously demonstrated that ACh variably influences voltage-dependent K (Kv) current in cat glomus cells (Shirahata et al., 2002). Excitability of glomus cells are regulated by several types of K channels including voltage-dependent (Kv) channels (Buckler, 1999; Shirahata and Sham, 1999). Kv channels are activated with membrane depolarization, and play an essential role for repolarizing the cell membrane. Several studies have shown that the activity of Kv channels are modulated by neurotransmitters (Brown et al., 1997; Dong and White, 2003; Fukuda et al., 1988; Huang et al., 1993; Shi et al., 1999, 2004), and this type of modification may be important for fine tuning of the excitability of glomus cells. Recent studies have shown that the carotid body of DBA/2J inbred strain of mice demonstrates morphological similarities to cat glomus cells (Yamaguchi et al., 2003). Glomus cells of these mice responds to ACh (Yamaguchi et al., 2003) and express 4-aminopyridine sensitive and charybdotoxin sensitive components of Kv channels (Yamaguchi et al., 2004). In this study, we have investigated whether Kv channels in glomus cells of DBA/2J mice are also modified by ACh. Further, some underlying mechanisms of Kv current modulation by ACh was also investigated.

2. MATERIALS AND METHODS

An undissociated carotid body from a DBA/2J mouse was used for this study as described before (Yamaguchi et al., 2004). Mice (3-6 weeks old; either sex) were deeply anesthetized by peritoneal injection of 50 mg/kg ketamine and 100 mg/kg sodium pentobarbital. After the heart was removed to avoid bleeding in the neck, the carotid bodies were quickly harvested together with the carotid bifurcation and immersed in ice-cold modified Krebs solution (in mM : NaCl, 118; KCl, 4.7; $CaCl_2$, 1.8; KH_2PO_4, 1.2; $MgSO_4 \cdot 7H_2O$, 1.2; $NaHCO_3$, 25; glucose, 11.1; EDTA, 0.0016; pH, 7.4 with 5% CO_2/Air). Fat, connective tissue, and the sympathetic cervical ganglion were gently removed, and the carotid body was localized. Subsequently, the tissue was warmed in Krebs solution at 37 °C for 60 minutes, and then placed in a recording chamber which was attached to a stage of an upright microscope (Axioskop 2, Zeiss). After the tissue was superfused with Krebs solution containing 0.0375% collagenase (Type IX; Sigma) for 40 minutes, the carotid body was visualized using a water immersion lens (ACHROPLAN, 40X) combined with an infrared differential interference video camera (DAGE-MTI Inc.). All experiments were performed approximately at 37.0 °C while the tissue was continuously superfused with Krebs solution (1.22 mL/min) using a peristaltic pump (Minipuls 3, Gilson). A conventional tight-seal whole-cell recording was applied. The patch electrodes which had resistance of 4~6MΩ were filled the internal solution (in mM: K gluconate, 90; KCl, 33; NaCl, 10; $CaCl_2$, 1; EGTA, 10; MgATP, 5; HEPES, 10; pH 7.2 with KOH). Voltage-dependent whole cell current was evoked by a voltage clamp pulse (from 80 mV of holding potential to +20 mV) for 100 ms. The current was amplified using an Axopatch 200B patch-clamp amplifier (Axon Instruments). The signals were acquired and procecced with a Digidata 1320A and pCLAMP 8.1 (Axon Instrument). ACh and nicotine were prepared in a modified Tyrode solution (in mM): NaCl 143, KCl 4.7, $MgSO_4 \cdot 7H_2O$ 1.2, KH_2PO_4 1.2, $CaCl_2$ 1.8, EDTA 0.0016, HEPES 10 adjusted to pH 7.4 with NaOH.). A micro glass pipette was set close to a patched cell for focal application of ACh, nicotine or Tyrode solution (vehicle control) as described before (Higashi et al., 2003). To determine significant difference, Student's t-test or ANOVA was used. Values were considered significant if $p<0.05$.

3. RESULTS

The application of Tyrode solution (vehicle control) on Kv current was tested in 15 cells. Although some variations from the control values were observed, Kv current was generally stable during 10-15 minutes of recordings. Based on these data, we considered that more than 15% changes from control values were the effects of pharmacological agents. We categorize the changes in Kv current in three classes (increase, no change, decrease). In contrast to Tyrode application, ACh induced variable effects on Kv current (Table 1).

Table 1. Effects of ACh on Kv current.

	ACh		
	10 nM	1 µM	100 µM
Increase	139±7% (n=7)*	151±18% (n=4)*	133±5% (n=9)*
No Change	100±1% (n=7)	107% (n=1)	93±7% (n=4)
Decrease	none	65±4% (n=5)*	50±6% (n=8)*

More than 15% changes from the control were considered pharmacological effects, which were categorized in three classes. The size of Kv current at 97 msec after the beginning of the step pulse was expressed as % of the control current. *, significantly different from control current.

Since ACh has both nicotinic and muscarinic actions, we examined whether these changes were via nicotinic or muscarinic receptors. First, we tested if nicotinic receptor activation modulate Kv channels, using atropine, a non-specific muscarinic receptor blocker, and nicotine, a nicotinic receptor agonist (Table 2).

Table 2. Nicotinic modulation of Kv current.

	Nicotine				ACh (100 μM) with atropine (1 μM)
	10 nM	1 μM	10 μM	100 μM	
Increase	none	none	150±13% (n=4)*	141±7% (n=7)*	135±13% (n=2)
No Change	none	none	115% (n=1)	110±10% (n=3)	96±2% (n=7)
Decrease	52±2% (n=7)*	67±4% (n=7)*	58±3% (n=5)*	64±6% (n=8)*	64% (n=1)

Nicotine induced variable effects on Kv current. Inhibition is more apparent in lower doses of nicotine. Atropine did not change the pattern of ACh-induced modulation of Kv current. *, significantly different from control.

Because neuronal nicotinic ACh receptors are highly permeable to Ca^{2+} (McGehee and Role, 1995), Ca^{2+} may be a mediator for Kv channel modulation by nicotinic receptor activation. When the carotid body was superfused with Ca^{2+}-free Krebs, Kv current significantly decreased. During Ca^{2+}-free Krebs superfusion, nicotine application did not further change Kv current (data not shown). We further tested whether nicotinic receptor-mediated modulation on Kv channels was mediated via the changes in the activity of voltage-dependent large conductance Ca^{2+}-activated K channels (BK channels). Superfusion of the carotid body with Krebs including iberiotoxin (200 nM), a specific antagonist for BK channels, significantly decreased Kv current. Nicotine application did not produce no further changes in Kv current (Fig. 1).

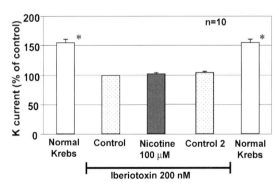

Figure 1. Elimination of nicotinic modulation of Kv current by iberiotoxin. Superfusion of the carotid body with Krebs solution containing iberiotoxin, decreased Kv current of the glomus cell. Under this condition, nicotine did not influence the current, suggesting that nicotine mainly affects BK channels. The current was evoked by a voltage step from -80 mV to +20 mV for 100 msec. K current at 97 msec after the beginning of the step pulse was expressed as % of the control current. Nicotine was applied to the patched cell via a puff pipette. *, significantly different from control.

Subsequently, cholinergic modulation of Kv channels via muscarinic receptors was evaluated after blocking nicotinic receptors. The carotid body was superfused with Krebs solution including hexamethonium (300-600 µM), and ACh was applied to the cell. ACh 100 µM induced only enhancement of Kv current with the presence of hexamethonium (Fig. 2). The enhancement of Kv current by muscarinic receptor activation does not seem to be mediated by Ca^{2+}. When ACh was applied during Ca^{2+}-free Krebs superfusion, Kv current was still either increased (148±4 % of control; n=7) or unchanged (100±1% of control; n=10).

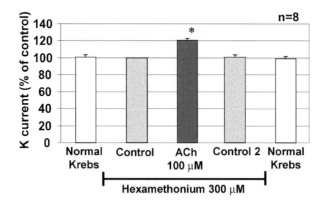

Figure 2. The carotid body was superfused with solution containing hexamethonium. Under this condition, ACh increased Kv current. *, significantly different from control.

4. DISCUSSION

The major finding of this study is that ACh modifies Kv current in glomus cells of the DBA/2J strain of mice. The effects of ACh are dose-related. With 10 nM of ACh, Kv current was enhanced or unchanged, but with increasing concentrations inhibition of Kv current was also observed. The effects of ACh are mediated by both nicotinic and muscarinic receptors. Variable effects of ACh on Kv current were observed with the presence of atropine (Table 2). Further, nicotine administration also showed variable effects on Kv current. Thus, it appears that nicotinic receptor activation increases or decreases Kv current. When both mechanisms are activated in one cell, no apparent changes would be observed. When the carotid body was superfused by Ca^{2+}-free Krebs solution, Kv current was not affected by nicotine, suggesting that intracellular Ca^{2+} mediates the responses induced by nicotine. Iberiotoxin, a specific BK channel blocker, significantly reduced Kv current in glomus cells. Further, with the presence of iberiotoxin, nicotine did not influence Kv current. The results suggest that nicotinic receptor activation modifies mainly BK channel activity. BK channels in the rat glomus cells, as well as recombinant human BKαβ1 channels expressed in HEK or CHO cells, are sensitive to hypoxia (Williams et al., 2004). BK channels in DBA mice are also oxygen sensitive (Otsubo et al, in this book). These results together with the current study indicate that BK channels in glomus cells are modulated by multiple factors.

With the presence of a nicotinic receptor blocker, hexamethonium, ACh only increased Kv current (Fig. 2), suggesting activation of muscarinic receptors enhances Kv current. In contrast to the nicotinic effects, muscarinic modulation of Kv current does not appear to be mediated Ca^{2+}-dependent pathways.

The presence and release of ACh from glomus cells has been shown in several species with biochemical and pharmacological techniques (Fitzgerald, 2000; Fitzgerald et al., 1997, 1999; Fitzgerald and Shirahata 1994; Kim et al., 2004). The release of ACh increases when the carotid body is stimulated by hypoxia or hypercapnea in the cat (Fitzgerald et al., 1999) and the rabbit (Kim et al., 2004). If this is also true in the mouse carotid body, low concentrations of ACh at rest (normocapnic normoxia) would enhance Kv current activity and decrease the excitability of glomus cells. With increased ACh release during the carotid body excitation such as hypoxia or hypercapnia, Kv current in some glomus cells is inhibited by ACh. In these cells, excitation of the cells would continue because Kv current inhibiton would interfere in repolarization processes. On the other hand, higher doses of ACh enhance Kv current in other glomus cells, in which repolarization of the cell membrane would be facilitated. Thus, modulation of Kv current by ACh may play a role for fine tuning the excitability of glomus cells.

ACKNOWLEDGEMENT

This study was supported by AHA0255358N and NHLBI 72293.

REFERENCES

Brown D.A., Abogadie F.C., Allen T.G., Buckley N.J., Caulfield M.P., Delmas P., Haley,J.E., Lamas,J.A., Selyanko,A.A., 1997. Muscarinic mechanisms in nerve cells. *Life Sci.* **60:** 1137-1144.

Buckler K.J., 1999. Background leak K^+-currents and oxygen sensing in carotid body type 1 cells. *Respir. Physiol.* **115:** 179-187.

Dong Y., White F.J., 2003. Dopamine D1-class receptors selectively modulate a slowly inactivating potassium current in rat medial prefrontal cortex pyramidal neurons. *J. Neurosci.* **23:** 2686-2695.

Fitzgerald R.S., 2000. Oxygen and carotid body chemotransduction: the cholinergic hypothesis - a brief history and new evaluation. *Respir. Physiol.* **120:** 89-104.

Fitzgerald R.S., Shirahata M., Ide T., 1997. Further cholinergic aspects of carotid body chemotransduction of hypoxia in cats. *J. Appl. Physiol.* **82:** 819-827.

Fitzgerald R.S., Shirahata M., 1994. Acetylcholine and carotid body excitation during hypoxia in the cat. *J. Appl. Physiol.* **76:** 1566-1574.

Fitzgerald R.S., Shirahata M., Wang H.Y., 1999. Acetylcholine release from cat carotid bodies. *Brain Res.* **841:** 53-61.

Fukuda K., Higashida H., Kubo T., Maeda A., Akiba I., Bujo H., Mishima M., Numa S., 1988. Selective coupling with K^+ currents of muscarinic acetylcholine receptor subtypes in NG108-15 cells. *Nature* **335:** 355-358.

Higashi T., McIntosh J.M., Shirahata M., 2003. Characterization of nicotinic acetylcholine receptors in cultured arterial chemoreceptor cells of the cat. *Brain Res.,* **974:** 167-175.

Huang X.Y., Morielli A.D., Peralta E.G., 1993. Tyrosine kinase-dependent suppression of a potassium channel by the G protein-coupled m1 muscarinic acetylcholine receptor. *Cell,* **75:** 1145-1156.

Kim D.K., Prabhakar N.R., Kumar G.K., 2004. Acetylcholine release from the carotid body by hypoxia: evidence for the involvement of autoinhibitory receptors. *J. Appl. Physiol.,* **96:** 376-383.

McGehee D.S., Role L.W., 1995. Physiological diversity of nicotinic acetylcholine receptors expressed by vertebrate neurons. *Annu. Rev. Physiol.*, **57**: 521-546.

Shi H., Wang H., Wang Z., 1999. M3 muscarinic receptor activation of a delayed rectifier potassium current in canine atrial myocytes. *Life Sci.*, **64**: L251-L257.

Shi H., Wang H., Yang B., Xu D., Wang Z., 2004. The M3 receptor-mediated K^+ current (IKM3), a Gq protein-coupled K^+ channel. *J. Biol. Chem.*, **279**: 21774-21778.

Shirahata M., Higashi T., Hirasawa S., Yamaguchi S., Fitzgerald R.S., Lande B., 2002. Excitation of Glomus Cells: Interaction between Voltage-gated K^+ Channels and Cholinergic Receptors. In: Oxygen Sensing: Responses and Adaptaion to Hypoxia. (Lahiri S., Semenza G.L., Prabhakar N.R., eds), pp365-379. New York: Marcel Dekker.

Shirahata M., Sham J.S., 1999. Roles of ion channels in carotid body chemotransmission of acute hypoxia. *Jpn. J. Physiol.* **49**: 213-228.

Williams S.E., Wootton P., Mason H.S., Bould J., Iles D.E., Riccardi D., Peers C., Kemp P.J., 2004. Hemoxygenase-2 is an oxygen sensor for a calcium-sensitive potassium channel. *Science*, **306**: 2093-2097.

Yamaguchi S., Balbir A., Schofield B., Coram J., Tankersley C.G., Fitzgerald R.S., O'Donnell C.P., Shirahata M., 2003. Structural and functional differences of the carotid body between DBA/2J and A/J strains of mice. *J. Appl. Physiol.* **94**: 1536-1542.

Yamaguchi S., Lande B., Kitajima T., Hori Y., Shirahata M., 2004. Patch clamp study of mouse glomus cells using a whole carotid body. *Neurosc. Lett.* **357**: 155-157.

Zhang M., Fearon I.M., Zhong H., Nurse C.A. 2003. Presynaptic modulation of rat arterial chemoreceptor function by 5-HT: role of K^+ channel inhibition via protein kinase C. *J. Physiol.* **551**: 825-842.

Modification of the Glutathione Redox Environment and Chemoreceptor Cell Responses

A. GÓMEZ-NIÑO*, M.T. AGAPITO, A. OBESO AND C. GONZÁLEZ

*Departamento de Bioquímica y Biología Molecular y Fisiología/IBGM. *Departamento de Biología Celular, Histología y Farmacología. Universidad de Valladolid / CSIC. Facultad de Medicina. 47005 Valladolid. Spain.*

1. INTRODUCTION

Carotid body (CB) chemoreceptor cells (CBCC) are involved in maintaining the homeostasis of O_2 by detecting arterial blood PO_2 and become activated when arterial PO_2 decreases. In response to hypoxia, CBCC release neurotransmitters which excite the adjacent afferent nerve terminals of the carotid sinus nerve, increase their action potential and, via the central projections of the nerve to the brain stem, activate ventilation. The proposed cascade of transduction of hypoxia requires the presence of an oxygen sensor that is coupled to specific K^+ channels. The decrease in the opening probability of these, leads to CBCC depolarization followed by Ca^{2+} entry via voltage-dependent Ca^{2+} channels and consequently, the release of excitatory transmitters (5). However, the molecular identity of the O_2 sensor and the mechanism coupling the oxygen sensor to the exocytotic machinery has thus far remained unclear.

Reactive oxygen species (ROS) have been proposed as mediators, either triggers or modulators, of the hypoxic responses in specialized O_2-sensing cell types including CBCC, erythropoietin producing cells and pulmonary artery smooth muscle cells (6). The superoxide radical, $O_2^{\bullet-}$, is quantitatively the most relevant primary ROS, capable of reacting with many molecules to generate altered structures and new secondary ROS. One of the major sources of ROS is the mitochondrial respiratory chain, where univalent reduction of O_2 may take place in some steps of the overall transport system. There are two places where the superoxide radical can be produced: the complex I NADPH-Coenzyme Q Oxido-Reductase and the ubiquinone pool. In both places there are electron shuttle molecules (flavin adenin mononucleotide and ubiquinone) that can accept electrons. When they accept a single electron they become free radical molecules themselves, capable of slipping the electron to molecular O_2. The mitochondrial respiratory chain is also the main place of ATP synthesis.

Two hypotheses postulate that ROS participate in the control of the CBCC activity. The first hypothesis developed by Acker and colleagues (3) proposed a decrease in ROS levels due to the hypoxic inhibition of a phagocyte-like NADPH oxidase present in CBCC as the signal triggering O_2-sensitive K^+ channel inhibition. The falling levels of ROS, or the concomitant increase in the reduced to oxidized glutathione ratio (GSH/GSSG), would cause a reduction of the disulfide bonds in the proteins forming the K^+ channels, thereby leading to a decrease in their opening probability, followed by cell depolarization (1). Opposed to that, Schumacker and coworkers found that hypoxia increased the rate of ROS production at mitochondrial level, this increase being the signal triggering chemoreception (2). According to these authors the production of ROS takes place mainly in the quinone pool.

Controversy remains as to whether hypoxia decreases or increases ROS, and also unresolved is which electron transport chain complex generates ROS and what might be the function of mitochondrial ROS in regulating the activity of oxygen-sensing cells. Because the GSH/GSSG system is quantitatively the most important mechanism to dispose ROS and to maintain the overall redox environment in mammalian cells, glutathione redox potential can be used as an indicator of the rate of ROS production in cells. Our laboratory has previously observed that different reducing agents which increase the glutathione redox potential (GSH/GSSG) affect CBCC normoxic and hypoxic activity in a variable manner (7). In a further attempt to define the significance of ROS in CBCC function we have manipulated the redox potential of cells with reducing and oxidizing agents and with inhibitors of the mitochondrial electron transport chain in rat diaphragm. Selected concentrations of the drugs, capable of altering the glutathione redox potential, were subsequently used to study their effects on the activity of CBCC measured as the rate of release of catecholamines (CA) (5).

2. METHODS

Surgical procedures. *Removal and dissection of rat diaphragm.* Wistar rats of both sexes and 250-300 g of body weight were anaesthetized by intraperitoneal injection of sodium pentobarbital (60mg/Kg) dissolved in physiological saline. The diaphragm was entirely removed, then brought to a Lucite chamber filled with ice-cold 100% O_2 saturated Tyrode where was cleaned and cut in four quadrants of comparable size. Each piece constituted an experimental sample.

Removal and dissection of rat CB. The same type of animals were anaesthetized and an incision in the neck allowed the dissection of the carotid artery bifurcation. With the help of a dissecting microscope the CB were cleaned of surrounding connective tissue.

Incubation of the tissues. *Incubation of the diaphragm to measure glutathione.* Each of the four pieces of a diaphragm was transferred to individual glass scintillation vials contained 10 ml of prewarmed bicarbonate-buffered Tyrode (in mM: NaCl, 116; KCl, 5; $CaCl_2$, 2; $MgCl_2$, 1.1; glucose, 5.5; $NaHCO_3$, 24; HEPES, 10) and maintained in a shaker bath at 37°C for a period of 30 min. while balancing the solution with 21%O_2/ 5% CO_2/ 75 N_2 saturated with water vapor. This initial incubation provided uniformity to all pieces of tissue. Afterwards, the incubation solutions contained the agents specified and were incubated for periods of time specified below. The inhibitors of the mitochondrial electron transport chain, rotenone (1 μM) and sodium azide (5 mM) were applied for 10 min and in another experiments we incubated the diaphragms during the 10 min period prior to and during the application of mitochondrial inhibitors with 2 mM NAC; the reducing agents N-acetylcysteine (NAC; 2 mM) and N-mercaptopropionylglycine (NMPG; 5 mM) were applied for 40 min; the oxidizing agents carmustine (1mM), buthionine-sulfoximine (250 μM), aminotriazol (10 mM), 2,2′-dithiodipiridine (0.1 mM), diamide (0.2 mM) and GSSG (1mM) were also applied for 40 min. At the end of the incubation the tissues were transferred to new vials containing 10 ml of (ice-cold) Tyrode, weighed and placed in eppendorf tubes containing 5-sulfosalicylic acid at 5% and 0,25mM EDTA. Tissues were homogenized at 0-4°C, centrifuged and the supernatant was used to measure GSH/GSSG.

Incubation of the CBs to label the (CA) deposits of CBCC. The deposits of CA were labeled by incubating the CBs 2 h at 37°C in a shaking bath, in a solution containing 3,5-^{3}H-tyrosine (30uM; 40-50Ci/mmol).Thereafter, CBs were transferred to glass scintillation vials containing precursor-free bicarbonate-buffered Tyrode solution balanced with 21%O_2 / 5% CO_2 / N_2 saturated with water vapor. Incubating solutions were collected every 10 min and saved for analysis in their ^{3}H-CA content. In the drug treated CBs the drug was added at the required concentration to the incubation solution (see Results). At the end of the experiments, the CB tissues were homogenized, centrifuged and the supernatant stored for analysis in ^{3}H-CA (4).

Analytical procedures. *Measurement of GSH and GSSG.* GSH and GSSG were measured with the method of Griffth (9) as recently described in detail by González et al. (8). By measuring the levels of reduced and oxidized glutathione (GSH and GSSG respectively) the ratio GSH/GSSG was calculated.

Measurement of ^{3}H-CA The analysis of ^{3}H-CA present in the solution collected every 10 min is based in the ability of alumina to absorb ^{3}H-CA at alkaline pH and their elution at acidic pH. Quantification of ^{3}H-CA was made by liquid scintillation counting.

3. RESULTS

Effects of inhibitors of mitochondrial electron transport chain. Rotenone a blocker of Complex I (NADH CoQ1 oxido-reductase) at 1 μM, but not at lower concentrations, significantly decreased the GSH/GSSG ratio, the effect being completely reversed by the addition of NAC (Figure 1). Yet, rotenone (0.25 μM) elicited release of ^{3}H-CA that was not modified by the presence of NAC (Figure 2). Sodium-Azide, blocker of Complex IV (Cytocrome c oxydase), significantly decreased the GSH/GSSG ratio, the effect being also reversed by the addition of NAC (Figure 1). At the same concentration sodium-azide strongly activated the normoxic ^{3}H-CA release response and the release was not modified by the presence of NAC (Figure 2).

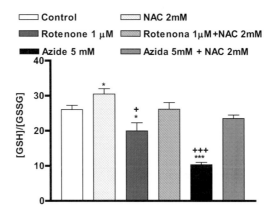

Figure 1. Effect of NAC (2mM) significantly increasing GSH/GSSG ratio, rotenone 1μM and Na-azide (5mM) significantly decreasing GSH/GSSG ratio. The effects of the metabolic poisons are reversed by the addition of NAC. Data are means + SEM with n = 6-8 individual values. * p< 0.05 vs control. *** p < 0.001 vs control. + p < 0.05 vs rotenone + NAC. +++ p < 0.001 vs azide + NAC.

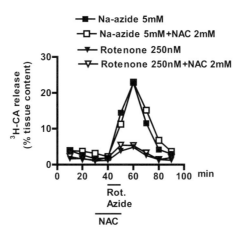

Figure 2. Rotenone (250nM), blocker of Complex I and sodium azide (5mM), blocker of Complex IV, induced an increase on ^{3}H-CA release from CBCC. The secretory response elicited by the mitochondrial poisons was nt modified by the presence of the antioxidant NAC.

Effect of reducing agents. NMPG is a thiol compound used as a ROS scavenger. NMPG produced a decrease of GSSG with minor modification of GSH and as a consequence significantly increased the GSH/GSSG ratio (figure 3). Incubation of CB for 40 min with the same concentration of scavenger, increased GSH level then GSH/GSSG ratio and did not affect the normoxic release of ^{3}H-CA. NMPG (5mM) produced an increase in the normoxic release of ^{3}H-CA that outlasted the application of the drug (Figure 4). NAC (2mM), a known precursor for glutathione synthesis and ROS.

Glutathione Redox Environment

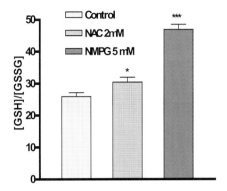

Figure 3. The reducing agents NAC (2mM) and NMPG (5mM) significantly increased the GSH/GSSG ratio. Data are means + SEM with n = 8 individual values. * p< 0.05; ***p < 0.001 vs control.

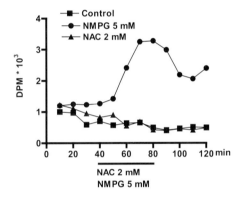

Figure 4. Time course of the ^{3}H-CA release from rat CBs incubated in the presence of NAC and NMPG in normoxic conditions. The basal release of ^{3}H-CA was not affected by NAC but NMPG induced release outlast the application of the drug.

Effect of oxidazing agents. The oxidizing agents tested 3-amino-1,2,4-triazol (10mM), inhibitor of catalase, carmustine (1mM), inhibitor of glutathione reductase, L-buthionine-sulfoximine (0.250mM), inhibitor of glutathione synthesis, and the permeable oxidizing compounds diamide (0.4mM), 2,2'-dithiopiridine (0.1mM) and GSSG (1mM), decreased the GSH/GSSG ratio but the effects on the release response of CBCC were very variable, increasing, decreasing or not altering the 3H-CA normoxic or hypoxic induced release (data not shown).

4. DISCUSSION

The pharmacological manipulation of the redox status of cells showed that there is not a unique relationship between the glutathion redox potential and the activity of CBCC, measured as the neurosecretory response. The metabolic poisons activated chemoreceptor cells independently of the redox environment, as the modification they produced on the GSH/GSSG level was reversed by NAC and the antioxidant did not alter the CBCC activation induced by these mitochondrial blockers. The reducing and oxidizing agents tested showed the same lack of correlation This discrepancy suggests that the general redox potential of cells is not a critical signal to trigger the activation of CBCC. These experimental findings however, do not exclude the possibility that in restricted cellular compartments the redox status changes locally. This local change could be able of modulating the functionality of CBCC as shown in the article by He et al. (this book).

ACKNOWLEDGEMENTS

Supported by Spanish DGICYT grant (BFU2004-06394) and ICiii-FISS (Red Respira and Grant PI042462).

REFERENCES

1) Acker H. and Xue D. Mechanisms of O2 sensing in the carotid body in comparison with other O2-sensing cells. *News Physiol. Sci.* 10, 211-215, 1995.
2) Chandel NS and Schumacker PT. Cellular oxygen sensing by mitochondria: old questions, new insight. *J. Appl. Physiol.* 88, 1880-1889, 2000.
3) Cross AR, Henderson L, Jones OTG, Delpiano MA, Hentschel J and Acker H. Involvement on NAD(P)H oxidase as a PO2 sensor protein in rat carotid body. *Biochem. J.* 272, 743-747, 1990.
4) Fidone S., González C. And Yoshizaki K. Effects of low oxygen on the release of dopamine from the rabbit carotid body in vitro. *J. Physiol.* 333, 93-110, 1982.
5) González C., Almaraz L., Obeso A. and Rigual R. Oxygen and acid chemoreception in the carotid body chemoreceptors. *Trends Neurosci.* 15, 156-153, 1992.
6) González C., Sanz-Alfayete G., Agapito M.T., Gómez-Niño A., Rocher A., and Obeso A. Significance of ROS in oxygen sensing in cell systems with sensitivity to physiological hypoxia. *Respir. Physiol. Neurobiol.* 132, 17-41, 2002.
7) González C, Sanz-Alfayete G., Agapito MT: and Obeso A. Effects of reducing agents on glutathione metabolism and the function of carotid bodies cgemoreceptor cells. *Biol. Chem.* 385 (3-4): 265-274, 2004.
8) González C., Sanz-Alfayete G., Obeso A., and Agapito MT. Role of glutathione redox state in oxygen sensing by carotid chemoreceptor cells. *Methods Enzymol.* 381, 40-71, 2004.
9) Griffith OW. Determination of glutathione and glutathione disulfide using glutathione reductase and 2-vinylpyridine. *Anal. Biochem.* 106, 207-212, 1980.
10) He L., Dinger B., González C., Obeso A. and Fidone S. Function of NADPH oxidase and signaling by reactive oxygen species in rat carotid body type I cells.

Carotid Body Transmitters Actions on Rabbit Petrosal Ganglion *in Vitro*

JULIO ALCAYAGA, CAROLINA R. SOTO, ROMINA V. VARGAS, FERNANDO C. ORTIZ, JORGE ARROYO, AND *RODRIGO ITURRIAGA

*Laboratorio de Fisiologia Celular, Facultad de Ciencias, Universidad de Chile y *Laboratorio de Neurobiología, Facultad de Ciencias Biológicas, P. Universidad Católica de Chile Santiago, Chile*

1. INTRODUCTION

The petrosal ganglion (PG) that innervate the carotid body (CB) are activated by transmitters released from the CB receptor (type I, glomus) cells, resulting in the carotid nerve (CN) chemosensory activity. Although the exact transmitter(s) of this synapse is unknown, several molecules present in the CB may play a role in the generation and/or maintenance of the afferent activity (Eyzaguirre & Zapata, 1984; González et al. 1994).

Acetylcholine (ACh) is present and released from the CB during stimulation (Eyzaguirre & Zapata 1968). ACh applied to the PG and its neurons, acutely isolated or in tissue culture, modify their membrane potential, induce inward currents, and/or increase their firing rate in a dose-dependent manner (Alcayaga et al. 1998; Varas et al. 2000; Zhong & Nurse 1997). Moreover, nicotinic cholinergic receptor antagonists reduce chemosensory responses in rat reconstituted chemoreceptor units in tissue culture (Zhang et al. 2000; Zhong et al. 1997), and in acutely isolated chemoreceptor units *in vitro* (Varas et al. 2003). Thus, ACh appears to play a role in the generation of chemosensory activity in PG neurons.

Adenosine 5´-triphosphate (ATP) increases cat chemosensory discharge (Ribeiro & McQueen 1984; Spergel & Lahiri 1993), and is released from the rat CB by hypoxia (Buttigieg & Nurse 2004). The PG and its isolated neurons respond to ATP with increased discharge (Alcayaga et al. 2000), and inward currents (Zhang et al., 2000). Chemosensory activity recorded from reconstituted chemosensory units is partly blocked by suramin (Prasad et al., 2001; Zhang et al. 2000), a P2X receptor antagonist. $P2X_2$ and $P2X_3$ receptor subunits are present in the rat PG neurons (Prasad et al. 2001), and the deletion of their encoding genes results in reduced chemosensory and ventilatory responses in the mouse (Rong et al. 2003). These data suggest that ATP receptors on PG neurons may participate in the generation of chemosensory activity.

Dopamine (DA) is synthesized and stored type-I cells, released during hyoxia (Gonzalez et al. 1994), while D2 receptors are present in PG neurons (Bairam et al. 1996). DA inhibits chemosensory activity in most species, but excitation or dual effects have been reported (Eyzaguirre & Zapata 1984). DA has no direct effect on isolated rat petrosal-jugular neurons, and on CN activity when applied to the cat PG *in vitro* (Alcayaga et al. 1998) but modulates responses induced by ACh and ATP in the latter preparation (Alcayaga et al. 1999, Alcayaga et al. 2003). Thus, DA receptors in PG neurons appear to modulate the actions of other transmitters.

Because the evidence for the role of these transmitters has been obtained from different species, we studied the effects of these transmitters on the rabbit PG and the activity conducted through the CN.

2. METHODS

The PGs were obtained from adult male White New Zealand rabbits anaesthetised with ketamine/xylazine (75/7.5 mg/kg, im); ⅓ of the initial dose was applied when necessary to maintain the anaesthetic level. The carotid bifurcation was exposed in the neck, and the CN cut close to the CB. The glossopharyngeal nerve, cut distally to origin of the CN, was followed into the cranium, the PG completely exposed and the central roots severed. The tissue obtained was placed in ice-chilled Hanks' solution, and the ganglion and nerves cleaned from surrounding connective tissue. The PG was pined to the bottom of a 0.2 ml chamber, and superfused with Hanks' solution supplemented with 5 mM Hepes buffer, pH 7.4 at $38 \pm 0.5°C$, flowing at 1.5 ml/min. The CN was placed on paired electrodes, and lifted into an upper compartment of the chamber filled with mineral oil. The electrodes were connected to an AC-preamplifier, and the electroneurogram amplified, displayed in oscilloscope, recorded, and fed to a spike amplitude discriminator, whose output pulses were digitally counted, to assess the CN frequency discharge (f), in Hz. The temperature of the chamber and f were acquired, displayed and recorded through a data acquisition system, at 1 Hz. ACh, ATP, and DA, in doses of 1 to 3000 µg in 10 µl boluses, were applie over the ganglion, and antagonists in the medium. The change in f (Δf) was calculated as the difference between the maximal frequency (max f) achieved during a response and the mean basal activity (bas f), computed 30 s before the response. The relation between the responses (Y) and the doses of any of the drugs used (X) was assessed fitting data to a sigmoid curve ($Y = \max Y / [1 + \{ED_{50} / X\}^S]$) characterized by its mean effective dose (ED_{50}) and slope (s). Correlation between variables was higher than 0.96 and significant ($P<0.05$; Student's t-test).

3. RESULTS

ACh applied to the PG produced a brief increase in f (Fig. 1A), which amplitude and duration were dose-dependent (Fig. 1B, C), and with little or no temporal desensitization. The responses presented a threshold between 0.1 and 0.5 µg and maximal responses for doses near 100 µg. The responses were mimicked by nicotine but not by bethanechol; the latter temporarily reduced the amplitude of ACh-induced responses. Moreover, during atropine superfusion the maximal responses induced by ACh were significantly increased (Fig. 1C). The ACh-induced responses were largely and reversibly reduced during the superfusion with hexamethonium 10 nM (Fig. 1D).

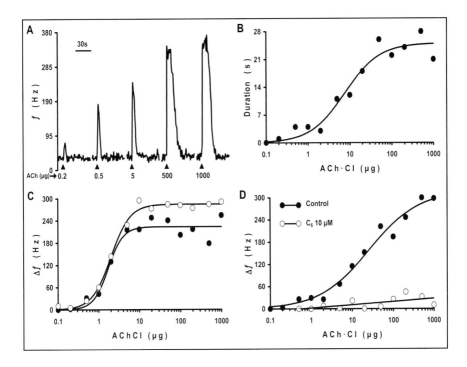

Figure 1. Effects of ACh on carotid nerve discharge. A. Changes in f induced by ACh. B. Dose-duration relationship. C. Dose-response relationships in control (filled dots) and during atropine (empty circles) superfusion. D. Dose-response relationships in control (filled dots) and during hexamethoium 10 nM (empty circles) superfusion. Representative recordings from single preparations.

ATP produced a brief increase in f (Fig. 2A), with no appreciable temporal desensitization, while adenosine 5′-monophosphate (AMP) was largely ineffective in inducing responses (Fig. 2B). The amplitude and duration of the ATP-induced responses were dose-dependent, with a threshold near 0.1 µg and maximal responses for doses over 100 µg (Fig. 2B, C). The ATP-induced responses were almost completely blocked when suramin 50 µM was present in the superfusion medium (Fig. 2D).

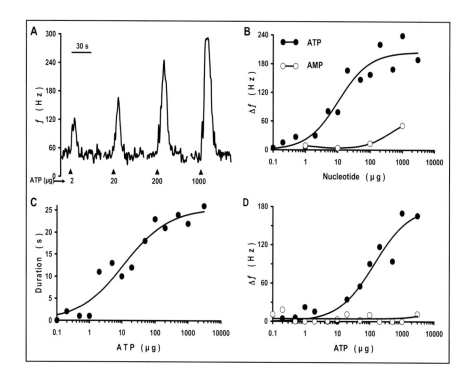

Figure 2. Effects of ATP on carotid nerve discharge. A. Changes in f induced by ATP. B. Dose-response relationships for the responses induced by ATP (filled dots) and AMP (empty circles). C. Dose-duration relationship. D. Dose-response relationships in control (filled dots) and during suramin 50 nM (empty circles) superfusion. Representative recordings from single preparations.

DA applied to the PG increased f (Fig. 3A), but the responses presented a large degree of temporal desensitization needing about 20 min between successive doses for responses to fully develop. The amplitude of the responses were dose-dependent (Fig. 3A), with an apparent threshold between 1 and 10 µg and maximal responses for doses over 200 µg (Fig. 3B). The duration of the response increased to a maximum for doses between 10 and 100 µg and then they were consistently reduced in duration (Fig. 3C). The responses induced by DA were largely reduced when spiperone 10 nM was present in the superfusion medium (Fig. 3D).

4. DISCUSSION

ACh applied to the rabbit PG increases f, response mimicked by nicotine and blocked by hexamethonium. Excitatory responses to ACh, with similar pharmacological properties, had been described in the cat isolated PG (Alcayaga et al. 1998) and in cultured cat and rat PG neurons (Varas et al. 2000, Zhong & Nurse 1997), suggesting that nicotinic cholinergic receptors (nAChR) are

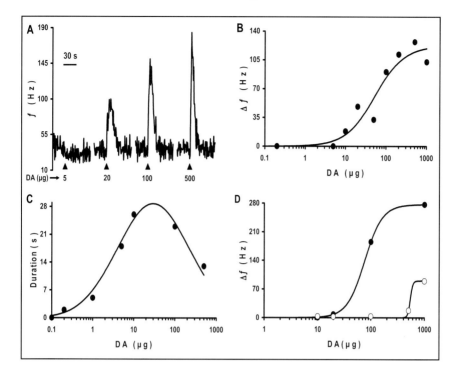

Figure 3. Effects of DA on carotid nerve discharge. A. Changes in f induced by DA. B. Dose-response relationship. C. Dose-duration relationship. D. Dose-response relationships in control (filled dots) and during spiperone 10 nM (empty circles) superfusion. Representative recordings from single preparations.

present in PG neurons. Moreover, chemosensory responses in rat reconstituted chemoreceptor units in tissue culture (Zhang et al. 2000, Zhong et al. 1997), and in acutely isolated chemoreceptor units *in vitro* (Varas et al. 2003) are largely reduced by nAChR antagonists. Bethanechol, applied to the rabbit PG, had no effect on f but transiently reduced ACh-induced responses, while inclusion of atropine in the medium increased the maximal responses induced by ACh. These data suggest that muscarinic receptors are also present in rabbit PG neurons, and their activation by ACh opposes the effects of concomitant nAChR activation. Similar inhibitory effects of ACh on cat PG neurons had been described (Shirahata et al. 2000). Our data indicate that rabbit PG neurons express both muscarinic and nicotinic receptors, the latter increasing the activity and the former reducing the maximal discharge without modifying the neuronal sensitivity to ACh.

ATP applied to the rabbit PG increases f, response blocked by suramin and marginally mimicked by AMP. Similar responses and pharmacological properties had been described for the cat PG (Alcayaga et al. 2000) and for rat PG neurons co-cultured with CB cells (Prasad et al. 2001; Zhang et al. 2000). Our data suggest that rabbit PG neurons and terminals are endowed with P2X receptors, as in the rat (Prasad et al. 2001) and mouse (Rong et al. 2003), and that their activation in the PG terminals increase the afferent discharge. Rabbit

ATP-induced responses did not desensitize as previously reported PG responses, suggesting that the receptor sub-unit composition may differ between different species (Prasad et al. 2001; Rong et al. 2003). Thus, rabbit PG neurons projecting through the CN express P2X receptors that may participate in the generation of chemosensory activity.

Application of DA to the rabbit PG induces a dose-dependent increase in f, effect that was partially blocked by the D2 receptor antagonist spiperone. DA applications were ineffective in modifying f in a similar cat PG preparation (Alcayaga et al., 1999), although they modulate the responses induced by ACh (Alcayaga et al., 1999) and ATP (Alcayaga et al., 2003). Similarly, chemosensory responses in rat reconstituted chemoreceptor units in tissue culture were unaffected by spiperone (Zhong et al. 1997), suggesting that activation of D2 receptors is not a key process in the generation of chemosensory activity in the PG neurons. Our results indicate that, in contrast with the effects in many species, DA can actually generate discharges in the PG neurons, and that the effect appears to be specifically mediated by D2 receptors. However, an unspecific excitatory effect of DA on chemosensory discharge has been described on the goat, mediated by $5HT_3$ receptors (Herman et al. 2003).

Our results indicate that rabbit PG neurons express nicotinic ACh, ATP, and DA receptors as in other species. However, muscarinic ACh receptors are also present, and activation of DA receptors increases CN activity instead of modulating the activity as in other species.

ACKNOWLEDGEMENTS

Work supported by FONDECYT 1040638.

REFERENCES

Alcayaga J, Cerpa V, Retamal M, Arroyo J, Iturriaga R & Zapata P (2000). Adenosine triphosphate-induced peripheral nerve discharges generated from the cat petrosal ganglion in vitro, *Neurosci Lett* **282**, 185-188.

Alcayaga J, Iturriaga R, Varas R, Arroyo J & Zapata P (1998). Selective activation of carotid nerve fibers by acetylcholine applied to the cat petrosal ganglion in vitro. *Brain Res* **786**, 47-54.

Alcayaga J, Retamal M, Cerpa V, Arroyo J & Zapata P (2003). Dopamine inhibits ATP-induced responses in the cat petrosal ganglion in vitro. *Brain Res* **966**, 283-287.

Alcayaga J, Varas R, Arroyo J, Iturriaga R & Zapata P (1999). Dopamine modulates carotid nerve responses induced by acetylcholine on cat petrosal ganglion in vitro. *Brain Res* **831**, 97-103.

Bairam A, Dauphin C, Rousseau F, Khandjian EW (1996). Dopamine D2 receptor mRNA isoforms expression in the carotid body and petrosal ganglion of developing rabbits. *Adv Exp Med Biol* **410**, 285-289.

Buttigieg J & Nurse CA (2004). Detection of hypoxia-evoked ATP release from chemoreceptors cells of the rat carotid body. *Biochem Biophys Res Comm* **322**, 82-87.

Eyzaguirre C & Zapata P (1968). The release of acetylcholine from carotid body tissue. Further study on the effects of acetylcholine and cholinergic blocking agents on the chemosensory discharge., *J Physiol (London)* **195**, 589-607.

Eyzaguirre C & Zapata P (1984). Perspectives in carotid body research. *J Appl Physiol* **57**, 931-957.

González C, Almaraz L, Obeso A & Rigual R (1994). Carotid body chemoreceptors: from natural stimuli to sensory discharges. *Physiol Rev* **74**, 829-898.

Herman JK, O'Halloran KD, Janssen PL & Bisgard GE (2003). Dopaminergic excitation of the goat carotid body is mediated by the serotonin type 3 receptor subtype. *Respir Physiol Neurobiol* **136**, 1-12.

Prasad M, Fearon IM, Zhang M, Laing M, Vollmer C & Nurse CA (2001). Expression of P2X2 and P2X3 receptor subunits in rat carotid body afferent neurones: role in chemosensory signaling. *J Physiol (London)* **537**, 667-677.

Ribeiro JA & McQueen DS (1984). Effects of purines on carotid chemoreceptors. In: DJ Pallot (Ed), The peripheral arterial chemoreceptors. Croom Helm, London, 1984, pp 383-390.

Rong W, Gourine AV, Cockayne DA, Xiang Z, Ford APDW, Spyer KM & Burnstock G (2003). Pivotal role of nucleotide $P2X_2$ receptor subunit of the ATP-gated ion channel mediating ventilatory responses to hypoxia. *J Neurosci* **23**, 11315-11321.

Shirahata M, Ishizawa Y, Rudisill M, Sham JS, Schofield B & Fitzgerald RS (2000). Acetylcholine sensitivity of cat petrosal ganglion neurons. *Adv Exp Med Biol* **475**, 377-387.

Spergel D & Lahiri S (1993). Differential modulation by extracellular ATP of carotid chemosensory responses. *J Appl Physiol* **74**, 3052-3056.

Varas R, Alcayaga J & Zapata P (2000). Acetylcholine sensitivity in sensory neurons dissociated from the cat petrosal ganglion. *Brain Res* **882**, 201-205.

Varas R, Alcayaga J, Iturriaga R (2003). ACh and ATP mediate excitatory transmission in identified cat carotid body chemoreceptor units in vitro. *Brain Res* **988**, 154-163.

Zhang M, Zhong H, Vollmer C & Nurse CA (2000). Co-release of ATP and ACh mediates hypoxic signalling at rat carotid body chemoreceptors, *J Physiol (London)* **525**, 143-158

Zhong H & Nurse CA (1997). Nicotinic acetylcholine sensitivity of rat petrosal sensory neurons in dissociated cell culture. *Brain Res* **766**, 153-161.

Zhong H, Zhang M & Nurse CA (1997). Synapse formation and hypoxic signalling in co-cultures of rat petrosal neurons and carotid body type 1 cells. *J Physiol (London)* **503**, 599-612.

Potassium Channels in the Central Control of Breathing

YOSHITAKA OYAMADA, KAZUHIRO YAMAGUCHI, MICHIE MURAI, AKITOSHI ISHIZAKA, YASUMASA OKADA*

*Department of Pulmonary Medicine, School of Medicine, Keio University, Shinjuku, Tokyo, Japan; *Department of Medicine, KeioUniversity Tsukigase Rehabilitation Center, Izu City, Shizuoka, Japan*

1. INTRODUCTION

Ventilation is closely tied to $PaCO_2$ and PaO_2 via a feedback control mechanism. Most importantly, changes in PCO_2 and/or H+ (pH) are sensed by the central nervous system (CNS) resulting in directly related changes in ventilation. The magnitude of this response relative to the stimulus is known as, "central ventilatory chemosensitivity." Several electrophysiological studies have demonstrated that chemosensitive neurons – i.e. neurons that change their electrical activity in response to a shift of PCO_2/pH – are widely distributed throughout the brainstem (Bernard and Nattie, 1996; Coates and Nattie, 1993; Dean et al., 1989, 1990; Kawai et al., 1996; Oyamada et al., 1998, 1999; Richerson, 1995), suggesting that they could be the central CO_2/pH sensors that regulate ventilation. In addition, PaO_2 levels are inversely related to ventilation. The type I cells of the carotid body serve as the main sensors stimulating ventilation when PaO_2 declines ("peripheral ventilatory chemosensitivity"). The afferent inputs from these PCO_2/pH- or PO_2-sensitive cells converge at the respiratory center located in the brainstem, where the generation of respiratory neural activity is integrated.

Potassium channels control the excitability of cells by setting its membrane potential. The importance of potassium channels in peripheral arterial chemoreception was first demonstrated by López-Barneo et al. (1988). They showed, in the rabbit, that the potassium current was reduced by hypoxia in type I carotid body cells. More recently it has been observed, in the rat, that members of the tandem pore domain family, TASK-1 and TASK-3, are also inhibited by hypoxia in type I cells (Buckler et al., 2000) and H146 cells (Hartness et al., 2001). It has also been suggested that the Kv4 is the main subfamily involved in the O_2-sensitive potassium current in type I cells in the rabbit (Perez-Garcia et al., 2000). Some potassium channels are also sensitive to CO_2/pH (Buckler et al., 2000; Chapman et al., 2000; Kim et al., 2000; Lesage and Lazdunski, 2000; Rajan et al., 2000; Xu et al., 2000a; Zhu et al., 2000). This suggests that they may play a key role in central ventilatory chemosensitivity, as they apparently do in the case of peripheral arterial chemoreception. We review the observations related to the roles of potassium channels in central ventilatory chemosensitivity. Their role in O_2-sensing of the CNS will also be summarized.

2. POTASSIUM CHANNELS IN CENTRAL VENTILATORY CHEMOSENSITIVITY

This section reviews studies relevant to central ventilatory chemosensitivity according to the type of experimental preparations.

2.1 *Ex Vivo* Preparations (Isolated Brainstem-spinal Cord & Slice Preparations)

Dean et al. (1989), using brainstem slices, first demonstrated a depolarizing response to hypercapnic acidosis from neurons in the solitary tract nucleus. This response was accompanied by an increase in input resistance. Kawai et al. (1996) showed the depolarization, along with an increase in input resistance, of certain types of respiratory neurons (e.g., inspiratory and respiration-modulated neurons) on the ventral aspect of isolated brainstem-spinal cord preparation in rat neonates. They also demonstrated that hypercapnic acidosis elicited a hyperpolarizing response and a decrease in input resistance in expiratory and post-inspiratory neurons. These observations suggested that potassium channels might be involved in the chemosensitivity of these respiratory neurons. This hypothesis was supported by Pineda and Aghajanian's experiments (1997), who measured the electrical activity of neurons located in the locus coeruleus (LC) in pontine slices. Oyamada et al. (1998) later proved that these neurons are related to ventilation, and Pineda and Aghajanian (1997) showed an increase in burst frequency of LC neurons induced by extracellular acidosis, with or without hypercapnia. This excitatory response was accompanied by a decrease in conductance of the outward portion of proton- and polyamine-sensitive inward rectifier potassium current.

In 2005, Okada et al. (2005) ascertained the important role played by extracellular potassium in the maintenance of respiratory rhythm and central chemosensitivity in isolated brainstem-spinal cord preparations of rat neonates.

2.2 *In Vitro* Preparations

The sensitivity of certain inwardly rectifying potassium channels (Kir) to CO_2/pH has been electrophysiologically studied in *Xenopus* oocytes, where they are prominently expressed (Xu et al., 2000a; Zhu et al., 2000). Among them, Kir2.3 exhibited a decrease in conductance in response to intra and extracellular acidosis, and the activity of Kir1.1 was suppressed by intracellular acidosis (Zhu et al., 2000). Kir4.1 was also inhibited by intracellular acidification, an inhibition augmented when it was co-expressed with Kir5.1 (Xu et al., 2000a). Attempts have been made to identify the molecular motif determining this pH sensitivity.

Qu et al. (1999) demonstrated that a critical motif responsible for proton gating of Kir2.3 is located in the N terminus, which contains approximately 10 residues centered by Thr^{53}. Several portions of Kir1.1 and Kir4.1 seem to control their pH sensitivity (Xu et al., 2000a, 2000b).

In addition, the tandem pore domain potassium channels, TASK-1 and TASK-3, are also inhibited by extracellular acidosis (Bayliss et al., 2001; Chapman et al., 2000; Kim et al., 2000; Lesage and Lazdunski, 2000; Rajan et al., 2000). Rajan et al. (2000) showed that the substitution of the histidine residue, His-98, by asparagine or tyrosine abolished the pH sensitivity of TASK-3. In contrast, K_{ATP}, an ATP-sensitive potassium channel consisting of Kir6.2 and sulfonylurea receptor (SUR) 1 subunits was activated by intracellular acidosis (Xu et al., 2001).

2.3 Experiments in Knockout Mice

Transgenic mice have been developed, in which a Kir channel (e.g., Kir1.1, Kir2.1, Kir2.2 or Kir4.1) is knocked-out (Kofuji et al., 2000; Lu et al., 2002; Zaritsky et al., 2001). A single study has related their phenotypes with ventilation. Oyamada et al. (2005) measured the hypercapnic ventilatory response in unanesthetized, unrestrained Kir2.2-knockout (Kir2.2-/-) mice, using pressure plethysmography on postnatal days 9-10, 14-15 and 18. The Kir2.2-/- mice developed a smaller increase in tidal volume and minute ventilation volume than their wild-type counterparts in response to exposure to hypercapnia on days 14-15. There was a transient increase in mRNA expression of Kir2.2 in the brainstem of the wild-type mice on days 14-15. These observations suggest that, during postnatal development, Kir2.2 in the brainstem plays an important role, albeit transient, in the hypercapnic ventilatory response, perhaps mediated by central ventilatory chemosensitivity.

2.4 Histological Studies

Using *in situ* hybridization techniques, Karschin et al. (1996) and Karschin and Karschin (1997) found temporal and spatial heterogeneities of the mRNA expression of Kir2.x and Kir3.x in the CNS of the rat. In their studies, several chemosensitive nuclei (e.g., LC and raphe nucleus) expressed Kir2.2 in an age-dependent manner. Washburn et al. (2002) demonstrated that TASK-1 and TASK-3 channels are expressed in serotonergic raphe neurons. Haller et al. (2001) showed that the Kir6.2 subunit was co-expressed with the SUR1 subunit in respiratory neurons. Wu et al. (2004), who similarly examined the mRNA expression of Kir1.1, Kir2.3, Kir4.1 and Kir5.1, showed that the mRNAs found in these channels were expressed in several brainstem nuclei of the rat, particularly those involved in the cardio-respiratory control (e.g., LC, ventrolateral medulla, parabrachial-Kolliker-Fuse nuclei, solitary tract nucleus). The expression of Kir5.1 and Kir4.1 was notably more prominent than Kir1.1 and Kir2.3.

3. POTASSIUM CHANNELS IN O_2-SENSING BY THE CENTRAL NERVOUS SYSTEM

In mammals, acute hypoxia triggers a biphasic ventilatory response with first an augmentation followed by depression. The cellular mechanisms behind the inhibitory phase, known as "hypoxic ventilatory decline" (HVD), have not been fully clarified, although it is believed to be a CNS-derived phenomenon. A hypoxia-induced loss of neuronal excitability is one possible mechanism of HVD. In contrast to TASK-1, TASK-3 or Kv4, the heteromultimeric Kir6.x/SUR complexes of K_{ATP} are activated by hypoxia, subsequently hyperpolarizing the cell. Hypoxia-induced activation of K_{ATP} has been documented in LC (Nieber et al. 1995), medullary inspiratory neurons (Mironov et al., 1998), rat midbrain dopaminergic neurons (Guatteo et al. 1998) and in dorsal vagal neurons (Kulik et al. 2002). Therefore, Kir6.x might be involved in the development of HVD. The diffuse expression of Kir6.2 in the CNS (Karschin et al. 1997a), including dorsal vagal (Karschin et al. 1998) and respiratory neurons (Haller et al. 2001) where functional K_{ATP} channels are constituted by the formation of Kir6.2 subunits with SUR1 receptors, suggests that Kir6.2 is involved. Actually, in a preliminary study, we observed that Kir6.2-/- mice developed a weaker HVD than the wild type mice (unpublished data).

Several territories of the CNS may be activated by hypoxia (Neubauer et al., 2004). These regions perhaps play an important role in the initial augmentation of the hypoxic ventilatory response. While the cellular mechanisms of hypoxia-induced activation remain to be clarified, it seems possible that O_2-sensitive potassium channels are involved, as in the case of peripheral arterial chemoreception.

4. CONCLUSIONS

Several specific potassium channels are sensitive to CO_2/pH or O_2. These channels are likely to play essential roles in CO_2/pH or O_2-sensing *in vivo*. Their contributions to the chemical control of breathing are probably heterogeneous both temporally and anatomically.

REFERENCES

Bayliss D.A., Talley E.M., Sirois J.E., and Lei Q., 2001. TASK-1 is a highly modulated pH-sensitive 'leak' K^+ channel expressed in brainstem respiratory neurons. *Respir Physiol.* 129: 159-174.

Bernard D.G., Li A., and Nattie, E.E., 1996. Evidence for central chemoreception in the midline raphe. *J. Appl. Physiol.* 80: 108-115.

Buckler K.J., Williams B.A., and Honoré E., 2000. An oxygen-, acid- and anaesthetic-sensitive TASK-like background potassium channel in rat arterial chemoreceptor cells. *J. Physiol.* 525 (Pt 1): 135-142.

Chapman C.G., Meadows H.J., Godden R.J., Campbell D.A., Duckworth M., Kelsell R.E., Murdock P.R., Randall A.D., Rennie G.I., and Gloger I.S., 2000. Cloning, localisation and functional expression of a novel human, cerebellum specific, two pore domain potassium channel. *Brain Res. Mol. Brain Res.* 82: 74-83.

Coates E.L., Li A., and Nattie E.E., 1993. Widespread sites of brain stem ventilatory chemoreceptors. *J. Appl. Physiol.* 75: 5-14.

Dean J.B., Lawing W.L., and Millhorn D.E., 1989. CO_2 decreases membrane conductance and depolarizes neurons in the nucleus tractus solitarii. *Exp. Brain Res.* 76: 656-661.

Dean J.B., Bayliss D.A., Erickson J.T., Lawing W.L., and Millhorn D.E., 1990. Depolarization and stimulation of neurons in nucleus tractus solitarii by carbon dioxide does not require chemical synaptic input. *Neuroscience* 36: 207-216.

Guatteo E., Federici M., Siniscalchi A., Knopfel T., Mercuri N.B., and Bernardi G., 1998. Whole cell patch-clamp recordings of rat midbrain dopaminergic neurons isolate a sulphonylurea- and ATP-sensitive component of potassium currents activated by hypoxia. *J. Neurophysiol.* 79: 1239-1245.

Haller M., Mironov S.L., Karschin A., and Richter D.W., 2001. Dynamic activation of K_{ATP} channels in rhythmically active neurons. *J. Physiol.* 537: 69-81.

Hartness M.E., Lewis A., Searle G.J., O'Kelly I., Peers C., and Kemp P.J., 2001. Combined antisense and pharmacological approaches implicate hTASK as an airway O_2 sensing K^+ channel. *J. Biol. Chem.* 276: 26499-26508.

Karschin C., Dißmann E., Stühmer W., and Karschin A., 1996. IRK(1-3) and GIRK(1-4) inwardly rectifying K^+ channel mRNAs are differentially expressed in the adult rat brain. *J. Neurosci.* 16: 3559-3570.

Karschin C., Ecke C., Ashcroft F.M., and Karschin A., 1997. Overlapping distribution of K_{ATP} channel-forming Kir6.2 subunit and the sulfonylurea receptor SUR1 in rodent brain. *FEBS Lett.* 401: 59-64.

Karschin C., and Karschin A., 1997. Ontogeny of gene expression of Kir channel subunits in the rat. *Mol. Cell. Neurosci.* 10: 131-148.

Karschin A., Brockhaus J., and Ballanyi K., 1998. K_{ATP} channel formation by the sulphonylurea receptors SUR1 with Kir6.2 subunits in rat dorsal vagal neurons in situ. *J. Physiol.* 509: 339-346.

Kawai A., Ballantyne D., Mückenhoff K., and Scheid P., 1996. Chemosensitive medullary neurones in the brainstem-spinal cord preparation of the neonatal rat. *J. Physiol.* 492: 277-292.

Kim Y., Bang H., and Kim D., 2000. TASK-3, a new member of the tandem pore K^+ channel family. *J. Biol. Chem.* 275: 9340-9347.

Kofuji P., Ceelen P., Zahs K.R., Surbeck L.W., Lester H.A., and Newman E.A., 2000. Genetic inactivation of an inwardly rectifying potassium channel (Kir4.1 subunit) in mice: phenotypic impact in retina. *J. Neurosci.* 20: 5733-5740.

Kulik A., Brockhaus J., Pedarzani P., and Ballanyi K., 2002. Chemical anoxia activates ATP-sensitive and blocks Ca^{2+}-dependent K^+ channels in rat dorsal vagal neurons in situ. *Neuroscience* 110: 541-554.

Lesage F., and Lazdunski M., 2000. Molecular and functional properties of two-pore-domain potassium channels. *Am. J. Physiol. Renal Physiol.* 279: F793-F801.

López-Barneo J., López- López J.R., Urena J., and González C., 1988. Chemotransduction in the carotid body: K^+ current modulated by PO_2 in type I chemoreceptor cells. *Science* 241: 580-582.

Lu M., Wang T., Yan Q., Yang X., Dong K., Knepper M.A., Wang W., Giebisch G., Shull G.E., and Hebert S.C., 2002. Absence of small conductance K^+ channel (SK) activity in apical membranes of thick ascending limb and cortical collecting duct in ROMK (Bartter's) knockout mice. *J. Biol. Chem.* 277: 37881-37887.

Mironov S.L., Langohr K., Haller M., and Richter D.W., 1998. Hypoxia activates ATP-dependent potassium channels in inspiratory neurones of neonatal mice. *J. Physiol.* 509: 755-766.

Neubauer J.A., and Sunderram J., 2004. Oxygen-sensing neurons in the central nervous system. *J. Appl. Physiol.* 96: 367-374.

Nieber K., Sevcik J., and Illes P., 1995. Hypoxic changes in rat locus coeruleus neurons in vitro. *J. Physiol.* 486: 33-46.

Okada Y., Kuwana S., Kawai A., Mückenhoff K., and Scheid, P., 2005. Significance of extracellular potassium in central respiratory control studied in the isolated brainstem-spinal cord preparation of the neonatal rat. *Respir. Physiol. Neurobiol.* 146: 21-32.

Oyamada Y., Ballantyne D., Mückenhoff K., and Scheid P., 1998. Respiration-modulated membrane potential and chemosensitivity of locus coeruleus neurones in the in vitro brainstem-spinal cord of the neonatal rat. *J. Physiol.* 513: 381-398.

Oyamada Y., Andrzejewski M., Mückenhoff K., Scheid P., and Ballantyne D., 1999. Locus coeruleus neurones in vitro: pH-sensitive oscillations of membrane potential in an electrically coupled network. *Respir. Physiol.* 118: 131-147.

Oyamada Y., Yamaguchi K., Murai M., Hakuno H., and Ishizaka A., 2005. Role of Kir2.2 in hypercapnic ventilatory response during postnatal development of mouse. *Respir. Physiol. Neurobiol.* 145: 143-151.

Perez-Garcia M.T., Lopez-Lopez J.R., Riesco A.M., Hoppe U.C., Marban E., Gonzalez, C., and Johns D.C., 2000. Viral gene transfer of dominant-negative Kv4 construct suppresses an O_2-sensitive K^+ current in chemoreceptor cells. *J. Neurosci.* 20: 5689-5695.

Pineda J., and Aghajanian G.K., 1997. Carbon dioxide regulates the tonic activity of locus coeruleus neurons by modulating a proton- and polyamine-sensitive inward rectifier potassium current. *Neuroscience* 77: 723-743.

Qu Z., Zhu G., Yang Z., Cui N., Li Y., Chanchevalap S., Sulaiman S., Haynie H., and Jiang C., 1999. Identification of a critical motif responsible for gating of Kir2.3 channel by intracellular protons. *J. Biol. Chem.* 274: 13783-13789.

Rajan S., Wischmeyer E., Xin Liu G., Preisig-Müller R., Daut J., Karschin A., and Derst C., 2000. TASK-3, a novel tandem pore domain acid-sensitive K^+ channel. An extracellular histidine as pH sensor. *J. Biol. Chem.* 275: 16650-16657.

Richerson G.B., 1995. Response to CO_2 of neurons in the rostral ventral medulla in vitro. J. Neurophysiol. 73, 933-944.

Washburn C.P., Sirois J.E., Talley E.M., Guyenet P.G., and Bayliss D.A., 2002. Serotonergic raphe neurons express TASK channel transcripts and a TASK-like pH- and halothane-sensitive K^+ conductance. *J. Neurosci.* 22: 1256-1265.

Wu J., Xu H., Shen W., and Jiang C., 2004, Expression and coexpression of CO_2-sensitive Kir channels in brainstem neurons of rats. *J. Membr. Biol.* 197: 179-191.

Xu H., Cui N., Yang Z., Qu Z., and Jiang C., 2000a, Modulation of Kir4.1 and Kir5.1 by hypercapnia and intracellular acidosis. *J. Physiol.* 524: 725-735.

Xu H., Yang Z., Cui N., Giwa L.R., Abdulkadir L., Patel M., Sharma P., Shan G., Shen W., and Jiang C., 2000b, Molecular determinants for the distinct pH sensitivity of Kir1.1 and Kir4.1 channels. *Am. J. Physiol. Cell Physiol.* 279: C1464-C1471.

Xu H., Cui N., Yang Z., Wu J., Giwa L.R., Abdulkadir L., Sharma P., and Jiang C., 2001. Direct activation of cloned K_{ATP} channels by intracellular acidosis. *J. Biol. Chem.* 276: 12898-12902.

Zaritsky J.J., Redell J.B., Tempel B.L., and Schwarz T.L., 2001. The consequences of disrupting cardiac inwardly rectifying K^+ current (I_{K1}) as revealed by the targeted deletion of the murine Kir2.1 and Kir2.2 genes. *J. Physiol.* 533: 697-710.

Zhu G., Liu C., Qu Z., Chanchevalap S., Xu H., Jiang C., 2000. CO_2 inhibits specific inward rectifier K^+ channels by decreases in intra- and extracellular pH. *J. Cell. Physiol.* 183, 53-64.

Role of Endothelin-1 on the Enhanced Carotid Body Activity Induced by Chronic Intermittent Hypoxia

SERGIO REY, RODRIGO DEL RIO AND RODRIGO ITURRIAGA

Laboratorio de Neurobiología, Facultad de Ciencias Biológicas, P. Universidad Católica de Chile, Alameda 340, Santiago, Chile

1. INTRODUCTION

The systemic ventilatory and cardiovascular adjustments to hypoxia depend on the activation of peripheral chemoreceptors, mainly the carotid bodies (CBs). Intermittent hypoxia produces CB chemosensory excitation, which may participate in the pathogenesis of obstructive sleep apnea (OSA) (Narkiewicz et al., 1999). Recently, we found that chronic intermittent hypoxia (CIH) for four days enhances cat ventilatory and chemosensory responses to acute hypoxia (Rey et al., 2004). We hypothesized that the enhanced chemosensory response to hypoxia induced by CIH would be the result of an increased effect of excitatory modulators, like endothelin (ETs) peptides, and/or a decreased effect of inhibitory molecules within the CB. Thus, we studied the contribution of ETs on the enhanced chemosensory response to hypoxia in CIH-treated cats.

Endothelin-1 (ET-1) a 21 aminoacid peptide with potent vasoconstrictor activity, is the only peptide with a chemoexcitatory effect in the CB (Rey & Iturriaga, 2004) that is upregulated during chronic hypoxia (Chen et al., 2002a; Chen et al., 2002b; Rey & Iturriaga, 2004). The excitatory effect of ET-1 was first described by McQueen and colleagues, who found that ET-1 injected into the common carotid artery increases the carotid chemosensory discharge and the respiratory frequency (Spyer et al., 1991; McQueen et al., 1994). Chen et al. (2002a) using an isolated superfused preparation of the rat CB devoid of vascular effects, found that ET-1 enhances the chemosensory response to hypoxia, suggesting that ET-1 may act upon glomus cells or the sensory endings of petrosal neurons. However, Rey & Iturriaga (2004) using a similar superfused preparation of the cat CB, found that ET-1 *perse* does not increase basal chemosensory discharges in doses up to 10 µg, but the same ET-1 dose applied intravascularly to the perfused cat CB produces a persistent chemosensory excitation, indicating that the excitatory effect of ET-1 is mainly due to an increase in CB vascular tone. To study the contribution of ET-1 to the enhanced chemosensory response to acute hypoxia in CIH-treated cats and to avoid the potent hypertensive effect of ET-1 *in situ*, we used a perfused *in vitro* preparation of the cat CB, which conserves its local vasomotor regulation. We studied the local ET-1 expression in control and CIH-exposed CBs with immunohistochemistry. Since plasma ET-1 levels are increased in CIH-exposed rats (Kanagy et al., 2001) and OSA patients (Phillips et al., 1999), we evaluated the plasma levels of ET-1 in control and CIH-treated cats.

2. METHODS

Experiments were performed on 12 male adult cats (3.4 ± 0.5 kg, mean ± SD). The animals were anaesthetized with sodium pentobarbitone (40 mg kg^{-1}, I.P.) additional doses (8-12 mg, I.V.) were given when necessary to maintain a level of surgical anaesthesia. The carotid bifurcation including the CB and the carotid sinus nerve (CSN), was excised and perfused *in vitro* with a Tyrode solution at 38.5± 0.5°C and pH 7.40 as previously described (Iturriaga et al., 1991). The chemosensory discharges were recorded from the CSN placed on a pair of platinum electrodes and lifted into mineral oil. The neural signals were preamplified, amplified, filtered and fed to an electronic amplitude discriminator. The selected chemosensory impulses were counted with a frequency meter to measure the frequency of chemosensory discharges (f_x), expressed in Hz. The f_x signal was digitized with an analog-digital board (Digidata 1200A, Axon Instruments, USA) for later analysis.

The CBs obtained from anesthetized cats were fixed in neutral buffered formalin and processed for immunohistochemistry. Deparaffinized 5 μm histological sections were rehydrated through ethanol and treated with 10% H_2O_2 for 10 minutes to block endogenous peroxidase activity. After rinsing with PBS-Tris 50 mM buffer for 5 minutes at pH 7.8, the slides were incubated in a humid chamber with protein block solution (DakoCytomation, Ely, UK) for 30 min. After washing, the slides were incubated for 18 h at 4°C with a rabbit anti-ET-1 (1 : 600) antibody directed against the complete ET-1 peptide (IHC-6901, Peninsula Labs, San Carlos, CA, USA). Following washing and rinsing with Tween-20 0.05% buffer, sections were immunostained using the biotin-streptavidin-peroxidase system from DakoCytomation (Ely, UK). Finally, the samples were treated for 15 min with 0.1% (w/v) 3-3´-diaminobenzidine in buffer containing 0.05% H_2O_2. The slides were counterstained with Harris' hematoxylin and mounted. We used femoral artery and myocardial tissue as positive controls, and negative controls without the primary antibody. Digital microphotographs were obtained at 400x magnification and analyzed with the ImageJ software (NIH, Bethesda, MD, USA) to calculate the immunoreactivity area with a color deconvolution algorithm (Ruifrok & Johnston, 2001). We averaged the percentage of immunoreactivity in three 400x fields from each CB, analyzing a total of three control and CIH-treated CBs.

Plasma-EDTA samples were collected from six control and six CIH cats and stored at -20 °C. After extraction, concentration and reconstitution, the samples were assayed immediately with a commercial ET-1 immunoassay (BBE5, R&D Systems, Minneapolis, USA), according to manufacturer instructions. The crossreactivity of the assay was: Big Endothelin < 1%; ET-2, 45% and ET-3, 14%.

Bosentan was dissolved in DMSO 0.01% v/v. and control Tyrode solutions contained the same DMSO concentration. Statistical analysis was done with the Mann-Whitney Test. The level of statistical significance was $p<0.05$. Data is expressed as mean ± SEM.

3. RESULTS

3.1 Endothelin Expression is Enhanced in the Carotid Body of CIH Cats

Positive immunohistochemical staining was mostly found between the glomic cell clusters with weak staining within the clusters (Fig. 1A, B). The percentage of endothelin-like immunoreactivity (ET-ir) in the CBs from control animals was $0.7 \pm 0.2\%$, whereas the ET-ir in CBs from CIH-treated cats was $10.0 \pm 0.8\%$ (Fig. 1C). This increase in the area of ET-ir in the CIH group was statistically significant ($p<0.001$). Negative controls showed no staining, while myocardial and femoral artery samples (positive controls) showed endothelial staining, which is consistent with ET-ir localization in other species. The cross-sectional area of the analyzed samples was not different between control and CIH groups ($44,037 \pm 469$ vs. $40,120 \pm 1,850$ μm^2, $p>0.05$).

3.2 Chronic Intermittent Hypoxia does not Modify ET-1 Plasma Levels in Cats

Plasma ET-1 levels in control animals was 2.83 ± 0.26 fmol ml^{-1} and 2.60 ± 0.28 fmol ml^{-1} in CIH-treated animals ($p>0.05$, Fig. 1D). The calculated limit of detection of the assay was 0.77 fmol ml^{-1}.

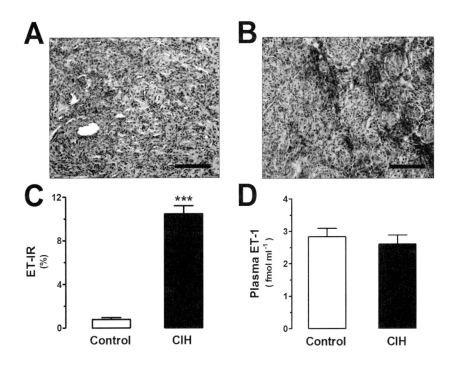

Figure 1. Endothelin-1 expression in the cat CB. Microphotographs from **A**, control CB; **B**, CIH-treated CB (bar = 100 μm). **C**, Endothelin-like immunoreactivity (ET-IR) in three CBs from control and CIH-treated cats. **D**, Plasma ET-1 levels in six control and six CIH-treated cats. *White bars*, control; *black bars*, CIH-treated. *** $p<0.001$.

3.3 Non-selective Endothelin Receptor Blockade Reduces the Hypoxic Chemosensory Response in the CB from Cats Exposed to CIH, but not in Control Conditions

Basal f_x was significantly higher in CIH-treated CBs (125 ± 13 Hz) compared to control CBs (69 ± 10 Hz, n=6, p<0.05). The chemosensory response to severe hypoxia (PO_2 ~30 Torr) was also higher in CIH-treated CBs. Indeed, in CIH-treated CBs, the maximal f_x attained during hypoxia was 403 ± 18 Hz, whereas in control CBs was 328 ± 24 Hz (n=7, p<0.05). Application of ET-1 to the perfused CB produces a long-lasting, dose-dependent chemoexcitation.

To assess the role of endogenous endothelin receptor activation upon the hypoxia-induced chemosensory response, we studied the effect of bosentan (50 µM), a non-selective $ET_{A/B}$ antagonist in CBs from control and CIH-treated cats. $ET_{A/B}$ blockade did not have any effect in the maximal f_x induced by severe hypoxia in control and CIH-treated CB (data not shown). However, bosentan reduced the chemosensory response evoked by mild hypoxia (PO_2 ~ 80 Torr) only in CBs from CIH-treated cats, leaving the chemosensory response of control CBs unchanged (Fig. 2). Bosentan had an inhibitory effect over basal f_x in approximately half of the CBs from both experimental groups.

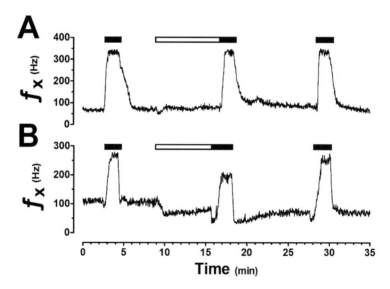

Figure 2. Effect of $ET_{A/B}$ blockade on the hypoxia-induced chemosensory response in perfused CBs from control and CIH-treated cats. Representative f_x tracings from **A**, Control; **B**, CIH-treated CB. f_x, chemosensory discharge; *filled bars*, hypoxic perfusion (PO_2 ~ 80 Torr); *empty bars*, bosentan (50 µM) perfusion.

4. DISCUSSION

In a previous work from our laboratory, we found that exposure of cats to CIH for four days enhanced ventilatory and chemosensory responses to acute hypoxia (Rey et al., 2004). Present results show that **i)** CIH increased basal f_x in the *in vitro* preparation of the cat CB and enhanced the chemosensory response to hypoxia. **ii)** ET-ir was markedly increased in CBs from CIH cats compared to the control group, but plasma ET-1 levels remained within the control range in CIH cats. **iii)** Bosentan reduced the chemosensory response to moderate hypoxia ($PO_2 \sim 80$ Torr), but not to severe hypoxia ($PO_2 \sim 30$ Torr) in CBs from CIH cats. These results suggest that the augmented expression of ET-1 within the CB may contribute to the enhanced chemosensory response to acute hypoxia in CIH. Our results confirm and extend previous findings showing that CIH enhances ventilatory and chemosensory responses to hypoxia (Peng et al., 2001; Peng & Prabhakar, 2004). Present results suggest that ET-1 mediates the enhanced chemosensory response to acute hypoxia after exposure to CIH, similarly to what is observed during chronic sustained hypoxia (Chen et al., 2002a; Chen et al., 2002b). The immunohistochemical data shows a marked increase in ET-1 expression locally in the CIH-treated CB, which is not paralleled by an increase in ET-1 plasma levels. Therefore, present results suggest that ET-1 and probably other endothelin isoforms are locally upregulated in the CB before any increase in the plasma levels. Other researchers have found increased plasma ET-1 levels and hypertension in a rat model of CIH exposure for 10 days (Kanagy et al., 2001). Current evidence shows that ET-1 primarily acts as a paracrine modulator of vascular tone, cell proliferation and fibrosis in various tissues (Rubanyi & Polokoff, 1994). However, only an increase of ET-1 levels within the CB can produce an enhancement of the systemic ventilatory and cardiovascular adjustments to hypoxia in CIH animals and OSA patients. Thus, the augmented expression of ET-1 locally in the CB may be relevant in the pathophysiology of sleep-disordered breathing. The effects of bosentan in the perfused CB preparation suggest a role of endogenous ET-1 in the enhanced hypoxic chemosensory response in CBs from animals exposed to CIH.

ACKNOWLEDGEMENTS

We would like to thank Mrs. Paulina Arias and Mrs. Carmen Gloria León for technical assistance in the experiments, and Mrs. Cecilia Chacon and Mrs. Jenny Corthorn for their help with the immunohistochemical techniques. Bosentan was a gift of Dr. Aquiles Jara, P.U.C. This work is supported by FONDECYT 1030330 (R.I.) and PG-22/04 (S.R.)

REFERENCES

Chen J, He L, Dinger B, Stensaas L & Fidone S. (2002a). Role of endothelin and endothelin A-type receptor in adaptation of the carotid body to chronic hypoxia. *Am J Physiol Lung Cell Mol Physiol* **282,** L1314-1323.

Chen Y, Tipoe GL, Liong E, Leung S, Lam SY, Iwase R, Tjong YW & Fung ML. (2002b). Chronic hypoxia enhances endothelin-1-induced intracellular calcium elevation in rat carotid body chemoreceptors and up-regulates ETA receptor expression. *Pflugers Arch* **443,** 565-573.

Iturriaga R, Rumsey WL, Mokashi A, Spergel D, Wilson DF & Lahiri S. (1991). In vitro perfused-superfused cat carotid body for physiological and pharmacological studies. *J Appl Physiol* **70,** 1393-1400.

Kanagy NL, Walker BR & Nelin LD. (2001). Role of endothelin in intermittent hypoxia-induced hypertension. *Hypertension* **37,** 511-515.

McQueen DS, Dashwood MR, Cobb VJ & Marr CG. (1994). Effects of endothelins on respiration and arterial chemoreceptor activity in anaesthetised rats. *Adv Exp Med Biol* **360,** 289-291.

Narkiewicz K, van de Borne PJ, Pesek CA, Dyken ME, Montano N & Somers VK. (1999). Selective potentiation of peripheral chemoreflex sensitivity in obstructive sleep apnea. *Circulation* **99,** 1183-1189.

Peng Y, Kline DD, Dick TE & Prabhakar NR. (2001). Chronic intermittent hypoxia enhances carotid body chemoreceptor response to low oxygen. *Adv Exp Med Biol* **499,** 33-38.

Peng YJ & Prabhakar NR. (2004). Effect of two paradigms of chronic intermittent hypoxia on carotid body sensory activity. *J Appl Physiol* **96,** 1236-1242; discussion 1196.

Phillips BG, Narkiewicz K, Pesek CA, Haynes WG, Dyken ME & Somers VK. (1999). Effects of obstructive sleep apnea on endothelin-1 and blood pressure. *J Hypertens* **17,** 61-66.

Rey S, Del Rio R, Alcayaga J & Iturriaga R. (2004). Chronic intermittent hypoxia enhances cat chemosensory and ventilatory responses to hypoxia. *J Physiol* **560,** 577-586.

Rey S & Iturriaga R. (2004). Endothelins and nitric oxide: Vasoactive modulators of carotid body chemoreception. *Curr Neurovasc Res* **1,** 465-473.

Rubanyi GM & Polokoff MA. (1994). Endothelins: molecular biology, biochemistry, pharmacology, physiology, and pathophysiology. *Pharmacol Rev* **46,** 325-415.

Ruifrok AC & Johnston DA. (2001). Quantification of histochemical staining by color deconvolution. *Anal Quant Cytol Histol* **23,** 291-299.

Spyer KM, McQueen DS, Dashwood MR, Sykes RM, Daly MB & Muddle JR. (1991). Localization of [125I]endothelin binding sites in the region of the carotid bifurcation and brainstem of the cat: possible baro- and chemoreceptor involvement. *J Cardiovasc Pharmacol* **17,** S385-389.

Concluding Remarks

C. Gonzalez
President of ISAC

Departamento de Bioquímica y Biología Molecular y Fisiología/IBGM Facultad de Medicina. Universidad de Valladolid/CSIC. 47005Valladolid. Spain

At the outset of these remarks I want to acknowledge Prof. Hisatake Kondo, past President of ISAC and President of XVIth ISAC Symposium for his excellent job as a host. He and his team of coworkers showed to all of us what the efficiency of organization in every regard means. I want to underline the two aspects that impressed me more: first, it was the *Shinkansen* type of punctuality in the initiation and completion of every event during the Meeting days; second, it was the sided projection that allowed every participant to be aware of coming events ahead of time. That virtue and this organizational strategy were dressed with a sincere expression of warmth that made the nearly four days of Symposium pleasant and substantial; the substance of the Meeting was provided by all symposiasts.

First session started on time. **Rodrigo Iturriaga** from Santiago de Chile was the Chairman. Two speakers in this session (**P.Kemp** and **Y. Zhang**) underscored the potential significance of heme oxygenase-2 (HO-2) in controlling the O_2-sensing machinery in carotid body chemoreceptor cells (CBCC), in optimizing the ventilation-perfusion relationship and hypoxic ventilatory response. However, there was a tinge of discrepancy in their findings. From the molecular and electrophysiological study of Kemp it should be expected that KO mice for HO-2 should be hyper-reactive to hypoxia, an expectation supported by previous studies with inhibitors of the enzyme, and yet, the KO animals showed a blunted ventilatory response to hypoxia. Could it be that a compensatory over-expression of HO-1 accounts for the discrepancy? Important was the demonstration of exaggerated ventilation-perfusion mismatching in the HO-2 KO mice, as it can lead to suggest that this enzyme is involved in the hypoxic transduction cascade of chemoreceptor cells and pulmonary artery smooth muscle cells (PASMC).

M. Evans study on AMP-activated protein kinase provided clues that may be crucial to understand how metabolic poisons excite CBCC. The most classical activators of the CB chemoreceptors keep activating ventilation since Heymans discovered the dyspnoeic effect of cyanide in 1931 and the Russian School of Chemoreception at St Petersburg tested all families of metabolic poisons, and here we are 75 years later asking how mitochondrial poisons act. It might be that the key is AMP kinase. All mitochondrial poisons should be expected to increase AMP levels, even if ATP is maintained, and this would lead to activation of the kinase. Pharmacological activation of the kinase produces PASMC contraction and activation of the sensory activity in the carotid sinus nerve. It is well known that the enzymes controlling basically every metabolic route are targets of AMP-kinase: *is it possible that some of the ion channels known to participate in the O_2 transduction cascade are also substrates for this kinase?*

Bruce Dinger presented sound evidence that hypoxia activates NADPH-oxidase in CBCC of normal mice, but not in CBCC from $p47^{phox}$ knockout mice. Even further, in the recordings he showed, obtained in cells form the KO mice it was also evident that bath PO_2 of 20-25 mmHg did not increase their production of $O_2^{\bullet-}$, implying that mitochondria are not sources of reactive oxygen species (ROS) at PO_2 strongly activating CBCC. He was further in showing that ROS act as negative modulator of the O_2-transduction cascade; thereby NADPH system would constitute a negative feedback controlling chemoreception. Additionally he showed that Nox-4 (an isoform of $gp91^{phox}$) and $p47^{phox}$ are induced in chronic hypoxia, implying that the system is able to control the sensitization status of the CB seen in these situations. *It probably would be nice to know if some of the enzymes of the hexose monophosphate shunt are also induced in chronic hypoxia.*

Second session was chaired by **Machicko Shirahata** from Baltimore. In the first paper presented by **R. Varas**, awarded third Prize Heymans-De Castro-Neil Award we learned that ATP and non-hydrolysable analogs inhibit O_2-sensitive leaky currents in inside-out isolated patches of rat CBCC. The study was neat, and in the context of the "metabolic hypothesis" put forward by the Russian School the data would provide, once again (see M.Evans), a link between mitochondria and O_2-sensing. *It would remain to define the domains in this leaky channel where ATP acts to increase their opening probability.*

MT Agapito presented an ample study carried out with different groups of oxidizing agents. The aim was to correlate the redox environment of CBCC with their activity in normoxia and hypoxia. Some members of every group of the oxidizing agents were able of activating CBCC neurosecretory activity in normoxia and/or hypoxia, and an equal numbers of agents did not produce any effect. Very intense hypoxia (PO_2 <8 mmHg) and hyperoxia (PO_2 around 630 mmHg) caused an oxidized redox environment in the cells, and hypoxia produced an intense neurosecretory activity in the cells while hyperoxia depressed the activity below the normoxic ($PO_2 \approx 150$ mmHg). The conclusion was that the activity of CBCC is independent of the general redox status of the cells. *In conjunction with the article by B. Dinger I consider important to define possible targets of ROS in the transduction cascade of hypoxia.*

M. Pokorki presented some data on the CB uptake of acyldopamine, more specifically of N-oleoyl-dopamine. These types of compounds are known to modulate the activity the TRP channels and thereby appear to mediate their biological activity. Since the CB is capable of accumulating important amounts of injected N-oleoyl-dopamine the possibility exists that it might control CBCC functionality via some member of the TRP family of channels expressed in the cells. Even further, anandamide that is a powerful inhibitor of some members of the K^+ leaky channels expressed in CBCC is another acylated amine, N-arachidonoyl-ethanolamine, making it plausible that N-oleoyl-dopamine modulates CBCC regulating the activity of this family of channels. *Therefore, it remains to be established the endogenous level of N-oleoyl-dopamine and to search for adequate targets for it.*

S.V. Conde showed data on the release of ATP by the intact rat CB and in parallel experiments on the release of adenosine. Basal release of both neurotransmitters was independent of the presence of extracellular Ca^{2+} in normoxia; however, it was Ca^{2+} dependent the release induced by hypoxia. There was a somehow inverse relationship between the release of both neurotransmitters: hypoxic stimuli of mild intensity release preferentially adenosine while intense hypoxic stimuli release preferentially ATP. The response to moderate hypoxic stimulus was specific to the CB. There are some very nice experiments that can be pursued in this study. For example, *it would be interesting to establish a molar relationship between ATP and DA release trying to correlate this ratio with the known stoichiometries of both neurotransmitters in dense-core vesicles of dopaminergic structures, ideally of the CB.*

Third session was chaired by **Ana Obeso** from Valladolid. **Chris Peers** presented an ample series of experiments showing that hypoxia (20-30 mmHg) produces modest increases in intracellular free Ca^+ that comes from intracellular deposits in endothelial cells and astrocytes. Both cell types expressed a capacitative Ca^{2+} entry mechanism. The Ca^{2+} rise induced by hypoxia was prevented by prior treatment with antioxidants indicating that ROS are the mediators of the responses. Although the origin of ROS could not be established, it was shown that uncoupling mitochondria produced opposite effects in both cell types, potentiating the hypoxic effect in astrocytes and abolishing it in endothelial cells. *The data open a window to explore in further detail the potential significance of the findings to processes such as atherogenesis and hypoxia induced brain damage.*

C. Wyatt presented the first study in literature dealing with the participation of AMP-kinase on the CB chemoreception. He was awarded with the First Prize Heymans-De Castro-Neil Award of the XVIth ISAC Symposium. This study was the other half of M. Evans paper. Activators of the kinase induced an increase in chemoreceptor discharges in intact preparations and in intracellular free Ca^{2+} in CBCC dispersed cells; last response being dependent of the presence of extracellular Ca^{2+}. The suggestions placed above on commenting M. Evans paper gain additional significance here.

I. M. Fearon presented a challenging paper relating hypoxia and production of amyloid peptides and mitochondrial ROS production. The notion is that Alzheimer disease incidence increases in subjects that had have episodes of brain hypoxia-ischemia. Since in this devastating disease an increase in amyloid peptides is pathogenic, the challenge is to define possible mechanisms leading to the increased levels of amyloids in hypoxia and the mechanisms triggered by the amyloids to produce neuronal death. Main findings included the observation that hypoxia-induced over expression of Ca^{2+} currents in HEK cells transfected with the α_{1C} subunit of the L-type Ca^{2+} channels was aborted by inhibitors of the enzyme generating the amyloid, that the expression of the Ca^{2+} currents was absent in cells depleted of mitochondria (ρ^o cells) and that a superoxide generating system restored the capacity of ρ^o cells to respond to hypoxia with over expression of Ca^{2+} currents. *In the context of the CBCC and physiological O_2-sensing it will be nice to know if hypoxia induces the expression of the amyloid peptides in these cells in chronic hypoxia and what is the effect of the inhibition of the enzymes processing the amyloid precursors on chemoreception.*

Fourth session was chaired by Dr. **Prem Kummar** from Birmingham. The first paper of this session was also presented by **Fearon**. In this study they have expressed THIK-1 K^+ channels in HEK cells and have studied in this heterologous system the effect of hypoxia. This leaky channel has previously been shown to be expressed in the parasympathetic neurons dispersed along the glossopharyngeal-carotid sinus nerve, that *bona fide,* can be defined as the mediators of the efferent pathway. In the natural preparation THIK-1 K^+ channels are inhibited by the same hypoxic stimuli capable of activating CBCC, and in the heterologous system they are also sensitive to hypoxia. However, and contrary to the situation with the TASK channels in CBCC rotenone did not affect the hypoxic inhibition of THIK-1 K^+ channels in the transfected system. *Although the findings are clear-cut, I would put a word of caution, as the possibility exists that the transfected cells do not express some piece of machinery needed for the expression of the inhibitory action of rotenone.*

T. Otsubo presented a very nice study in which it was correlated the density of maxi-K^+ currents and the levels of mRNA for the maxi-K^+ α and β subunits in the CBs of DBA/2J and A/J mice with the capacity of these two strains of mice to generate a hypoxic ventilatory response. It was found a positive correlation between the hypoxic ventilatory response and expression of the maxi-K^+ α subunit and an inverse relationship with the expression of maxi-K^+ β subunits; thus the DBA/2J strain that showed a strong hypoxic ventilatory response, expressed high levels of maxi K^+ α subunit and low level of expression of the β subunits. The opposite was true in the A/J mice. This elegant study that deserved the second Prize of Heymans-De Castro-Neil Award of the XVIth ISAC Symposium showed, I believe, *unequivocally that maxi-K^+ channels are involved in the genesis of the hypoxic responses, thereby closing an open debate in the literature but opening a new question: how it does so?*

A. Obeso's presentation dealt with some unexpected effects of caffeine on CBCC function. In her study on the potential significance of intracellular Ca^{2+} stores she tested caffeine as a well-known activator of Ca^{2+} release from intracellular stores, expecting that it would promote an increase in intracellular free Ca^{2+} and promote an increase in the exocytosis of catecholamines from CBCC. Contrary to that, it was found that caffeine markedly inhibited basal as well as the release evoked by low intensity hypoxic and high external K^+ stimulation. Since this inhibitory effect was seen in Ca^{2+} free solutions and had an IC_{50} 0.4 mM it was evident that was not related to the Ca^{2+} stores. A search for the mechanisms involved lead to discover that the effect was mediated by adenosine A2b receptors, described by the first time in CBCC, coupled to Dopamine D2 receptors. *Future experiments should be directed to explore the functional arrangement between the two sets of receptors and to disclose possible intracellular signaling pathways.*

Fifth session was chaired by **Chris Peers** from Leeds. **Y. Kameda** presented a neat study on the development of the CB and adjacent structures in two models of mice lacking HOXA3 and Mash1 genes. The study struck me firstly, by the superb images that she showed to the audience. In the notes that I took at the bench, during the presentation, I wrote: "muy bonito". Second, because we finally learnt the real embryological origin of the CB elements: CBCC derive from neuroblasts associated to superior cervical ganglion and their differentiation is under the control of Mash1 genes; sustentacular cells are under the control of HOXA3 and have their origin in the third pharyngeal pouch. Thirdly, and this represents a naïf view on the embryogenesis process, that if a gene is disrupted there is nothing we can do to rescue at least partially the normal phenotype. *My question could be formulated in these terms: knowing that the disruption of a morphogenetic gene causes a given array of malformations, is it possible to back regulate pharmacologically the signaling pathways that have generated the malformations?*

M. Shirahata and Yamaguchi papers dealt with additional differences between DBA/2J and A/J mice. In a morphometric and molecular biology study Shirahata showed that glial cell-derived neurotrophic factor controls the maintenance of CBCC mass in early postnatal periods as the CB from A/J mice showed lower levels of expression of this neurotrophic factor and a continuous decay in the CBCC number even to adult age. This decay in CBCC number and decrease in CB size was related to the phenotypic hypo responsiveness to hypoxia described before by T. Otsubo. This paper added a further value to the previous one of this group, as it made obvious that the scarcity of CBCC in the adult A/J mice would pose additional difficulties to the patch clamp studies. In the study by Yamaguchi they crossed both strains of mice to determine the patterns of inheritance of their specific phenotypes and found that the size of the CB, the organization of the CB glomeruli and the number of CBCC are genetically determined. Further, in the first generation of DBA/2J and A/J hybrids, the three parameters studied were intermediate between those of parents. *My enthusiastic invitation to the Baltimore laboratory to pursue exploiting these strain differences. It should be expected that additional differences would be discovered and no doubt they would be ideal paths to clarify the functioning mechanisms in CBCC.*

Next presentation was mine. We have tried to exploit previous observations on the deleterious effects of perinatal hyperoxia on the CB function, and extend them to other O_2-sensitive cell lineages. We observed that perinatal hyperoxia does not alter the ability of EPO-producing cells to respond to hypoxia generating normal plasma levels of the hormone. However, perinatal hyperoxia damaged PASMC rendering them unable to respond to hypoxia (i.e., they did not hypoxic pulmonary vasoconstriction), while the vasoconstrictor effect of depolarizing agents remained intact. *Our immediate interests are focused on the dissection of the molecular machinery presumably involved in O_2-sensing in PASMC aiming to disclose the "missing element" (to use Evans words) in PASMC of hyperoxic animals.*

Sixth session was chaired by **Yasumasa Okada** from Izu City, and the first presentation was given by **H.Kazemi**. He presented an integrated view on the role of brainstem neurotransmitter systems in the genesis of the hypoxic ventilation. He analyzed microdyalysates collected from cannulas placed in the nucleus tractus solitarius and ventral medullary surface and correlated the levels on neurotransmitters found with the phrenic activity recorded. In the initial phase of the hypoxic response there was a progressive increase in glutamate in the NTS and VMS coincident with an augmentation of phrenic activity; glutamate output would result from the augmented release of CSN terminals in the brainstem. After 3-5 min of hypoxia GABA started to increase coincident with the roll-off phase of the phrenic activity; this increase in GABA was suggested to be due to a direct effect of hypoxia in brain tissue. Taurine levels decreased in the initial phase of hypoxia and increased in the second phase, contributing to the roll-off effect. *This step of analysis of neurotransmitter dynamics would help to direct future studies focused on specific synapses of the circuits connecting the nucleus tractus solitarius to other nuclei of the central controller of respiration.*

R. Iturriaga made nice comparisons between the episodes of hypoxic sleep apnea and the ascent to the Chilean Andes: people suffering sleep apnea would climb at 4-5 Km of altitude up to 30-40 times/hour during the 8 hours of sleep. As expected from this comparison, the CBs of his cats exposed to intermittent hypoxia mimicking the episodes of sleep apnea were hyper-reactive in test to acute hypoxia. He also found that the ratio low/high frequency heart rate was increased in the experimental animals, reflecting changes in the overall activity of the autonomic nervous system. *I consider that this type of experimental models that so clearly demonstrate sensitization of the CB chemoreceptors constitute ideal preparations to search for up regulation of the molecular machinery involved in the genesis of the hypoxic response in CBCC.*

Y. Okada presented a beautiful model on the location and organization of central chemoreceptors. As a professor of physiology non-expert in central chemoreception, I would consider solved the nearly 60 years elusive problem of identification of the primary central chemoreceptor elements. The combination of methodologies used has allowed the authors to demonstrate that the primary CO_2 chemoresponsive elements (Type I cells) are small sized neurons that surround the vessels of the ventral medulla surface and are clustered in specific areas. Type II cells, located deeper in the medulla are larger and receive a cholinergic input from type I cells. The arrangement of central chemoreceptors recalls me the studies of Fernando de Castro in the CB: *"the relationship of epithelioid cells (CBCC) in the glomus caroticum with blood vessels is intimate, suggesting that they are specialized in sensing qualitative changes in the composition of the blood..."* At last!

P. Kumar study addressed systemic aspects of the CB function. He has described previously that insulin-induced hypoglycemia produces an increase in the CB drive of ventilation via an, as yet, unidentified mechanism that should be generated systemically because hypoglycemia, even 0 glucose superfusion, does not modify the behavior of the isolated CB. Now he took these previous findings to compare the status of hypermetabolism produced by insulin-induced hypoglycemia with that encountered in physical exercise. He measured the gain of the CB chemoreflex in response to altered arterial PCO_2 as the delta in the phrenic output or as ventilation in control, insulin-hypoglycemic and insulin-hypoglycemic CSN denervated rats. Insulin hypoglycemia doubled metabolic rate in all animals but only in those with the intact CSN showed a parallel increase in ventilatory response/phrenic activity to CO_2; CSN denervation resulted in the loss of this extra sensitivity to CO_2 with the result of a retention of CO_2 that elevated arterial PCO_2 by about 6 mmHg during hypoglycemia, while the animals with intact CSN remained isocapnic. Comparisons with data obtained in isocapnic physical exercise yielded superimposed curves of responsiveness to CO_2. Thanks Prem, because studies like yours make meaningful the efforts made by us people working in minute things.

H. Kiwull-Schöne paper represented a further step in letting us feel that the work at the cellular and molecular level is worthwhile. To begin with, I want to thank Prof. Heidrum Kiwull-Schöne. She is a faithful attendant to the ISAC Symposia and together with Prof. Robert Fitzgerald and Prof. Katsushide they were the most senior symposiasts, having participate in the Bristol, Delhi or Dortmund Symposia. Prof. Kiwull-Schöne defined the steady-sate gain of the respiratory loop control system as the point of intersection between the nearly lineal respiratory response to CO_2 and the hyperbolic respiratory response to metabolism. Her purpose was to make evident that the instability of the respiration in situations with a high gain in the entire control system can be attenuated by reducing the gain of the system. Experiments with inhibitors of carbonic anhydrase and of the isoform 3 of the Na^+/H^+ exchangers the predictions and revealed different effects on the CO_2 and metabolic responses.

Seventh session was chaired by **Colin Nurse** from McMaster University. The first paper was presented by **D.J. Kwak**. They tested the hypothesis that the deleterious effect of hyperoxia on the CB structure and function seen in neonatal, but not in adult animals, is related to the higher production of ROS in neonates in comparison to adults. He made vibratome sections of CB and petrosal-nodose ganglia and incubated them during four hours in hypoxia (8% O_2), normoxia (21% O_2) and hyperoxia (95% O_2) in the presence of a 2',7'-dichlorodihydrofluorescein a fluorescent indicator of ROS. The hypothesis was correct, at least in part, as he found that the rise in fluorescence in going from hypoxia to normoxia and hyperoxia was significantly greater in neonate tissues in comparison to young (18 days old). *To really satisfy the hypothesis it would be important to know if we are dealing with an increase production of ROS or undeveloped defense mechanisms, and most important of all, if the signal got* *with the fluorescent indicator used is reliable. I recommend the authors to have a look to Gonzalez et al. Methods in Enzymology* **381:** *40-71, 2004.*

C. **Saiki** tested a simple hypothesis: whether or not an episode of neonatal anoxia interferes with the development of the CB. The episode of anoxia was rather severe (exposure to N_2 for 20 min), but nonetheless the rat pups were successfully autoresuscitated. At the age of 25 days a morphometric analysis (CB volume and CBCC/CB ratio) revealed the absence of differences between experimental correspondent controls. The simplicity of the question asked and the adequacy of the tools used to answer the question are remarkable. *Yet, I would suggest Dr. Saiki, and the old friend S. Matsumoto, to repeat a comparable study in animal models mimicking more closely the situations of hypoxia encountered in infants whether normal or with different types of hypoxia producing pathologies. In addition, a simple functional test (e.g., whole body plethysmography at different ages) would provide a superb study to hear in the next ISAC Symposium in Valladolid.*

E. **Gauda** presented the next paper dealing with adenosine receptors in the CB and petrosal ganglion neurons. In presenting her data, once again she exhibited the enviable vitality and enthusiasm she always shows. A combination of real time PCR, immunocytochemistry and electrophysiological recordings evidenced the presence of A_1 and A_{2A} receptors in the rat CB, the former located at mRNA level in petrosal neurons and immunocytochemically in the endings of the CSN, the second present both in petrosal neurons, where it increased with postnatal age, and in the CBCC. Electrophysiological data showed that A_1 receptors are inhibitory, as their inhibition increased CSN activity. Being A_{2A} excitatory, she concluded that the predominance of A_1 receptors in early postnatal days could contribute to the hyporesponsiveness of the CB to hypoxia in the initial periods of extrauterine life. Nice work. *However, it might be worth thinking within the frame that there are important species differences as in the rabbit A_1 receptors are also present in CBCC. These differences can be very important for a neonatologist pediatrician.*

R. **Rigual** made a straightforward comparison of the hypoxic responses encountered in adrenomedullary cells and CBCC. The main point in the comparison is that, even though both cell types apparently express the same basic transduction cascade (O_2-sensitive K^+ channels, voltage operated Ca^{2+} channels...), the gain of the overall transduction cascade of the hypoxic stimulus into a secretory response is approximately one hundred more efficient in the CBCC. The gain of a depolarizing stimulus (high extracellular K^+) is identical. Additionally he showed that the adrenomedullary cells are also responsive to hypercapnic/acidic stimuli, existing a positive interaction between hypoxic and acidic stimuli. This finding is highly meaningful: the surge of plasma catecholamines seen in infants vaginally delivered is in all likelihood due to this interaction, because the low gain of the hypoxic transduction cascade would be unable of producing such intense adrenomedullary response. *My suggestion would be to follow a step-by-step approach to identify the steps or mechanisms that are poorly expressed in adrenomedullary cells in comparison to CBCC.*

J.A. **Mendoza** presented the next paper. In fact, the program was altered and this paper was presented earlier in the same session. They tested the hypothesis that hypoxia augments the entry of Ca^{2+} in CBCC via nicotinic receptors by preventing the desensitization observed on continuous or repeated activation of the receptors. The experimental data supported the hypothesis. In a search for the mechanism involved, they found that antioxidant agents have mixed effects with dithiothreitol mimicking hypoxia but GSH having no effect. They further showed that probably coactivation of muscarinic receptors would contribute to decrease the level of desensitization on repeated administration of acetylcholine. In my notes, I have that additional observations presented seem to suggest that polyamines might play a significant role in the control of desensitization. *Considering that polyamines are strong modulators of ion channels, including ionotropic receptors, in many invertebrate and vertebrate neurons, and that this area has not been explored in the CBCC I would encourage Dr. Mendoza to go deep in this area.*

Eight session was chaired by professor **Yoshizaki Hayashida** from Osaka. First paper in the session was presented by **L.Y.Lee**. His paper was a classical study of electrophysiology, well in the line of those performed back in the sixties and seventies by renowned physiologists like Paintal, Iggo or Perl to quote some. His study was dealing with C fibers located in the broncopulmonary tissue, and the effect of transient alveolar hypercapnia in their response. In normal condition hypercapnia increased activity in only 10% of the C fibers studied, fulfilling all criteria of polymodal nociceptors, but after the applications of broncopulmonary irritants most/all C fibers were intensely activated by hypercapnia. *The molecular mechanisms of activation (involvement of transient receptor potential cationic channels,) and the physiological significance (for example in terms of ventilation or activity of bronchial muscle...) should ideally follow the identification of the responses.*

S.A. Montero has followed the path of his mentor Ramón Alvarez-Buylla in the study of the role of the CB chemoreflex in the utilization of glucose by the brain. He provided data supporting a complex interaction between the CB nitrergic system, chemoreflex activation with cyanide and cerebrospinal fluid perfusion with NO donors and NOS inhibitors alter the level of glucose utilization by brain tissue; however the pathways of the interaction cannot be easily traced with the data presented. *Likely, less ambitious experiments, aimed to answer one question at a time, would help to understand how the CB could modulate the rate of glucose utilization by the brain. For example, isolating the carotid bifurcation and perfusing it in situ at normal blood pressure would allow to activate the chemoreflex while keeping the entire body, including the brain, at normal PO_2. In these conditions, it can be dissected the real effect of the chemoreflex on glucose utilization. This preparation would allow manipulating a CB neurotransmitter system at a time and probably obtaining palatable data.*

J. Yu experimental proposal was that pulmonary nociceptors might constitute a target for the neuroactive substances (mainly serotonin) released by neuroepithelial bodies. His morphological studies were beautiful and convincing: high threshold A-δ receptors and C-fiber receptors, functionally nociceptors, penetrate or reach the close neighborhood of neuroepithelial bodies. His electrophysiological data showed that microinjections of serotonin in the receptive field of the A-δ receptors strongly, but transiently, activated their activity. *Probably next question to explore would be the function of this afferent activity: does it alter the production of mucus? does it produce bronchial spasm? Does it modulate ventilation?*

Ninth session was chaired by **Bruce Dinger** from Salt Lake City and had **A.V. Gourin** as the first speaker. He presented a nearly completed history demonstrating a prominent role for ATP as neurotransmitter in the CB and central chemoreceptors. Although with some minor critiques regarding the methods used to detect acetylcholine (I do believe the method was measuring choline instead), the combination of morphological, immunocytochemical, neurochemical, electrophysiological and measurement of ventilatory responses, both in normal and ATP-receptors knockout mice, generated well tied data to support the message given in the title "ATP is a common mediator of central and peripheral chemosensory transduction". Keep doing this kind of elegant experiments!

R.S. Fitzgerald presented an overview on "Modulators of cat carotid body chemotransduction". Bob, this was difficult task that might let many people happy and some unhappy, but nonetheless the effort was worthwhile. Particularly interesting were the interactions between adenosine and cholinergic and dopaminergic transmission: the excitatory action of adenosine, both at the level of the CSN and at the level of ventilation correlated with its ability to increase acetylcholine release and to decrease dopamine release. Similarly, NO-donors and precursors (which act as inhibitors of the CSN output) reduced the release of acetylcholine induced by hypoxia. These, and additional data presented, certainly support the contention sustained by his laboratory, namely, that acetylcholine is an excitatory, probably "the" excitatory, neurotransmitter in the cat CB and that dopamine role fits better with the notion of an inhibitory modulator. Yet, the colleagues from Santiago de Chile have recently found that dopamine excites petrosal ganglion neurons in the rabbit.

Last presentation in the Symposium was made by **Colin Nurse,** next ISAC President and host for the Symposium of 2011. The contention of his presentation was that 0.1 mM glucose causes release of ATP and acetylcholine from CBCC, because the electrical activity elicited by low glucose recorded from petrosal neurons co-cultured with CBCC was inhibited by P2X and nicotinic blockers. The poof for that release was obtained by direct measurement of ATP in saline bathing fresh slices of the CB. In this preparation it was further shown that the release was sensitive to Cd^{2+}. Interestingly, he found that the effect of low glucose and hypoxia were additive, suggesting different transduction pathways for hypoxic and hypoglycemic stimuli. In a sense, these findings are provocative because in the intact preparation even 0 glucose does not activate chemosensory discharges either in the rat or in the cat. Sincerely, I cannot envision the reasons for differences between preparations.

Obvious constrains of space force me to stop here this overview of the XVIth ISAC Symposium. However, this overview would be lame in one foot if I do not state openly that an equal number of papers used the poster format for their presentation. The topics covered were the same and the quality of the works as good as oral presentations. In many occasions, it has been seniority what has decided one format of presentation or the other; in others, it has been language difficulties for the native of non-English speaking countries. Once again, I want to express my personal gratitude for the creditors of half of the success of the symposium.

The last word of these closing remarks is one of invitation to the XVIIth ISAC Symposium. My colleagues in Valladolid and myself will organize it for all of you and for any other scientist working in allied fields with enthusiasm. However, as our predecessors have shown us, we are aware that in addition to enthusiasm we need to be efficient. We will do our best.

ACKNOWLEDGEMENTS

Supported by Spanish DGICYT grant (BFU2004-06394) and ICiii-FISS (Red Respira and Grant PI042462).

Valladolid 14 of June 2005.

Index

Acetylcholine, 275, 307, 319, 331
Acidosis, 131, 267, 307
Acclimatization, 49
Adenosine, 179, 307
   receptor, 121, 215, 307
Adrenal medulla, 121
Adrenaline, 245
Afferent activity, 191
   Vagal, 301
Anoxia,
   Neonatal, 115
Antioxidant protein SP-22, 73
Apnea,
   Obstructive sleep, 227, 345
Astrocyte, 185
ATP, 167, 179, 185, 307, 331

Breathing, 257
   Control of, 339

Caffeine, 215
Calcium channel, 197, 267
Calcium ion, 215
   Intracellular, 185, 191, 275
CaM kinase I, 87
Carbonic anhydrase, 239
Carotid body, 9, 15, 21, 29, 37, 43, 49, 55, 73, 87, 93, 99, 105, 115, 131, 167, 173, 179, 191, 209, 215, 227, 245, 275, 293, 307, 319, 331, 345
Cat, 227, 257, 275, 345
Cell,
   HEK293, 197, 203
Chemoreceptor, 43, 251, 263
   Airway, 301
   Cell, 215, 325
   Central cell, 233
Chemosensitivity,
   Central, 233, 239, 339
Chromaffin cells, 79

Chipmunk, 73
CO_2,
   Sensitivity, 245
   Sensor, 281
Comperative study, 131
Cyanide, 293

Deacclimatization, 49
Development, 43, 93, 99, 111, 121, 131
Differential display, 267
Dihydroethidium, 155
Dog, 223, 281
Dopamine, 307, 331
   N-Oleoyl-, 173

Endothelin, 21, 345
Endothelial cell, 185
Exercise, 251

GABA, 223
Gene expression, 21, 29, 37, 43
Genetics, 93, 99, 105, 121
Glucose
   Homeostasis, 267
   Brain retention, 293
Glutamate, 223
Glutathione,
   Redox potential, 325
Guinea pig, 281

Hem oxygenase, 137, 161
Hibernation, 73
H^+ sensitivity, 267
Human, 63, 185, 251, 263
Hypoxia inducible factor (HIF), 21, 29, 63
Hydrogen peroxide, 155
Hypercapnia, 55, 263, 281
   Induced gene, 267

Hypoxia, 93, 99, 147, 161, 185, 191, 209, 263, 275, 307
  Acute, 203, 223
  Chronic, 29, 155, 197
  Chronic hypocapnic, 49
  Chronic hypercapnic, 55
  Intermittent, 15, 21, 227, 345
Hyperoxia, 111, 137
  Chronic, 37
Hyperpnea, 251

Iberiotoxin, 209
Immunochemistry, 9, 49, 55, 73, 93, 99, 105, 115, 191
Inflamation,
  Airway, 281

L-Arginine, 307
L-NAME, 293
Loop gain, 239
Lung, 281
  Noniceptor, 301

Methazolamide, 257
Midbrain, 223
  Ventral surface, 233, 267
Mitochondria, 79, 147
Morphometry, 93, 99, 105, 115
Mouse, 87, 93, 99, 105, 161, 209, 319, 349
Muscarine, 275, 319

NADPH, 155
Nuclear transcription factor, 267
Neural crest cell, 93
Neuroepithelial body, 301, 313
Neurotransmitter, 223
Neuroglobin, 15
Neurotransmitters, 319
Nicotine, 275, 319
Nitric oxide(NO), 293
Nodose ganglion cell, 111
Norepinephrine, 307

O_2-sensing, 79, 161

Paraganglia, 9
Patch clamp, 209, 319
Peptidergic Innervation, 49, 55

Petrosa ganglion cell, 111, 331
pO_2, 179
Potassium channels, 137, 155, 203, 209, 313, 319, 339
  TASK, 9, 37, 43, 167
  TREK, 9
  THIK, 203
Phrenic output, 223
Phosphorylation,
  Oxydative, 147, 191
Plethysmography, 339
Protein Kinase, 87, 191
  AMP-activated, 147
Pulmonary
  C-fiber, 281
  Endothelial cells, 63
  Vasculature, 147, 161
Proteomics, 137

Rabbit, 215, 239, 301, 313, 331
Rat, 9, 15, 21, 29, 37, 43, 49, 55, 79, 87, 111, 115, 121, 131, 147, 155, 167, 173, 179, 185, 191, 215, 223, 245, 267, 281, 293, 325
Reactive oxygen species (ROS), 79, 111, 155
Receptors, 319
Recombinant expression, 203
Repolarization, 155
Respiratory
  Control, 239
  Response, 223
RNA, 137
RT-PCR, 209

Serotonin, 331
SIF cells, 87
S-Nitrosoglutathione (SNOG), 63
Sodium nitroprusside, 307
Sodium-proton exchanger, 239
Stimulus interaction, 263

Taurine, 223
Transmitter, 341
Tyrosine hydroxylase, 93, 99, 105

Ventilation, 93, 99
  Perfusion maching, 161
  Hypoxic response, 339